MW00345968

Running Li.....

Edip Yuksel

v.s.

Abu Ameenah Bilal Philips
Carl Sagan
Daniel Abdurrahman Lomax
Ayman Mutaakher
James (Amazing) Randi
Michael Shermer
… and other Ingrates

Brainbowpress

Running Like Zebras

A message for those who prefer reason over blind faith. For those who seek ultimate peace and freedom by submitting themselves to the Truth alone.

www.19.org
www.yuksel.org
www.quranix.com
www.islamicreform.org
www.brainbowpress.com

ISBN 978-0-9825867-3-0

Cover design: Edip Yuksel, Aykut Aydoğdu, Uğur Şahin

Printed in the United States of America

10 9 8 7 6 5 4 3 2

A cartoon of mine, which was published in the first edition of my Turkish book: *Üzerinde 19 Var* in 1997 by Ad (Milliyet) publishing house, Istanbul.

Sunni and Shiite scholars, mullahs, and sheiks are in agreement that *Basmalah* (i.e.*Bismillahi-Rahmaanir-Raheem)* does not consist of 19 letters.

The one with Abacus: I counted the letters of Basmalah, and they are always 18.
The one with beans: According to my beans, it is sometimes 20 and sometimes 21.
The one with rosary: You are both correct. I am sure it is not 19.

3

EDIP YUKSEL, American-Turkish-Kurdish author and activist, was born in Turkey in 1957 the son of late Sadreddin Yuksel, a prominent Sunni scholar. During his youth he was an internationally active Islamist who spent over four years in Turkish prisons in the 1980's for his political writings and activities that promoted an Islamic revolution in Turkey. While a popular Islamist author and youth leader, Edip adopted the Quran Alone philosophy after corresponding with Rashad Khalifa and reading his landmark book, *Quran, Hadith and Islam*. This led him to experience a paradigm change on 1st of July, 1986, transforming him from a Sunni polytheist to a reformed muslim, a peacemaker, and a rational monotheist. As a consequence, he was excommunicated and declared to be a heretic, an apostate deserving the death penalty. In 1989, he was sponsored for immigration to the USA by Rashad Khalifa and worked together for a year in Masjid Tucson, becoming a prominent member of the United Submitters International. However, shortly after the assassination of Rashad by a group of Sunni terrorists, he was *excommunicated* from the Submitters as he objected to an idol-carving gang that had infected the group with cult mentality and false ideas about Rashad.

Edip Yuksel is the author of over twenty books and hundreds of articles on religion, politics, philosophy and law in Turkish and English. His English books are recently published by Brainbow Press, which include *Quran: a Reformist Translation, Manifesto for Islamic Reform, Peacmaker's Guide to Warmongers*, and *NINETEEN: God's Signature in Nature and Scripture*. Edip is the co-editor of the annual anthology, *Critical Thinkers for Islamic Reform*.

After receiving his bachelor degrees from the University of Arizona in Philosophy and Near Eastern Studies, Edip received his law degree from the same university. Besides writing and lecturing, Edip works as an Adjunct Philosophy professor at Pima Community College. Edip is fluent in Turkish, English and Classic Arabic; proficient in Persian, and barely conversant in Kurdish, his mother tongue. Edip is the founder of Islamic Reform and co-founder of MPJP organizations. His online books, interviews, and articles are published at various Internet sites, including:

www.19.org
www.yuksel.org
www.quranix.com
www.islamicreform.org
www.quranmiracles.com
www.free-minds.org
www.quranic.org
www.quran.org
www.mpjp.org

Edip's books can be ordered at: www.amazon.com ● www.ozanyayincilik.com

Edip Yuksel

nineteen

God's Signature in Nature and Scripture

BrainbowPress

6

Table of Contents

Edip Yuksel and Rashad Khalifa in front of Masjid Tucson in 1988. Because of the implications of his discovery of the Secret, as well as his strong criticism of the sectarian teachings based on Hadith and Sunna, Rashad was declared a heretic/apostate by leading Sunni scholars from 38 countries who held an emergency conference in Saudi Arabia in February 27, 1989 to discuss the Salman Rushdie controversy. While Rushdie survived the fatwa, Rashad was assassinated in this Masjid in January 31, 1990, by a terrorist group linked to al-Qaeda. The author of this book also received similar fatwas, yet he has escaped several assassination attempts, so far.

1

The Prime Argument

NINETEEN: God's Signature in Nature and Scripture

Edip Yuksel vs. Carl Sagan

For my honor thesis as an undergrad student in philosophy, I used code 19 as a philosophical argument for the existence of God. Before submitting my paper, I decided to discuss the issue with late professor Carl Sagan. As an agnostic astronomer, Sagan expressed his doubts about such an inference. The two-round short argument provides a different perspective on the "Miraculous Code 19" of the Quran. It is a prime argument on a mysterious prime number. I updated the section where I presented the Code 19 so that those readers who have not yet read my book *NINETEEN: God's Signature in Nature and Scripture* have better idea about issue.

In the name of God, Gracious, Merciful

12/20/1993

Dear Carl Sagan,

Since I read your novel "Contact" three years ago I have been thinking to "contact" you. Recently, your Parading articles[1] forced me to write this letter and send you my first books in English.

To create a mutual knowledge of each other I will give headline information about me with segregated phrases.

---

[1] [Here, I refer to Sagan's article "How Can Games Test Ethics? A New Way To Think About Rules To Live By" published in Parade Magazine, November 28, 1993. There Sagan used empirical evidence to compare some well-known ethical rules. He concluded that the Goldplated Brazen Rule is the most efficient rule. He referred to a Quranic "Brazen" verse quoted by President Clinton at the Israeli/Palestinian peace accords. "If the enemy inclines toward peace, do you also incline toward peace." (8:61; 4:90). Nevertheless, other verses encouraging forgiveness in the practice of retaliation (such as 2:178) makes the Quranic rule a "Goldplated Brazen Rule".]

The following unsolicited information of snap-shot moments of my life with a grammatically handicapped English (or "literary challenged", if you like!), may send twinkling silly messages for a world-wide famous scientist. But, I bet you like silly things. I know from the conspicuous smile on your face.

[Personal information omitted]

Here I enclose two of my first booklets in English. I am planning to complete them with "19 Questions for atheists."

Please read them, at least the last question (which is almost identical in both). It is about the "miraculous" mathematical structure of the Quran. Please don't reject it without studying it, since it is very easy to confuse it with numerology. How will you react to someone who thinks that Astronomy is a branch of Astrology?

I would like to discuss with you on many issues. Just for this reason I'm applying to Cornell University for graduate program in Philosophy. We may end up co-authoring a book together, if you keep alive your diverse curiosity and sparkles in your eyes.

Edip Yuksel

PS: If you want, I can send some of my short arguments on paranormal phenomena. By the way, I strongly believe that there is life in other parts of the universe, with a probability of being a few intelligent kinds. My primary reason for this belief is very different than yours.

"THIS IS THE ONLY WAY"

"No, don't you see? This would be different. This isn't just starting the universe out with some precise mathematical laws that determine physics and chemistry. This is a message. Whoever makes the universe hides messages in transcendental numbers so they'll be read fifteen billion years later when intelligent life finally evolves. I criticized you and Rankin the time we first met for not understanding this. 'If God wanted us to know that he existed, why didn't he send us an unambiguous message?' I asked. Remember?"

"I remember very well. You think God is a mathematician."

"Something like that. If what we're told is true. If this isn't a wild-goose chase. If there's a message hiding in pi and not one of the infinity of other transcendental numbers. That's a lot of ifs."

10

"You're looking for Revelation in arithmetic. I know a better way."

"Palmer, this is the only way. This is the only thing that would convince a skeptic. Imagine we find something. It doesn't have to be tremendously complicated. Just something more orderly than could accumulate by chance that many digits into pi. That's all we need. Then mathematicians all over the world can find exactly the same pattern or message or whatever it proves to be. Then there are no sectarian divisions. Everybody begins reading the same Scripture. No one could then argue that the key miracle in the religion was some conjurer's trick, or that later historians had falsified the record, or that it's just hysteria or delusion or a substitute parent for when we grow up. Everyone could be a believer." (Sagan, Carl. Contact. Simon and Schuster. New York: 1985, p 418-419)

The above excerpts are quoted from CONTACT, a book by Dr. Carl Sagan the late astronomer who became popular with the TV series, Cosmos. Sagan's CONTACT is a novel expression of philosopher's prime dream: Mathematical evidence for God's existence.

Mathematics is considered a priori, knowledge gained independently of experience. Most of the philosophers highly relied on mathematics. Descartes who employed extreme doubt as a method to reach the knowledge (certainty) could not doubt from mathematics. The language of mathematics is universal.

The Most Controversial Concept

Hindus believe that he is incarnated in many human beings. Christians pontificate that he has multiple personalities, one of them being sacrificed for humanity. Jews assert that he is Jehovah. Muslims claim that he is Allah. Many question his gender. Millions die for him, millions fight for him, millions cry for him. Clergymen use his name as a trademark for their business, and the very same name motivates many devotees to give away their belongings as charity. Many joyfully sing songs for his love, and others outrageously declare dialectic or scientific wars against him. Some even exclaim that he is no longer alive.

Volume upon volume of books are published for and against him. Big lies are attributed to him while scientific hoaxes are arranged to deny him. He is in the courts, he is on the money, he is in the schools, he is in the mind of saints and in the mouth of hypocrites. Yes, he is everywhere. And yet, philosophers continuously question his existence. In fact, world religions, with numerous versions of odd gods, have not helped philosophers prove his existence. On the contrary, they created further intellectual problems and logical obstacles for questioning minds who try to reach him.

The Prime Evidence

The "prime" evidence comes in the form of a highly sophisticated mathematical code embedded in an ancient document. Computer decoding of this document was originally started by Dr. Rashad Khalifa, a biochemist, in 1969. In 1974, this study unveiled an intricate mathematical pattern based on a prime number. (Having interested with the subject the author, like many others, I examined Dr. Khalifa's findings and assisted him in his further research.)*

For more than 14 centuries it was a hidden secret in the most read yet one of the most ignored books, the Quran (The Book of Recitation), until 1974. The discovery of the code not only explained many verses, but it also exposed the diabolic nature of sectarian teachings and the work of clergymen.

With the computer decoding of the Quran, summarized below, the argument for the existence of God gained examinable physical evidence. Although the Quran had been in existence for fourteen centuries, its mathematical code remained a secret until computer decoding became possible. As it turned out, the code ranges from extreme simplicity to a complex, interlocking intricacy. Thus, it can be appreciated by persons with limited education, as well as scholars.

This ancient document is the Quran, revealed to Muhammad of Arabia early in the seventh century as The Final Testament. The following is a condensed summary of this unique literary code. Please note that one does not need to know Arabic, the original language of the Quran, to examine most of the evidences presented below. For some of them one may only need to recognize the 28 letters of the Arabic alphabet.

The Message for the Computer Generation

Chapter 74 of the Quran is dedicated to the PRIME number 19. This chapter is called "Al-Muddassir" (The Hidden Secret). The number 19 is specifically mentioned in that Chapter as a "punishment" for those who state that the scripture is human-made (74:25). This number is also called "One of the greatest portents" (74:35). In 74:31, the purpose of the number 19 is described: to remove all doubts regarding the authenticity of the Quran, to increase the appreciation of the believers, and to be a scientific punishment for hypocrites and disbelievers. However, the implication of this number as a proof for the authenticity of the Quran remained unknown for centuries. For fourteen

12

centuries, the commentators tried in vain to understand the function and fulfillment of the number 19.

Before The Secret Was Decoded

Before the discovery of the 19-based system, we were aware of a symmetrical mathematical wonder in the Quran. For example:

- The word "month" (ŞaHR) occurs 12 times.
- The word "day" (YaWM) occurs 365 times.
- The word "days" (ayyam, YaWMayn) occurs 30 times.
- The words "perverse/satan" (shaytan) and "controller/angel" (malak), each occur 88 times.
- The words "this world" (DuNYa) and "hereafter" (ahirah), each occur 115 times.
- The words "they said" (QALu) and "you say" (QUL), each occur 332 times.
- The proper names of 26 messengers mentioned in the Quran and the frequencies of all the derivatives of the root word RaSaLa (message/send messenger), each occur 512 times.

A Great Prophecy is Fulfilled and the Secret is Unveiled

The miraculous function of the number 19 prophesized in Chapter 74 was unveiled in 1974 through a computerized analysis of the Quran. Though, in retrospect, the implication of 19 in Chapter 74 traditionally called Hidden One, were obvious, it remained a secret for 1406 (19x74) lunar years after the revelation of the Quran. Ironically, the first words of the Chapter 74, The Hidden One, was revealing, yet the code was a divinely guarded gift allocated to the computer generation; they were the one who would need and appreciate it the most. As we have demonstrated in various books, hundreds of simple and complex algorithms, we witness the depth and breath of mathematical manipulation of Arabic, an arbitrary human language, to be profound and extraordinary.

This is the fulfillment of a Quranic challenge (17:88). While the meaning of the Quranic text and its literal excellence was kept, all its units, from chapters, verses, words to its letters were also assigned universally recognizable roles in creation of mathematical patterns. Since its discovery, the number 19 of the Quran and the Bible has increased the appreciation of many of those who acknowledge the truth, has removed doubts in the minds of many People of the Book, and has caused discord, controversy and chaos among those who have traded the Quran with men-made sectarian teachings. This is indeed a fulfillment of a Quranic prophecy (74:30-31).

Various verses of the Quran mention an important miracle that will appear after its revelation.

10:20   They say, "If only a sign was sent down to him from His Lord." Say, "The future is with God, so wait, and I will wait with you."

21:37   The human being is made of haste. I will show you My signs; do not be in a rush.

38:0   In the name of God, the Gracious, the Compassionate.

38:1   S90, and the Quran that contains the Reminder.

38:2   Indeed, those who have rejected are in false pride and defiance.

38:3   How many a generation have We destroyed before them. They called out when it was far too late.

38:4   They were surprised that a warner has come to them from among themselves. The ingrates said, "This is a magician, a liar."

38:5   "Has he made the gods into One god? This is indeed a strange thing!"

38:6   The leaders among them went out: "Walk away, and remain patient to your gods. This thing can be turned back."

38:7   "We never heard of this from the people before us. This is but an innovation."

38:8   "Has the remembrance been sent down to him, from between all of us?" Indeed, they are doubtful of My reminder. Indeed, they have not yet tasted My retribution.

38:87   "It is but a reminder for the worlds."

38:88   "You will come to know its news after awhile"

41:53   We will show them Our signs in the horizons, and within themselves, until it becomes clear to them that this is the truth. Is it not enough that your Lord is witness over all things?

72:28   … He has counted everything in numbers.

The beginning of Chapter 25 refers to the arguments of the opponents who denied the divine nature of the Quran:

25:4   Those who rejected said, "This is but a falsehood that he invented and other people have helped him with it; for they have come with what is wrong and fabricated."

14

25:5   They said, "Tales of the people of old, he wrote them down while they were being dictated to him morning and evening."

The awaited miracle, the hidden mathematical structure

The subsequent verse gives an enigmatic answer to the assertion of those who claimed that the Quran is manmade.

25:6   Say, "It was sent down by the One who knows the secrets in the heavens and the earth. He is always Forgiving, Compassionate."

How can "knowing the secret" constitute an answer for those who assert that the Quran is Muhammad's work? Will the proof or evidence of the divine authorship of the Quran remain a secret known by God alone? Or, will the antagonists be rebuffed by that divine mystery? If there should be a relationship between the objection and answer, then we can infer from the above verse that a SECRET will demonstrate the divine nature of the Quran.

The miracle promised throughout the Quran might have been destined to appear after Muhammad's death:

13:38   We have sent messengers before you and We have made for them mates and offspring. It was not for a messenger to come with any sign except by God's leave, but for every time there is a decree.

13:39   God erases what He wishes and affirms, and with Him is the source of the book.

13:40   If We show you some of what We promise them or if We let you pass away, for you is only to deliver, while for Us is the reckoning.

The last phrase of verse 13:40 quoted above is interesting, The meaning of this phrase becomes clearer after the discovery of Code 19, since the Arabic word "HeSaaB" refers to both the "day of judgment" and "numerical computation." The deliberate use of multi-meaning words is very common in the Quran. For instance, the word "AaYah" occurs 84 times in the Quran and in all occurrences it means "miracle," "sign" or "law." However, its plural form "Aayaat" also means "revelation" or "verses of the Quran." This unique usage equates minimum three verses of the Quran (in Arabic, a different form is used for duality) with miracle. It also pulls our attention to the parallelism between God's signs/laws in the universe and God's revelation in human language: they share the same source and the same truth. In the following verse the plural word

15

A*ayaat* is used not to mean revelation but "miracle, sign or physical manifestation":

> 6:158　Do they wait until the angles/controllers will come to them, or your Lord comes, or some signs from your Lord? The day some signs come from your Lord, it will do no good for any person to acknowledge if it did not acknowledge before, or it gained good through its acknowledgement. Say, "Wait, for we too are waiting."

The verse 30 of Chapter 74, Al-Muddassir (The Hidden/The Secret) of the Quran reads exactly, "Over it Nineteen." The entire chapter is about the number 19. Let's read the chapter from the beginning.

> 74:0　In the name of God, the Gracious, the Compassionate.
> 74:1　O you hidden one.
> 74:2　Stand and warn.
> 74:3　Your Lord glorify.
> 74:4　Your garments purify.
> 74:5　Abandon all that is vile.
> 74:6　Do not be greedy.
> 74:7　To your Lord be patient.
> 74:8　So when the trumpet is sounded.
> 74:9　That will be a very difficult day.
> 74:10　Upon the ingrates it will not be easy.
> 74:11　So leave Me alone with the one I have created.
> 74:12　I gave him abundant wealth.
> 74:13　Children to bear witness.
> 74:14　I made everything comfortable for him.
> 74:15　Then he wishes that I give more.
> 74:16　No. He was stubborn to Our signs.
> 74:17　I will exhaust him in climbing.
> 74:18　He thought and he analyzed.
> 74:19　So woes to him for how he analyzed.
> 74:20　Then woe to him for how he analyzed.
> 74:21　Then he looked.
> 74:22　Then he frowned and scowled.
> 74:23　Then he turned away in arrogance.
> 74:24　He said, "This is nothing but an impressive magic."
> 74:25　"This is nothing but the words of a human."
> 74:26　I will cast him in the *Saqar*.
> 74:27　Do you know what *Saqar* is?

74:28   It does not spare nor leave anything.

74:29   Manifest to all the people.

74:30   On it is nineteen.

The punishment issued for the opponent is very interesting: nineteen. Almost all numbers mentioned in the Quran is an adjective for a noun. Forty nights, seven heavens, four months, twelve leaders. But here the numerical function of nineteen is emphasized. Nineteen does not define or describe anything. The disbeliever will be subjected to the number nineteen itself. Then, what is the mission or function of this nineteen? Those who tended to understand the meaning of *Saqar* as "hell" naturally understood it as the number of guardians of hell. However, the punishment that is described with phrases such as, difficult task, precise, and universal manifestations, was an intellectual punishment; a mathematical challenge. Indeed, the following verse isolates the number nineteen from the number of controllers and lists five goals for it.

74:31   We have made the guardians of the fire to be angels/controllers; and We did not make their number except as a test for those who have rejected, to convince those who were given the book, to strengthen the acknowledgment of those who have acknowledged, so that those who have been given the book and those who acknowledge do not have doubt, and so that those who have a sickness in their hearts and the ingrates would say, "What did God mean by this example?" Thus God misguides whoever/whomever He wishes, and He guides whoever/whomever He wishes. None knows your Lord's soldiers except Him. It is but a reminder for people.

Traditional commentators of the Quran had justifiably grappled with understanding this verse. They thought that disbelievers would be punished by 19 guardians of hell. That was fine. But they could not explain how the number of guardians of hell would increase the appreciation of believers and convince the skeptical Christians and Jews regarding the divine nature of the Quran. Finding no answer to this question, they tried some explanations: the Christians and Jews would believe in the Quran since they would see that the number of guardians of hell is also nineteen in their scripture. Witnessing the conversion of Christians and Jews, the appreciation of Muslims would increase.

This orthodox commentary has three major problems. First, neither the Old, nor the New Testament mentions number nineteen as the guardians of hell. Second, even if there was such a similar statement, this would not remove their doubts but to the contrary; it would increase their doubts since they would consider it

17

one of the many evidences supporting their claim that the Quran plagiarized many stories from the Bible. Indeed, there are many Biblical events are told by the Quran, though occasionally with some differences. Third, none so far converted to Islam because of guardians of hell.

Some scholars noticed this flaw in traditional commentaries. For instance, Fahraddin el-Razi, in his classic commentary, Tafsyr al-Kabyr, offered many speculations, including that the number nineteen indicates the nineteen intellectual faculty of human being. Tough it is a clever interpretation, but it fails to explain the emphasis on the number nineteen itself and it also fails to substantiate the speculation.

The following verses emphasized the crucial function of number nineteen:

74:32	No, by the moon.
74:33	By the night when it passes.
74:34	By the morning when it shines.
74:35	It is one of the great ones.
74:36	A warning to people.
74:37	For any among you who wishes to progress or regress.

The purpose of "oath" statements in the Quran is not similar to their common usage. The Quran uses oath to pull our attention to a divine sign or a lesson in the subject matter. The Quran does not use the statements of oath to make us believe, but to make us think (see: 89:5). The passing of the night and the shining of the morning are obviously allegories used to indicate an intellectual enlightenment or salvation. But the expression "by the Moon" is literal and it pulls our attention to the relationship of the Moon and the number nineteen. The year Apollo 11 astronauts dug Moon's surface and brought a piece of the Moon to the Earth, the same year, a biochemist named Dr. Rashad Khalifa started feeding the Quran into a computer in St. Louis, which would end up in the discovery of Code 19. This might be considered a mere coincidence, but a Quranic verse implies the correlation between the two events.

54:1	The moment drew near, and the moon was split.
54:2	If they see a sign, they turn away and say, "Continuous magic!"
54:3	They rejected, and followed their desires, and every old tradition.
54:4	While the news had come to them in which there was sufficient warning.
54:5	A perfect wisdom; but the warnings are of no benefit.

18

Let's continue the reading of Chapter 74 (The Hidden):

74:38   Every person is held by what it earned;

74:39   Except for the people of the right.

74:40   In paradises, they will be asking

74:41   About the criminals.

74:42   "What has caused you to be in *Saqar*?"

74:43   They said, "We were not of those who offered support (or observed contact prayer)."

74:44   "We did not feed the poor."

74:45   "We used to participate with those who spoke falsehood."

74:46   "We used to deny the day of Judgment."

74:47   "Until the certainty came to us."

74:48   Thus, no intercession of intercessors could help them.

74:49   Why did they turn away from this reminder?

How can someone who does not believe in the day of judgment believe in intercession? Those who believe in the intercession of saints and prophets on their behalf surely believe or at least claim to believe in the Day of Judgment. Nevertheless, according to the Quranic definition of the "Day of Judgment" a person cannot simultaneously believe in both intercession and the Day of Judgment. The Quran defines the "Day of Judgment" as "the day that no person can help another person, and all decisions, on that day, will belong to God." (82:19). That day neither Muhammad nor Ahmad, neither Jesus nor Mary, neither Ali nor Wali can help those who reject the number nineteen, its role and implication in the Quran.

Simple to Understand, Impossible To Imitate

The mathematical structure of the Quran, or The Final Testament, is simple to understand, yet impossible to imitate. You do not need to know Arabic, the original language of the Quran to examine it for yourself. Basically, what you need is to be able to count until 19. It is a challenge for atheists, an invitation for agnostics and guidance for believers. It is a perpetual miracle for the computer generation. Dr. Rashad Khalifa introduces this supernatural message as follows:

"The Quran is characterized by a unique phenomenon never found in any human authored book. Every element of the Quran is mathematically composed-the chapters, the verses, the words, the number of certain letters, the number of words from the same root, the number and variety of divine names, the unique spelling of certain words, and many other elements of the Quran besides its content. There are two major facets of the Quran's mathematical system: (1) The mathematical literary composition, and (2) The mathematical structure involving the numbers of chapters and verses. Because of this comprehensive mathematical coding, the slightest distortion of the Quran's text or physical arrangement is immediately exposed".

Physical, Verifiable and Falsifiable Evidence

Here are some examples of this historical message:

- The first verse, i.e., the opening statement "*BiSM ALLaH AL-RaĤMaN AL-RaĤYM*", shortly "Bismillah," consists of 19 Arabic letters.
- The first word of Bismillah, Ism (name), without contraction, occurs in the Quran 19 times.
- The second word of Bismillah, Allah (God) occurs 2698 times, or 19x142.
- The third word of Bismillah, RaĤMaN (Gracious) occurs 57 times, or 19x3.
- The fourth word of Bismillah, RaĤYM (Compassionate) occurs 114 times, or 19x6.
- The multiplication factors of the words of the Bismillah (1+142+3+6) add up to 152 or 19x8.
- The Quran consists of 114 chapters, which is 19x6.
- The total number of verses in the Quran including all unnumbered Bismillahs is 6346, or 19x334. If you add the digits of that number, 6+3+4+6 equals 19.
- The Bismillah occurs 114 times, (despite its conspicuous absence from chapter 9, it occurs twice in chapter 27) and 114 is 19x6.
- From the missing Bismillah of chapter 9 to the extra Bismillah of chapter 27, there are precisely 19 chapters.
- The occurrence of the extra Bismillah is in 27:30. The number of the chapter and the verse add up to 57, or 19x3.
- Each letter of the Arabic alphabet corresponds to a number according to their original sequence in the alphabet. The Arabs were using this system for calculations. When the Quran was revealed 14 centuries ago, the numbers known today did not exist. A universal system was used where the letters of the Arabic, Hebrew, Aramaic, and Greek alphabets were used as numerals. The number assigned to each letter is its "Gematrical Value." The numerical values of the Arabic alphabet are shown below: [the table is omitted]
- There are exactly 114 (19x6) verses containing all these 14 letters.
- A study on the gematrical values of about 120 attributes of God which are mentioned in the Quran, shows that only four attributes have

20

gematrical values which are multiples of 19. These are "Wahid" (One), "Zul Fadl al Azim" (Possessor of Infinite Grace), "Majid" (Glorous), "Jaami" (Summoner). Their gematrical value are 19 , 2698, 57, and 114 respectively, which are all divisible by 19 and correspond exactly to the frequencies of occurrence of the Bismillah's four words.

- The total numbers of verses where the word "Allah" (God) occurs, add up to 118123, and is 19x6217.
- The total occurrences of the word Allah (God) in all the verses whose numbers are multiples of 19 is 133, or 19x7.
- The key commandment: "You shall devote your worship to God alone" (in Arabic "Wahdahu") occurs in 7:70; 39:45; 40:12,84; and 60:4. The total of these numbers adds up to 361, or 19x19.
- The Quran is characterized by a unique phenomenon that is not found in any other book: 29 chapters are prefixed with "Quranic Initials" which remained mysterious for 1406 lunar years. With the discovery of the code 19, we realized their major role in the Quran's mathematical structure. The initials occur in their respective chapters in multiples of 19. For example, Chapter 19 has five letters/numbers in its beginning, K20H8Y10A'70Ŝ90, and the total occurrence of these letters in this chapter is 798, or 19x42.
- For instance, seven chapters of the Quran starts with two letter/number combinations, Ĥ8M40, and the total occurrence of these letters in those chapters is 2347 (19x113). The details of the numerical patterns among the frequency of these two letters in the seven chapters they initialize follows a precise mathematical formula.
- To witness the details of the miracle of these initials, a short chapter which begins with one initial letter/number, Q100, will be a good example. The frequency of "Q" in chapter 50 is 57, or 19x3. The letter "Q" occurs in the other Q-initialed chapter, i.e., chapter 42, exactly the same number of times, 57. The total occurrence of the letter "Q" in the two Q-initialed chapters is 114, which equals the number of chapters in the Quran. The description of the Quran as "Majid" (Glorious) is correlated with the frequency of occurrence of the letter "Q" in each of the Q-initialed chapters. The word "Majid" has a gematrical value of 57. Chapter 42 consists of 53 verses, and 42+53 is 95, or 19x5. Chapter 50 consists of 45 verses, and 50+45 is 95, or 19x5.
- The Quran mentions 30 different cardinal numbers: 1, 2, 3, 4, 5, 6, 7, 8, 9, 10, 11, 12, 19, 20, 30, 40, 50, 60, 70, 80, 99, 100, 200, 300, 1000, 2000, 3000, 5000, 50000, & 100000. The sum of these numbers is

162146, which equals 19x8534. Interestingly, nineteen is mentioned the 30th verse of chapter 74 and the number 30 is 19th composite number.

- In addition to 30 cardinal numbers, the Quran contains 8 fractions: 1/10, 1/8, 1/6, 1/5, 1/4, 1/3, 1/2, 2/3. Thus, the Quran contains 38 (19x2) different numbers. The total of fractions is approximately 2.
- If we write down the number of each verse in the Quran, one next to the other, preceded by the number of verses in each chapter, the resulting long number consists of 12692 digits (19x668). Additionally, the huge number itself is also a multiple of 19.

Code 19: The Real Bible Code

It is significant that the same 19-based mathematical composition was discovered by Judah ben Samuel in the 12th century AD in a preserved part of the Old Testament. Below is a quote from Studies in Jewish Mysticism.

> "The people (Jews) in France made it a custom to add (in the morning prayer) the words: " 'Ashrei temimei derekh (blessed are those who walk the righteous way)," and our Rabbi, the Pious, of blessed memory, wrote that they were completely and utterly wrong. It is all gross falsehood, because there are only nineteen times that the Holy Name is mentioned (in that portion of the morning prayer), . . . and similarly you find the word Elohim nineteen times in the pericope of Ve-'elleh shemot . . . .

> "Similarly, you find that Israel were called "sons" nineteen times, and there are many other examples. All these sets of nineteen are intricately intertwined, and they contain many secrets and esoteric meanings, which are contained in more than eight volumes. Therefore, anyone who has the fear of God in him will not listen to the words of the Frenchmen who add the verse " 'Ashrei temimei derekh (blessed are those who walk in the paths of God's Torah, for according to their additions the Holy Name is mentioned twenty times . . . and this is a great mistake. Furthermore, in this section there are 152 words, but if you add " 'Ashrei temimei derekh" there are 158 words. This is nonsense, for it is a great and hidden secret why there should be 152 words . . ."

Running like Zebras

The last section of Chapter 74 (The Hidden) likens those who turn away from the message of nineteen to zebras running away from a lion.

74:49	Why did they turn away from this reminder?
74:50	Like fleeing zebras,
74:51	Running from the lion?
74:52	Alas, every one of them wants to be given separate manuscripts.
74:53	No, they do not fear the Hereafter.
74:54	No, it is a reminder.
74:55	Whosoever wishes will take heed.
74:56	None will take heed except if God wills. He is the source of righteousness and the source of forgiveness.

Numerous books and articles rejecting the importance of the number nineteen in the Quran have been published in many languages worldwide. Some of the publications were freely distributed by the support of petrol-rich countries, such as Kuwait and Saudi Arabia.

The Quran is the only miracle given to Muhammad (29:51). Muhammad's *mushrik* companions could not comprehend that a book could be a miracle, and they wanted miracles "similar" to the ones given to the previous prophets (11:12; 17:90-95; 25:7,8; 37:7-8). Modern *mushriks* also demonstrated a similar reaction when God unveiled the prophesied miracle in 1974. When the miracle demanded from them the dedication of the system to God alone, and the rejection of all other "holy" teachings they have associated with the Quran, they objected, "How can there be mathematics in the Quran; the Quran is not a book of mathematics" or, "How can there be such a miracle; no previous messenger came up with such a miracle!" When a monotheist who was selected to fulfill the prophecy and discover the code started inviting his people to give up polytheism and the worship of Muhammad and clerics, he was officially declared an apostate by Sunni scholars gathered in Saudi Arabia from 38 different countries in February 1989. Within less than a year he was assassinated by a group linked to al-Qaida in early 1990 in Tucson, Arizona. See 3:81; 40:28-38; 72:24-28; 74:1-56. For the prophetic use of the word reminder (ZKR), see 15:9; 21:2-3; 21:24,105; 26:5; 29:51; 38:1,8; 41:41; 44:13; 72:17; 74:31,49,54.

Was the Discovery of the Code 19 a Coincidence?

Dr. Rashad Khalifa did not have any knowledge that his curiosity regarding the meaning of the alphabet letters that initialize 29 chapters of the Quran would end up with the discovery of its mathematical system. His computerized study that started in 1969 gave its fruits in 1974 by the discovery of the 14 century old SECRET.

If the Code 19 was going to provide strong evidence for the existence of God and for the authenticity of the Quran, then it is reasonable to expect that the identity of the discoverer and the time of the discovery would not be coincidental. Indeed, the events have demonstrated a prophetic design in the timing of this miraculous mathematical design.

The number 19 is mentioned only in a chapter known "The Hidden," the 74th chapter of the Quran. Juxtaposing these two numbers yields 1974, exactly the year in which the code was deciphered. (Calendar based on the birth of Jesus and the solar year is accepted by the Quran as units of calculating time. See, 19:33; 43:61; 18:24. Besides, this is the most commonly used calendar in the world.) If we multiply these two numbers, 19x74, we end up with 1406, the exact number of lunar years between the revelation of the Quran and the discovery of the code.

Furthermore, the first statement expressed in the first two verses of Chapter 74 is about the unveiling of the secret. It is interesting that if we consider one version of spelling the first word, which contains three Alifs instead of two, there are 19 letters in the first statement of chapter 74. More interestingly, when we add the numerical values of each letter in these two verses the sum is a very familiar number. Here is the value of each letter in "Ya ayyuhal Muddassir; qum fa anzir" (O you Hidden one, stand and warn):

Y =	10
A =	1
A =	1
Y =	10
H =	5
A =	1
A =	1
L =	30
M =	40
D =	4
TH =	500
R =	200
Q =	100
M =	40
F =	80
A =	1
N =	50
Z =	700

R =   200

Total: 1974

And 1974 is the year when the hidden secret came out and warned us!

Adding to this prophetic mathematical design is the fact that the derivatives of the name of the discoverer, RShD (guidance), occurs in the Quran exactly 19 times. (See 2:186; 2:256; 4:6; 7:146; 11:78; 11:87; 11:97; 18:10; 18:17; 18:24; 18:66; 21:51; 40:29; 40:38; 49:7; 72:2; 72:10; 72:14; 72:21.). The exact form, RaSHAD, occurs twice and they sandwich the claim of unappreciative people who wish to end the messengership (4:28-38).

In sum, the relationship between the following seven elements is more than interesting:

- The mathematical code (19).
- The number of the chapter mentioning the code (74).
- The year of the discovery of the code (1974).
- The number of lunar years between the revelation of the Quran and the year of the discovery (19x74).
- The numerical value of the 19 letters comprising the first statement of the chapter 74 (1974).
- The frequency of derivatives of the discoverer's name (19).
- The context of the verses where the exact name of the discoverer is mentioned as an adjective (40:28-38).

In January 31, 1990, Rashad Khalifa was assassinated in Tucson, Arizona, by a terrorist group affiliated with al-Qaida. Ironically, soon after his departure, ignorant people started idolizing him and created a cult distorting his message of strict monotheism. Also See 27:82-85; 40:28-38; 72:19-28.

27:82   When the punishment has been deserved by them, We will bring out for them a creature made of earthly material, it will speak to them that the people have been unaware regarding Our signs.

27:83   The day We gather from every nation a party that denied Our signs, then they will be driven.

27:84   Until they have come, He will say, "Have you denied My signs while you had no explicit knowledge of them? What were you doing?"

27:85   The punishment was deserved by them for what they transgressed, for they did not speak.

How Can We Explain This Phenomenon?

There are basically four possible explanations:

1. Manipulation: One may be skeptical about our data regarding the mathematical structure of the Quran. However, one can eliminate this option by spending several hours of checking the data at random. (We recommend Quran: *The Visual Presentation of Miracle* by Rashad Khalifa and *Nineteen: God's Signature in Nature and Scripture* by Edip Yuksel). Sunni and Shiite scholars and clerics who have traded the Quran with primitive mediaeval fabrications, that is, Hadith and Sunna, strongly reject this mathematical system, since the mathematical system exposes the corruption of religions by clergymen.

2. Coincidence: This possibility is eliminated by the statistical probability laws. The consistency and frequency of the 19-based pattern is much too overwhelming to occur coincidentally.

3. Human fabrication: While fabricating a literary work that meets the criteria of the document summarized here is a stunning challenge for our computer generation, it is certainly even more improbable during the time of initiation of the document, namely, 610 AD. One more fact augments the improbability of human fabrication. If a certain person or persons had fabricated this literary work, they would want to reap the fruits of their efforts; they would have shown it to people to prove their cause. In view of the originality, complexity, and mathematical sophistication of this work, one has to admit that it is ingenious. However, no one has ever claimed credit for this unique literary code; the code was never known prior to the computer decoding accomplished by Dr. Khalifa. Therefore, it is reasonable to exclude the possibility of human fabrication.

The timing of the discovery may be considered another evidence for the existence and full control of the Supreme Being: The mystery of the number 19 which is mentioned as "one of the greatest events" in the chapter 74 (The Hidden Secret) was discovered by Dr. Khalifa in 1974, exactly 1406 (19x74) lunar years after the revelation of the Quran. The connection between 19 (the code) and 74 (the number of the chapter which this code is mentioned) is significant in the timing of the discovery.

4. Super Intelligent Source: The only remaining reasonable possibility is that a super intelligent source is responsible for this document; one who designed the work in this extraordinary manner, then managed to keep it a well guarded secret for 14 centuries, for a predetermined time. The mathematical code ensures that the source is super intelligent and also that the document is perfectly intact.

CORNELL UNIVERSITY

Center for Radiophysics and Space Research

SPACE SCIENCES BUILDING
Ithaca, New York 14853-6801

Telephone (607) 255-4971
Fax (607) 255-9888

Laboratory for Planetary Studies

January 11, 1994

Mr. Edip Yuksel
742 W. Wheatridge Drive
Tucson, Arizona 85704

Dear Mr. Yuksel:

Thanks for your recent letter. In matters of this sort, it is important to distinguish between *a priori* and *a posteriori* statistics; and also to remember that there were mathematicians in Muhammad's time and long before. As far as I could follow your argument, it does seem to me very close to numerology. Enclosed is an article I wrote on another bright person who thought he could find a mathematical proof of the existence of God in certain publications. Please tell me what you think.

With best wishes,

Cordially,

Carl Sagan

CS:lkp
Enclosure

CHAPTER 8

NORMAN BLOOM,
MESSENGER OF GOD

***

[The French encyclopedist] Diderot paid a visit to the Russian Court at the invitation of the Empress. He conversed very freely, and gave the younger members of the Court circle a good deal of lively atheism. The Empress was much amused, but some of her councillors suggested that it might be desirable to check these expositions of doctrine. The Empress did not like to put a direct muzzle on her guest's tongue, so the following plot was contrived. Diderot was informed that a learned mathematician was in possession of an algebraical demonstration of the existence of God, and would give it him before all the Court, if he desired to hear it. Diderot gladly consented though the name of the mathematician is not given, it was Euler. He advanced towards Diderot, and said gravely, and in a tone of perfect conviction: 'Monsieur, $(a + bn)/n = x$, donc Dieu existe; répondez' [Sir, $(a + bn)/n = x$. Therefore God exists; reply!] Diderot, to whom algebra was Hebrew, was embarrassed and disconcerted, while peals of laughter arose on all sides. He asked permission to return to France at once, which was granted.

AUGUSTUS DE MORGAN
A Budget of Paradoxes (1672)

THROUGHOUT human history there have been attempts to contrive rational arguments to convince skeptics of the existence of a God or gods. But most theologians have held that the ultimate reality of divine beings is a matter for faith alone and is inaccessible to rational endeavor. St. Anselm argued that since we can imagine a perfect being, he must exist—because he would not be perfect without the added perfection of existence. This so-called ontological argument was more or less promptly attacked on two grounds: (1) can we imagine a completely perfect being? (2) Is it obvious that perfection is augmented by existence? To the modern ear such pious arguments seem to be about words and definitions rather than about external reality.

More familiar is the argument from design, an approach that penetrates deeply into issues of fundamental scientific concern. This argument was admirably summarized by David Hume: "Look round the world contemplate the whole and every part of it; you will find it to be nothing but one great machine, subdivided into an infinite number of lesser machines. . . . All these various machines, even their most minute parts, are adjusted to each other with an accuracy which ravishes into admiration all men who have ever contemplated them. The curious adapting of means to ends, throughout all nature, resembles exactly, though it much exceeds, the production of human contrivance; of human design, thought, wisdom, and intelligence. Since therefore the effects resemble each other, we are led to infer, by all the rules of analogy, that the causes also resemble; and that the Author of Nature is somewhat similar to the mind of man; though possessed of much larger faculties proportioned to the grandure of the work which he has executed."

Hume then goes on to subject this argument, as did Immanuel Kant after him, to a devastating and compelling attack, notwithstanding which the argument from design continued to be immensely popular—as, for example, in the works of William Paley— through the early nineteenth century. A typical passage by Paley goes: "There cannot be a design without a designer; contrivance without a contrivor; order without choice; arrangement without anything capable of arranging; subserviency and relation to a purpose, without that which could intend a purpose; means suitable to an end, and executing their office and accomplishing that end, without the end ever having been contemplated, or the means accommodated to it. Arrangement, disposition of parts, subserviency of means to an end, relation of instruments to a use; imply the presence of intelligence and mind."

It was not until the development of modern science, but most particularly the brilliant formulation of the theory of evolution by natural selection, put forth by Charles Darwin and Alfred Russel Wallace in 1859 that these apparently plausible arguments were fatally undermined.

There can, of course, be no disproof of the existence of God— particularly a sufficiently subtle God. But it is a kindness neither to science nor religion to leave unchallenged inadequate arguments for the existence of God. Moreover, debates on such questions are good fun, and at the very least, hone the mind for useful work. Not much of this sort of disputation is in evidence today, perhaps because new arguments for the existence of God which can be understood at all are exceedingly rare. One recent and modern version of the argument from design was kindly sent to me by its author, perhaps to secure constructive criticism.

NORMAN BLOOM is a contemporary American who incidentally believes himself to be the Second Coming of Jesus Christ. Bloom observes in Scripture and everyday life numerical coincidences which anyone else would consider meaningless. But there are so many such coincidences that, Bloom believes, they can be due only to an unseen intelligence, and the fact that no one else seems to be able to find or appreciate such coincidences convinces Bloom that he has been chosen to reveal God's presence. Bloom has been a fixture at some scientific meetings where he harangues the hurrying, preoccupied crowds moving from session to session. Typical Bloom rhetoric is "And though you reject me, and scorn me, and deny me, YET ALL WILL BE BROUGHT ONLY BY ME. My will will be, because I have formed you out of the nothingness. You are the Creation of My Hands. And I will complete My Creation and Complete My Purpose that I have Purposed from of old. I AM THAT I AM. I AM THE LORD THY GOD IN TRUTH." He is nothing if not modest, and the capitalization conventions are entirely his.

Bloom has issued a fascinating pamphlet, which states: "The complete faculty of Princeton University (including its officers and its deans and the chairmen of the departments listed here) has agreed that it cannot refute, nor show in basic error the proof brought to it, in the book, The New World dated Sept. 1974. This faculty acknowledges as of June 1 1975 that it accepts as a proven truth THE IRREFUTABLE PROOF THAT AN ETERNAL MIND AND HAND HAS SHAPED AND CONTROLLED THE HISTORY OF THE WORLD THROUGH THOUSANDS OF YEARS." A closer reading shows that despite Bloom's distributing his proofs to over a thousand faculty members of Princeton University, and despite his offer of a $1,000 prize for the first individual to refute his proof, there was no response whatever. After six months he concluded that since Princeton did not answer, Princeton believed. Considering the ways of university faculty members, an alternative explanation has occurred to me. In any case, I do not think that the absence of a reply constitutes irrefutable support for Bloom's arguments.

Princeton has apparently not been alone in treating Bloom inhospitably: "Yes, times almost without number, I have been chased by police for bringing you the gift of my writing . . . Is it not so that professors at a university are supposed to have the maturity and judgment and wisdom to be able to read a writing and determine for themselves the value of its contents? Is it that they require THOUGHT CONTROL POLICE to tell them what they should or should not read or think about? But, even at the astronomy department of Harvard University, I have been chased by police for the crime of distributing that New World Lecture, an irrefutable proof that the earth-moon-sun system is shaped by a controlling mind and hand. Yes, and THREATENED WITH IMPRISONMENT, IF I DARE BESMIRCH THE HARVARD CAMPUS WITH MY PRESENCE ONCE MORE. . . . AND THIS IS THE UNIVERSITY THAT HAS UPON ITS SHIELD THE WORD VERITAS: VERITAS: VERITAS:—Truth, Truth, Truth. Ah, what hypocrites and mockers you are!

The supposed proofs are many and diverse, all involving numerical coincidences which Bloom believes could not be due to chance. Both in style and content, the arguments are reminiscent of Talmudic textual commentary and cabalistic lore of the Jewish Middle Ages: for example, the angular size of the Moon or the Sun as seen from the Earth is half a degree. This is just 1/720 of the circle (360°) of the sky. But $720 = 6! = 6 \times 5 \times 4 \times 3 \times 2 \times 1$. Therefore, God exists. It is an improvement on Euler's proof to Diderot, but the approach is familiar and infiltrates the entire history of religion. In 1658 Gaspar Schott, a Jesuit priest, announced in his Magia Universalis Naturae et Artis that the number of

31

degrees of grace of the Virgin Mary is 2256 = 228 H 1.2 X 1077 (which, by the by, is very roughly the number of elementary particles in the universe).

Another Bloomian argument is described as "irrefutable proof that the God of Scripture is he who has shaped and controlled the history of the world through thousands of years." The argument is this: according to Chapters 5 and 11 of Genesis, Abraham was born 1,948 years after Adam, at a time when Abraham's father, Terah, was seventy years old. But the Second Temple was destroyed by the Romans in A.D. 70, and the State of Israel was created in A.D. 1948 Q.E.D. It is hard to escape the impression that there may be a flaw in the argument somewhere. "Irrefutable" is, after all, a fairly strong word. But the argument is a refreshing diversion from St. Anselm.

Bloom's central argument, however, and the one that much of the rest is based upon, is the claimed astronomical coincidence that 235 new moons is, with spectacular accuracy, just as long as nineteen years. Whence: "Look, mankind, I say to you all, in essence you are living in a clock. The clock keeps perfect time, to an accuracy of one second/day! . . . How could such a clock in the heavens come to be without there being some being, who with perception and understanding, who, with a plan and with the power, could form that clock?"

A fair question. To pursue it we must realize that there are several different kinds of years and several different kinds of months in use in astronomy. The sidereal year is the period that the Earth takes to go once around the Sun with respect to the distant stars. It equals 365.2564 days. (The days we will use, as Norman Bloom does, are what astronomers call "mean solar days.") Then there is the tropical year. It is the period for the Earth to make one circuit about the Sun with respect to the seasons, and equals 365.242199 days. The tropical year is different from the sidereal year because of the precession of the equinoxes, the slow toplike movement of the Earth produced by the gravitational forces of the Sun and the Moon on its oblate shape. Finally, there is the so-called anomalistic year of 365.2596 days. It is the interval between two successive closest approaches of the Earth to the Sun, and is different from the sidereal year because of the slow movement of the Earth's elliptical orbit in its own plane, produced by gravitational tugs by the nearby planets.

Likewise, there are several different kinds of months. The word "month," of course, comes from "moon." The sidereal month is the time for the Moon to go once around the earth with respect to the distant stars and equals 27.32166 days. The synodic month, also called a lunation, is the time from new moon to new moon or full moon to full moon. It is 29.530588 days. The synodic month is different from the sidereal month because, in the course of one sidereal revolution of the Moon about the Earth, the Earth- Moon system has together revolved a little bit (about one-thirteenth) of the way around the Sun. Therefore the angle by which the Sun illuminates the Moon has changed from our terrestrial vantage point. Now, the plane of the Moon's orbit around the Earth intersects the plane of the Earth's orbit around the Sun at two places—opposite to each other— called the nodes of the Moon's orbit. A nodical or draconic month is the time for the Moon to move from one node back around again to the same node and equals 27.21220 days. These nodes move, completing one apparent circuit, in 18.6 years because of gravitational tugs, chiefly by the Sun. Finally, there is the anomalistic month of 27.55455 days, which is the time for he Moon to complete one circuit of the Earth with respect to the nearest point in its orbit. A little table on these various definitions of the year and the month is shown below.

KINDS OF YEARS AND MONTHS, EARTH-MOON SYSTEM

Years

Sidereal year	365.2564 mean solar days
Tropical year	365.242199 days
Anomalistic year	365.2596 days

Months

Sidereal month	27.3 2 1 66 days
Synodic month	29.530588 days
Nodical or draconic month	27.21220 days
Anomalistic month	27 55455 days

Now, Bloom's main proof of the existence of God depends upon choosing one of the sorts of years, multiplying it by 19 and then dividing by one of the sorts of months. Since the sidereal, tropical and anomalistic years are so close together in length, we get sensibly the same answer whichever one we choose. But the same is not true for the months. There are four different kinds of months, and each gives a different answer. If we ask how many synodic months there are in a sidereal year, we find the answer to be 235.00621 as advertised; and it is the closeness of this result to a whole number that Is the fundamental coincidence of Bloom's thesis. Bloom, of course, believes it to be no coincidence.

But if we were to ask instead how many sidereal months there are in nineteen sidereal years we would find the answer to be 254.00622; for nodical months, 255.02795; and for anomalistic months, 251.85937 251.85937. It is certainly true that the synodic month is the one most strikingly apparent to a naked-eye observer, but I nevertheless have the impression that one could construct equally elaborate theological speculations on 252, 254, or 255 as on 235

We must now ask where the number 19 comes from in this argument. Its only justification is David's lovely Nineteenth Psalm, which begins: "The heavens declare the glory of God, and the firmament sheweth his handiwork. Day unto day uttereth speech, and night unto night sheweth knowledge." This seems quite an appropriate quotation from which to find a hint of an astronomical proof for the existence of God. But the argument assumes what it intends to prove. The argument is also not unique. Consider, for example, the Eleventh Psalm, also written by David. In it we find the following words, which may equally well bear on this question: "The Lord is in his holy temple, the Lord's throne is in heaven: his eyes behold, his eyelids try, the children of men," which is followed in the following Psalm with "the children of men . . . speak vanity." Now, if we ask how many synodic months there are in eleven sidereal years (or 4017.8204 mean solar days), we find the answer to be 136.05623. Thus, just as there seems to be a connection between nineteen years and 235 new moons, there is a connection between eleven years and 136 new moons. Moreover, the famous British astronomer Sir Arthur Stanley Eddington believed that all of physics could be derived from the number 136 (I once suggested to Bloom that with the foregoing information and just a little intellectual fortitude it should be possible as well to reconstruct all of Bosnian history.)

One numerical coincidence of this sort, which is of deep significance, was well known to the Babylonians, contemporaries of the ancient Hebrews. It is called the Saros. It is the period between two successive similar cycles of eclipses. In a solar eclipse the Moon, which appears from the Earth just as large ($1/2°$) as the Sun, must pass in front of it. For a lunar eclipse, the Earth's shadow in space must intercept the Moon. For either kind of eclipse to occur, the Moon must, first of all, be either new or full—so that the Earth, the Moon and the Sun are in a straight line. Therefore the synodic month is obviously involved in the periodicity of eclipses. But for an eclipse to occur, the Moon must also be near one of the nodes of its orbit. Therefore the nodical month is involved. It turns out that 233 synodic months is equal to 241.9989 (or very close to 242) nodical months. This is the equivalent of a little over eighteen years and ten or eleven days (depending on the number of intervening leap days), and comprises the Saros. Coincidence?

Similar numerical coincidences are in fact common throughout the solar system. The ratio of spin period to orbital period on Mercury is 3 to 2. Venus manages to turn the same face to the Earth at its closest approach on each of its revolutions around the Sun. A particle in the gap between the two principal rings of Saturn, called the Cassini Division, would orbit Saturn in a period just half that of Mimas, its second satellite. Likewise, in the asteroid belt there are empty regions, known as the Kirkwood Gaps, which correspond to nonexistent asteroids with periods half that of Jupiter, one-third. two-fifths, three-fifths, and so on.

None of these numerical coincidences proves the existence of God—or if it does, the argument is subtle, because these effects are due to resonances. For example, an asteroid that strays into one of the Kirkwood Gaps experiences a periodic gravitational pumping by Jupiter. Every two times around the Sun for the asteroid, Jupiter makes exactly one circuit. There it is, tugging away at the same point in the asteroid's orbit every revolution. Soon the asteroid is persuaded to vacate the gap. Such incommensurable ratios of whole numbers are a general consequence of gravitational resonance in the solar system. It is a kind of perturbational natural selection. Given enough time—and time is what the solar system has a great deal of—such resonances will arise inevitably.

34

That the general result of planetary perturbations is stable resonances and not catastrophic collisions was first shown from Newtonian gravitational theory by Pierre Simon, Marquis de Laplace, who described the solar system as "a great pendulum of eternity, which beats ages as a pendulum beats seconds." Now, the elegance and simplicity of Newtonian gravitation might be used as an argument for the existence of God. We could imagine universes with other gravitational laws and much more chaotic planetary interactions. But in many of those universes we would not have evolved—precisely because of the chaos. Such gravitational resonances do not prove the existence of God, but if he does exist, they show, in the words of Einstein, that, while he may be subtle, he is not malicious.

BLOOM CONTINUES his work. He has, for example, demonstrated the preordination of the United States of America by the prominence of the number 13 in major league baseball scores on July 4, 1976. He has accepted my challenge and made an interesting attempt to derive some of Bosnian history from numerology—at least the assassination of Archduke Ferdinand at Sarajevo, the event that precipitated World War 1. One of his arguments involves the date on which Sir Arthur Stanley Eddington presented a talk on his mystical number 136 at Cornell University, where I teach. And he has even performed some numerical manipulations using my birth date to demonstrate that I also am part of the cosmic plan. These and similar cases convince me that Bloom can prove anything.

Norman Bloom is, in fact, a kind of genius. If enough independent phenomena are studied and correlations sought, some will of course be found. If we know only the coincidences and not the enormous effort and many unsuccessful trials that preceded their discovery, we might believe that an important finding has been made. Actually, it is only what statisticians call "the fallacy of the enumeration of favorable circumstances." But to kind as many coincidences as Norman Bloom has requires great skill and dedication. It is in a way a forlorn and perhaps even hopeless objective—to demonstrate the existence of God by numerical coincidences to an uninterested, to say nothing of a mathematically unenlightened public. It is easy to imagine the contributions Bloom's talents might have made in another field. But there is something a little glorious, I find, in his fierce dedication and very considerable arithmetic intuition. It is a combination of talents which is, one might almost say, God-given.

***

In the name of God, Gracious, Merciful

February 5, 1994

Dear Dr. Sagan,

Thank you very much for your letter. I hope we can continue this debate with better understanding of each other's position. As I have expected, you fell in the common trap and confused my argument with traditional numerology. I don't blame you, since you are too busy to study a new argument made by an ordinary person. Furthermore, all of us are, more or less, influenced by our preconceived ideas regardless of our age and education.

I see a big difference between Norman Bloom's argument and mine. As for Bloom's argument, I entirely agree with your criticism on it. I am familiar with many such claims. For instance, after the discovery of the code 19, some zealots started to play with numbers. Currently, I am trashing hundreds of those speculations made by enthusiastic "believers" who send their work for evaluation. Certainly, "the fallacy of the enumeration of favorable circumstances," and the "ignorance of mathematical properties and the law of probability" can mislead many naive people to confuse diamonds with pieces of glasses.

Enclosed are several short articles I wrote on paranormal phenomena, hoping that you will not classify me with those so-called "bright" persons.

Let me try to clarify the issue in several points:

1. I don't claim that the "miraculous mathematical structure" of the Quran "proves" the existence of God or the divine source of the book for sceptics. Rather, I claim that it provides a falsifiable strong evidence for it. The strength of this evidence entirely depends on the initial reaction, approach and objectivity of the audience. If God wanted to prove Himself, He could, by definition, have made everyone a believer instantly.

2. The purpose of the mathematical evidence is clearly mentioned in Chapter 74, verse 31 (74:31). The number 19 increases the faith of believers and removes their doubts regarding the authenticity of the Quran. Therefore, I don't expect from an atheist to become a deist by this evidence. Atheists, it seems to me, lack the required objectivity and the intellectual motivation to evaluate and appreciate the evidence. Can you see a 3-D holographic picture with naked eyes if you don't think the probability for its existence? In order to see them you need to follow the rules and concentrate on them and wait with patience until it appears. On my part, the number 19, as the code of the Quran's mathematical system, has intellectually convinced me, and related metaphysical experiences in my private life have spiritually satisfied me (Quran: 41:53).

3. The discovery of the mathematical structure of the Quran was made after a painstaking research without any pre-determined target. When Dr. Khalifa first published the computer data in 1973 he was not aware of the code, that is, the common denominator 19. He had found some interesting relations and correlation among the numbers of the frequency of certain letters. (We have national newspapers, magazines and books to witness this fact.) However, in the beginning of 1974 he discovered that most of those numbers were the multiple of 19 and then he realized the connection of Chapter 74 (Al-Muddathir, The Hidden One) with his discovery. So far, this was based on a priori statistics.

After the discovery of the code, quite a few people, including me, studied the claim and made further discoveries. However, this empirical study, later lead us to make few changes and modifications. For instance, we all came to the conclusion that the last two verses of Chapter 9 (The Ultimatum) are not from the original Quran. Therefore, both a priori statistics and a posteriori statistics have involved in our study. I should acknowledge that this study is not free of the philosophical problem, i.e., the theory-dependence of observation that stains every scientific research. However, the mathematical system of the Quran is in a unique way precise and objective that can lead independent researchers to the same conclusion. Besides, we have very interesting experiences regarding this research. If you want I can give you some examples.

4. I am aware of the history of mathematics and mathematicians. (My prime hobby is to read books on mathematics, solve puzzles and entertain with paradoxes.) If you could only spend an hour to study the examples of our argument you probably would not have reminded me of mathematicians in Muhammad's time.

This is because the mathematical structure of the Quran involves both mathematics and Arabic. The language of Mathematics is universal and deductive; however, the language of Arabic is special and arbitrary. The Quran integrates these two different languages in a very clear and extensive fashion. I am convinced that no human mathematician and men of literary can ever manipulate language to such extend without sacrificing the meaning. To reach this conviction we do not need to know the Arabic. Just get rid of our prejudices and start evaluating the argument itself.

5. It is very difficult, if not impossible, to "calculate" the mathematical probability of this kind of phenomena. But you may not need to calculate the statistical probability of it to reach a judgment. Your intuition, most likely, will find it silly to waste time on such a calculation. You will see the evidence so clear and overwhelming that you will become certain without mechanical calculations of probability.

After this reminder let's look at the argument in two steps:

A. Does the Quran has a mathematical structure or not?

We can discuss this issue. You may claim that our examples are all within the limit of coincidence. Then, I have to provide you with enough examples and reasons that should make you refrain from such a claim.

B. If we agree that these examples (in case that our counting is accurate) show a deliberate mathematical pattern in the Quran then we can continue and discuss the nature of it: whether is it human or divine.

If you want I can send a couple of books demonstrating the basic features of the argument. Nevertheless, here I will pick just a set of examples out of hundreds to discuss. Please notice that all the examples are coherent and integrated circuits of a "simple-to-understand-impossible-to-imitate" program.

The example is the first verse of the Quran, i.e., "Bismillahirrahmanirrahim," (In the Name of God, Most Gracious, Most Merciful) or shortly, "Basmalah." Please see how the code 19 is consistent in the structure of the verse and its relation with the whole body of the book. And please remember that the number 19 is explicitly mentioned in 74:30 as a response for disbeliever's claim that it is made up by Muhammad.

"On it is Nineteen"

- The first verse, i.e., the opening statement "Bismillahirrahmanirrahim," shortly "Basmalah," consists of 19 Arabic letters.
- The first word of Basmalah, Ism (name) occurs in the Quran 19 times.
- The second word of Basmalah, Allah (God) occurs 2698 times, or 19x142.
- The third word of Basmalah, Rahman (Gracious) occurs 57 times, or 19x3.

- The fourth word of Basmalah, Rahim (Merciful) occurs 114 times, 19x6.
- The multiplication factors of the words of the Basmalah mentioned above (1+142+3+6) add up to 152, or 19x8.
- The total number of verses where the word Allah (God) occurs add up to 118123, and is 19x6217.
- The Quran consists of 114 chapters, which is 19x6.
- The total number of verses in the Quran including all Basmalahs is 6346, or 19x334. If you add the digits of that number, 6+3+4+6 equals 19.
- The Basmalah opens all chapters in the Quran except chapter 9. Yet, despite its conspicuous absence from chapter 9, it occurs twice in chapter 27, so that its total occurrence is still 114, or 19x6.
- From the missing Basmalah of chapter 9 to the extra Basmalah of chapter 27, there are precisely 19 chapters.
- It follows that the sum of the chapter numbers from 9 to 27 (9+10+11+12...+26+27) is 342. This total (342) equals the number of words between the two Basmalahs of chapter 27, and 342 is 19x18.
- Each letter of Arabic alphabet corresponds to a number according to their original sequence in the alphabet. The Arabs used this system for calculations. When the Quran was revealed 14 centuries ago, the numbers as we know today did not exist. A universal system was used where the letters of the Arabic, Hebrew, Aramaic, and Greek alphabets were used as numerals. The number assigned to each letter is its "Gematrical Value."

A study on the gematrical values of more than 120 attributes of God that are mentioned in the Quran, shows that only four names have gematrical values that are multiples of 19. These are "Wahid" (One), "Zul Fadl al Azim" (Possessor of Infinite Grace), "Majid" (Glorious), "Jaami" (Summoner). Their gematrical values are 19 , 2698, 57, and 114 respectively, which are all divisible by 19 and correspond exactly to the frequencies of occurrence of the Basmalah's four words.

If you have confined your judgment with a preconceived assumption that either these are "interesting coincidences," or "the work of a medieval mathematician" then there will be no point to carry our discussion further. By this way, I believe, scientists can handicap themselves from hearing the message of the Supreme Creator of the Universe.

Sincerely, Edip Yuksel

PS: If you are interested, I can send you a list of Quranic verses about the creation and astronomy. You may appreciate them more than anyone else. Remember that the Quran may be the only book in the world where the word "Shahr (month)" is repeated 12 times and the word "Yawm (day)" 365 times.

CORNELL UNIVERSITY

Center for Radiophysics and Space Research

SPACE SCIENCES BUILDING
Ithaca, New York 14853-6801

Telephone (607) 255-4971
Fax (607) 255-9888

Laboratory for Planetary Studies

February 17, 1994

Mr. Edip Yuksel
742 West Wheatridge Drive
Tucson, Arizona 85704

Dear Mr. Yuksel:

Thanks for your recent letter, but I don't think you have understood the distinction I was making between *a priori* and *a posteriori* judgements. Let's take the example you've sent me. You have chosen the first verse of the Qur'an to do your numerology, but of course that is an arbitrary decision. You could, for example, have chosen the 19th *sura*, or the last one, or one of particular elegance or felicity or fame. You could have looked for any "code" index you wished and then tried to find it repeated throughout the Qur'an. I see many signs of arbitrary decisions in what you consider evidence; for example, why do you count from the missing "basmalah" to the extra "basmalah", instead of from Chapter 1 to Chapter 9 or from Chapter 27 to the end? With a sufficient computer data base and absolutely no *a priori* constraints a clever person should be able to find many regularities. If in addition the author and transcribers of the Qur'an consciously inserted a little numerology, I believe that the coincidences you describe can readily be understood. Of course I might be wrong.

With best wishes,

Cordially,

Carl Sagan

CS:lkp

40

29 March 1994

Dear Dr. Sagan,

The conclusion of your concise letter was a brilliant skeptical hypothesis: "With a sufficient computer data base and absolutely no a priori constraints a clever person should be able to find many regularities. If in addition the author and transcribers of the Qur'an consciously inserted a little numerology, I believe that the coincidences you describe can readily be understood." Nevertheless, you, as a cautious scientist, did not forget to add "Of course I might be wrong."

Unfortunately or fortunately, I can't reciprocate your humble statement by saying "of course I too might be wrong!" On the contrary, I am obliged to say "of course, Dr. Sagan, you are wrong." I am certain that the mathematical structure of the Quran is the work of the Supreme Mathematician, as you are certain that the planet earth is round and rotates around the sun. I hope you won't get offended if I remind you of the possibility of having some "a priori constraints" that may force you not to see the inimitable signature of the Most Wise.

Before starting my counter argument, I want to tell you that I am impressed by your meticulousness in spelling the word "Qura'n" (The Book of Recitation) with its accurate transliteration. For convenience I prefer to write "Quran." As for my misunderstanding of "a priori, and a posteriori statistics" I can blame the different implications of this terminology in different fields.

Arbitrary Decisions

In order to provide a concrete base for our argument, in my previous letter, I had picked some mathematical patterns related to "basmalah," the opening statement of the Quran. You claim that it is arbitrary to focus on "basmalah".

I disagree, since I think "basmalah" is the most appropriate candidate among all the options. If you ask people who are familiar with the Quran to choose an idiosyncratic or representative verse, I am sure that a great majority of them will pick the "basmalah" without hesitation. It is not only the first verse of the Quran, it is also the opening statement of every chapter except Chapter 9. Furthermore, it is the most repeated verse in the Quran and the most popular verse among

Muslims. We say "Bismi Allahi Rahmani Rahim" (In the name of God, Most Gracious, Most Merciful) before eating and drinking, or before starting our cars, etc. It is virtually a sacred password.

For the sake of argument, let's assume that "basmalah" is not special and we picked it arbitrarily. The probability of finding a verse in the Quran that exhibits ONLY six features of "basmalah" is very low. If you consider the astounding interlocking relation between the number of the occurrences of the words in "basmalah" and the gematrical values of the names of God, then the probability will diminish dramatically. Please note that the relation between "basmalah" and God's names is not arbitrary but a necessity coming from the very meaning of "basmalah." Besides, we have discovered many more mathematical facts just about "basmalah" which eventually reduce the probability to zero. Therefore, neither "arbitrary decisions" nor "a little numerology" nor "coincidences," nor the combination of the three can be candidates for a plausible explanation.

As another example of "arbitrary decisions" you mention the relation between the missing "basmalah" and the extra "basmalah." You ask: "why do you count from the missing 'basmalah' to the extra 'basmalah', instead of from Chapter 1 to Chapter 9 or from Chapter 27 to the end? "

I have at least four answers for this objection:

1. Just look at two adjectives of the "basmalah" you mention in your question: MISSING and EXTRA. Isn't it more logical to look for a relation between the MISSING and EXTRA, instead of the FIRST and MISSING, or the EXTRA and LAST?

2. The number of chapters from Chapter 1 to Chapter 9 AND from Chapter 27 to the end is also a multiple of 19: 95 (19x5), since the number of all Chapters are multiple of 19.

3. The author of the Quran obviously has willed to hide the implication of the code 19 in Chapter 74 (The Hidden Secret) until 1974. It is a message reserved for the computer generation. If the missing "basmalah" or the extra "basmalah" had been in Chapter 19, as you suggest, then the code of the Quran could be easily discovered prematurely, since many previous Muslim scholars were aware of the fact that "basmalah" consists of 19 letters. The relation of "basmalah" with another 19 could easily lead them to search for its mathematical function in the Quran. I believe that there are many reasons behind

42

the timing of this discovery. If God Almighty did not want the corrupt Muslim clergymen to discover and abuse this miraculous phenomena, then it is understandable to see why the simple facts of Quran's mathematical system are hidden from oblivious eyes.

4. Besides, if there was a conscious effort to insert a little numerology in the Quran, as you suspect, then satisfying your suggestions would be very easy. Arranging the Chapter with missing "basmalah" as the 19th chapter and the Chapter with the extra one as the 37th chapter would eliminate at least one objection of skeptics. Why should Muhammad ignore this very simple arrangement while wasting his time and energy on arranging more complicated ones? If you claim that he was not smart enough to do this simple task, you will loose your main argument regarding more clever and complex patterns; because then you cannot say that Muhammad was a clever mathematician etc.

From your criticism it seems you are not sure whether the claimed mathematical pattern is intentional or not. You seem to want to have it both ways. You have a two-sided judgment ready in your disposal: either it is ENTIRELY coincidence, OR it is semi-conscious and semi-coincidence. This is a very sure way of discarding anything you disagree with. You can jump between these two preconceived judgments whenever you want. For instance, you can refute my answer above in number 4, by claiming that the mathematical structure of the Quran is entirely coincidental. On the other hand, when the word "coincidence" is too improbable to reject my argument, you can seek refuge in the other side of the label by claiming that it is a little deliberate numerology. I think, for a healthier argument you should clarify your position. Refuting a thesis by oscillating between a contradictory disjunction indicates prejudice.

Is Every Scientific Experiment Arbitrary?

Narrow Inductive Model of Scientific Investigation requires the following:

> 1. Observe and record all facts.
> 2. Classify and analyze without prior hypothesis.
> 3. Induce generalizations from observed facts.

But scientists never follow this utopic principles when they conduct their experiments. They do not observe and record all facts. How can they? Their previous observations, reflections and expectations determine the relevant facts and experiments.

You must be familiar with Newton's "Experimentation Cruces" on the nature of light and colors. A skeptic could object to his findings by claiming that he started his experiment with an a priori hypothesis and with arbitrary decisions: "Why didn't Newton measure the temperature of the room? Why didn't he consider the distance of the source of light? Why did he ignore the role of the type of the mirror? etc...." As you know, Newton had a better intuition than any of his contemporary scientists about which phenomena were relevant and which were not in searching for the nature of light.

The point is, non-relevant phenomena may seem relevant for an outsider, but a scientist familiar with a particular subject will have clear or vague reasons to decide on the relevancy of observations. We can have a hypothesis as long as they guide our observation but not determine the result of our observation. Why don't you grant the following self-correcting scientific method for the study of the mathematical system of the Quran?:

Simple Hypothetico-Deductive Model:

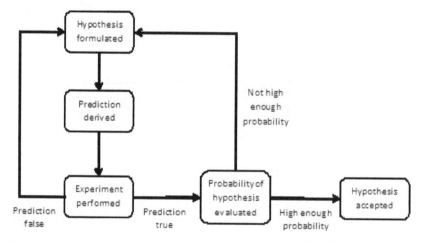

Fir Cone Argument

You know that each term of Fibonacci sequence which runs: 1, 1, 2, 3, 5, 8, 13, 21, 34 . . . , is the sum of the two preceding terms. This numerical sequence appears in nature and plants, such as reproduction of rabbits, fir cones and petals of certain flowers.

44

Let's take fir cones. Observers claim that the pattern of scales in fir cones follows the order of Fibonacci series. This claim or phenomenon can be evaluated in two ways. Let's have a hypothetical argument regarding the existence of Fibonacci pattern in fir cones. Simultaneously, we will consider it analogous to our argument on the mathematical pattern of the Quran.

PROPONENT: There are thousands of other plants where Fibonacci sequence cannot be observed. It is entirely an arbitrary decision to choose fir cones as an example of this so-called pattern. The appearance of Fibonacci sequence in fir cones is only an interesting coincidence. Of course, I might be wrong.

OPPONENT: Though when other plants are considered, the appearance of Fibonacci sequence in fir cones can be a coincidence but the very pattern in fir cones cannot be coincidence. There must be a reason behind this regularity in fir cones.

PROPONENT: I was just being sarcastic. I was trying to use your own argument against you. The status of "basmalah" in the Quran is much more prestigious and important than the status of fir cones in plant kingdom, let alone in all of nature. Second, there are many more examples of the 19-based mathematical pattern in the Quran, than the examples of Fibonacci Sequence in nature.

OPPONENT: You ignore something about fir cones. We can repeat our observation millions of times on millions of different fir cones and come to the same conclusion. The repeated observation of a similar pattern makes a scientist to accept the existence of that pattern. However, you have only one "basmalah" with a pattern.

PROPONENT: Well, how many fir cones are enough to convince you that Fibonacci sequence exists in the nature of fir cones?

OPPONENT: I don't know. Probably, ten or twenty observation would be enough.

PROPONENT: That means, after twenty observation you will generalize your conclusion on billions of fir cones that you haven't observed. Why do you "believe" that the other fir cones also will show the same property?

OPPONENT: For two reasons: First, all my twenty observations confirmed the pattern without a single exception. Indeed, after several observations we start predicting and each extra observations is a fulfillment of our prediction. Thus,

45

there is no reason to suspect the 21'st fir cone will spoil this coherent orchestra. Second, our previous observations on many other cases have created a very strong belief that apparently similar things share similar internal structures. I mean that our previous observations, say, on mangos or bananas have created a faith in determinism.

PROPONENT: What if, similarly, my observation on different elements of the Quran has created a faith in the intention and style of its designer.

OPPONENT: Well, you can select any thing that confirms your faith, and you can ignore or interpret any exception that contradict your expectation.

PROPONENT: Ironically, I can say the same thing about your scientific method. You observe and predict. If any observation contradict your prediction you will modify your theory or create a new formula. Your prediction can be "certain" only about the objects of your previous observations. However, on new cases you cannot guarantee your prediction, since history is full of failures of scientific predictions. Nevertheless, I believe that the predictive power of science increases with every new observation. It is the same with our observations on the mathematical structure of the Quran. With every new discovery or observation we get much more comprehensive understanding of it.

OPPONENT: You are undermining the predictive power of science. I can challenge you to bring as many as fir cones and examine their scales. You will find the Fibonacci sequence in all of them.

PROPONENT: Though I don't think that induction provides us with certainty, nevertheless, I believe that your prediction will be confirmed. But, my question will be this: Based on the observation on fir cones how accurately can you predict the pattern of fir branches, or sunflower petals or corn kernels?

OPPONENT: Certain things can be observed and therefore predicted for all material objects. However, there are details which we cannot predict with certainty without sufficient observation.

PROPONENT: That is exactly the same for our observation on the mathematical structure of the Quran. We observe and discover. Each discovery increases our knowledge of the big picture.

Here I want to give you two examples for the accuracy or the predictive nature of mathematical code:

Correcting a Scribal Error

Three chapters of the Quran, chapter 7, 19 and 38 contains letter "Saad" in their initial letter combination. Curiously, in verse 7:69 we see a word with a unique spelling: "Basstatan ." Over the letter "Saad" is written a small "Sin." This word occurs in the Quran with two different spellings and it makes no difference to the meaning. Just like the English words skeptic or sceptic. Commentaries of the Quran interpret it as an instruction on how to read the word. They claim that though it is written with "Saad" it should be read as if it is "Sin." They narrate three Hadiths (allegedly Muhammad's words) to support this interpretation .

The total number of "Saad" with this word "Basstatan" becomes 153, and it is not multiple of 19. Therefore, we concluded that the letter "Saad" in the word "Basstatan" is an ortographic error and should be corrected.

Indeed, when I checked one of the oldest manuscripts of the Quran, I found that our prediction was confirmed. Please see the document below:

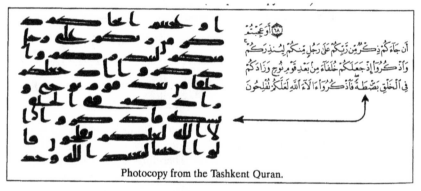

Photocopy from the Tashkent Quran.

Symmetry in the Table of God's Attributes

In my second letter I had attached a list of God's attributes with their numerical values and frequencies in the Quran, and the mathematical relation between these names and "basmalah."

Dr. Cezar Edib Majul, in his book "The Names of Allah in Relation To The Mathematical Structure of Quran" discovered the following two facts:

47

1. Only four attributes of God have gematrical (numerical) values that are multiple of 19.

2. The numerical values of these four attributes exactly correspond to the frequency of the four words of "basmalah," that is, 19, 2698, 57, and 114.

Here is the table of this observation:

Frequency of:	Gematrical Value of	Numerical Value of:
Ism (Name)	19 (19x1)	Wahed (One)
Allah (God)	2698 (19x142)	Zul fadlil'azeem (Possessor of Infinite Grace)
Rahman (Gracious)	57 (19x3)	Majeed (Glorious)
Rahim (Merciful)	114 (19x6)	Jami' (Editor; Gatherer)

While I studied this table I noticed an asymmetry. On the right side of the table there are four names of God, but on the left side there are only three. Obviously, "Ism" (Name) was not an attribute of God. Therefore, I made two predictions. There must be one attribute of God that must have a frequency of 19, and only four names of God must have frequencies of multiple of 19. When I examined the frequency of all the names of God, I found that ONLY four of them are repeated in the Quran as multiple of 19. We already had discovered three of them: Allah (God), Rahman (Gracious), and Raheem (Merciful). The fourth one was "Shaheed" (Witness). This attribute is mentioned in the Quran exactly 19 times and thus fills the empty space corresponding to the numerical value of "Wahed" (One) on the right side of the table. Now we can modify our tables of God's attributes according to the fulfillment of our prediction:

48

Frequency of:	Gematrical Value of	Numerical Value of:
Shaheed (Witness)	19 (19x1)	Wahed (One)
Allah (God)	2698 (19x142)	Zul fadlil'azeem (Possessor of Infinite Grace)
Rahman (Gracious)	57 (19x3)	Majeed (Glorious)
Rahim (Merciful)	114 (19x6)	Jami' (Editor; Gatherer)

Finally,

Dear Dr. Sagan, I'm enclosing a discovery made by Milan Sulc of Switzerland on the frequency of HH and M letters. They are the initials of 7 chapters. The frequency of HH and M letters in those 7 chapters are 2147, that is 19 x 113. The probability of this fact is 1/19. There are four subgroups whose combinations result in a multiple of 19. This is obviously coincidence, since there are 112 possible combinations. If you add the digits of each frequency (the absolute value of each number) you will end up with 113, which is exactly equal to the multiplication factor. This is valid also for all the subgroups. The probability of this phenomenon for each subgroup is 1/9. Therefore, the combined probability of these phenomena occurring altogether is 1/19 x 9 x 9 x 9 x 9 x 9, or 1/1,121,931. (FN1)Please let me know what you think.

- Is it merely a coincidence?
- Is it a mathematical property?
- Is it a little numerology?
- Is it a combination of the above?
or
- Is it a great mathematical art?

NINETEEN: God's Signature in Nature and Scripture. EXAMPLES:

The number of Arabic letters in the opening statement of the Quran, Bismillahirrahmanirrahym (1:1)	19	x 1
Every word in Bismillah… is found in the Quran in multiples of	19	
The frequency of the first word (Ism = Name)	19	x 1
The frequency of the second word (Allah = God)	19	x 142
The frequency of the third word (Rahman = Gracious)	19	x 3
The fourth word (Rahym = Compassionate) occurs	19	x 6
Out of more than hundred attributes of God, only four has numerical values of multiple of	19	
One (WAHiD)	19	x 1
Possessor of Great Bounties (ŻuW ALFaDL ALÂŻYM)	19	x 142
Glorious (MaJYD)	19	x 3
Summoner/Editor (JAMeÂ)	19	x 6
The number of chapters in the Quran	19	x 6
Despite its conspicuous absence from Chapter 9, Bismillah occurs twice in Chapter 27, making its frequency in the Quran	19	x 6
Number of chapters from the missing Ch. 9 to the extra in Ch. 27.	19	x 1
The total number of all verses in the Quran, including the 112 unnumbered Bismillah	19	x 334
29 Chapters of the Quran starts with 14 different combinations of 14 different letters. Their frequencies in the chapters that they initialize demonstrate an interlocking numerical pattern based on	19	
The number of all verses containing all those 14 letters.	19	x 6
Frequency of the letter Q in two chapters it initializes	19	x 6
Frequency of the letter Ŝ in three chapters it initializes	19	x 8
Frequency of the letters Y.S. in the chapter they initialize	19	x 15
Frequency of the letters K.H.Y.A.Ŝ in chapter they initialize	19	x 42
Frequency of the letters H.M. in seven chapters they initialize	19	x 113
The number of all different numbers mentioned in the Quran	19	x 2
The number of all numbers repeated in the Quran	19	x 16
The sum of all whole numbers mentioned in the Quran	19	x 8534
The number of lunar years in which the meaning of 19 remained hidden in Chapter 74, known al-Muddathir (The Hidden)	19	x 74
The year meaning of 19 in Ch. 74 (The Hidden) was unveiled	19	74
The number of letters in the first statement of Chapter 74	19	x 1
The numerical value of those first 19 letters in Chapter 74	19	74
The numerical value of the main message of the Quran, WAHiD	19	x 1

2

The Gullible, the Blind, and Boxes of Diamonds and Glass

The following two articles were published in the end of "Running Like Zebras," my debate with a Sunni on the mathematical miracle of the Quran. Because of their import, I decided to publish them here separately.

In the first part of this article my audience is the gullible people who get easily excited by the news of mathematical miracle discoveries. Like some confuse astronomy with astrology, they confuse math with esoteric numerology. A few, whom I call the "numerologist prophets of doom," have also announced many specific dates for doom; such as, devastating meteors, earthquakes, wars, etc. These people end up finding many miracles about themselves or their cult leaders. The prophets of doom are diabolic and manipulative fame-seekers who mutated from gullible and sincere miracle-hunters. Soon after each of their doomsday prophecy fails to fulfill, they come up with a new prophecy. Though they lack critical thinking or common sense, they have plenty of silly rationalization, megalomania, and enough numbers to play with.

In the second part of this article, my audience is the dogmatic people who have developed a bigoted allergy against the mathematical miracle of the Quran. For various reasons, they have problem in objectively evaluating the facts, despite the overwhelming evidence and the unveiled divine prophecy in chapter 74. I had specifically exposed their bias or hypocrisy in article titled, "Which One Do You See: Hell or Miracle?" which is published at 19.org.

How to Distinguish a Good Scientific Theory from a Bad One?

According to the majority of scientists, a good scientific theory should demonstrate three qualifications:

1. It should have an explanatory power.
2. It should be able to make accurate prediction.
3. It should provide economy in thought.

When scientists are going to choose among alternative theories they act according to these three criteria. The advance in science and technology is the result of using these criteria. For instance, before the Copernican model of solar system was proved, scientists preferred Copernicus' sun-centered model to Ptolemaic earth-centered model by using these intuitive and practical criteria. History of science is full of positive results of these criteria.

In order to help those who consider the mathematical structure of the Quran as a "theory," we will evaluate the mathematical system according to the first two criteria. You think on the third one!

1. Explanatory Power

- The challenge of the Quran regarding the impossibility of imitating the unique and superhuman nature of the Quran gains an objective (mathematical) criterion with the code 19. The Quranic challenge (2:23-24) is provided a meaningful and universal arena, instead of practically meaningless (or, subjective) and culturally limited "literary miracle."

- The evidence for a divine authorship being in a mathematical design in physical structure rather than being in eloquent use of Arabic is in harmony with the universal appeal and message of the Quran.

- Initial Letters (Huruf-u Muqattaa) that prefixes 29 chapters obtains a unique meaning and purpose. Numerous speculations made by interpretators for fourteen centuries ending up or starting with the confession of "we really do not know their meaning" finally are replaced with a clear message.

- The reason why the Quranic expression "These are the miracles (ayaat) of this book" follow the Initial Letters in all eight occurrences is understood.

- The discussion regarding different spellings gains a new dimension. For instance, why the first word of BismillahirRaĤMaNirrahim "Bsm" (in the name) is written without "A" (Alif) and why the "Bism" in the first verse of Chapter Alaq (Embryo) is written with an "A" (Alif)?

- The mathematical system puts an end to the chronic arguments among various sects whether Bismillah is the first verse of Chapter Al-Fatiha or not. Now, it is clear that the Bismillah in the beginning of Chapter Al-Fatiha (The Opener) is the first verse while other 112 Bismillahs crowning the other chapters are un-numbered verses.

- It is not a question anymore why the Chapter 9, *Baraah* (Ultimatum), does not start with Bismillah and why Chapter 27, Al-Naml (The Ant) contains an extra Bismillah in verse 30.

- The prophecy of Chapter 74, *Al-Mudathir* (The Hidden One), is unveiled and especially 74:31 is being fulfilled.

- We have learned another reason why some numbers are mentioned in an unusual way. For instance, the Quran gives Noah's age with a subtraction, 1000 - 50. The number of years which the young monotheists spent in a cave is expressed with an addition, 300 + 9. There is a clear relation between these interesting unusual expressions and the mathematical code.

- The question "Why the Quran consists of 114 chapters?" receives several meaningful answers.

- The divine guarantee regarding the preservation of the Quran is confirmed by the discovery of the code. Thus, a skeptical argument that casts doubt on authenticity of the verses, including the ones that guarantee preservation of the Quran is refuted.

- The historical speculations about what is *Ism-i Azam*, the Greatest Name of God, are ended.

- It becomes more evident that Prophet Muhammad was a literate *ummi* (gentile).

- The identity of the witness mentioned in verse 46:10 is unveiled as Judah ben Samuel of the eleventh century. The identity of the curious creature made of earthly material that was prophesied in verse 27:82 is unveiled as the "computer."

- The description of *Kitabun Marqum* (Numerically Coded Book) in verses 83:9 and 20 is clearly understood.

- The question "Why Jonah (Yunus) is referred as *Sahibul Hut* (The friend of fish) in verse 68:48 and referred as 'Zannoon' (Possessor of Noon)?" receives a meaningful explanation.

- Quranic verses states that those who follow their parents and leaders blindly or those who rejects the truth with arrogance and ignorance are not able to see the clear miracles and signs. The mathematical miracle of the Quran proves to us this incredible fact.

2. The accuracy and predictive nature of the Code

Here I want to give two examples. First a spelling correction, second a prediction.

a. Correcting a scribal error

Three chapters of the Quran, chapter 7, 19 and 38 contains letter *Saad* in their initial letter combination. Curiously, in verse 7:69 we see a word with a unique spelling: *BaŜTatan*. Over the letter *Saad* is written a small *Sin*. This word occurs in the Quran with two different spellings and it makes no difference to the meaning. Just like the English words skeptic or skeptic. Commentaries of the Quran interpret it as an instruction on how to read the word. They claim that though it is written with *Saad* it should be read as if it is *Sin*. They narrate three Hadiths (allegedly Muhammad's words) to support this interpretation.

The total number of *Saad* with this word *BaŜTatan* becomes 153, and it is not multiple of 19. Therefore, we concluded that the letter *Saad* in the word *BaŜTatan* is an orthographic error and should be corrected.

Indeed, when I checked one of the oldest manuscripts of the Quran, I found that our prediction was confirmed. Please see the document below:

b. Symmetry in the table of God's names

In my second letter I had attached a list of God's names with their numerical values and frequencies in the Quran, and the mathematical relation between these names and *Bismillah.*.

Dr. Cezar Edip Majul, in his book, *The Names of Allah in Relation To the Mathematical Structure of Quran* discovered the following two facts:

1. Only four names (adjectives) of God have Gematrical (numerical) values that are a multiple of 19.

2. The numerical values of these four names exactly correspond to the frequency of the four words of "Basmala," that is, 19, 2698, 57, and 114.

While I studied this table I noticed an asymmetry. On the right side of the table there are four names of God, but on the left side there are only three. Obviously, "Ism" (Name) was not a name of God. Therefore, I made two predictions. There must be one name of God that must have a frequency of 19, and only four names of God must have frequencies of multiple of 19. When I examined the frequency of all the names of God, I found that ONLY four of them are repeated in the Quran as multiple of 19. We already had discovered three of them: *ALLaH* (God), *RaĤMaN* (Gracious), and *RaĤyM* (Merciful). The fourth one was ŞaHYD (Witness). This name is mentioned in the Quran exactly 19 times and thus fills the empty space corresponding to the numerical value of *WAĤiD*(One) on the right side of the table.

How can one Distinguish Diamonds from Pieces of Glass?

It needs a good mathematical intuition and experience in order to distinguish arbitrary and selective calculations from a systematic and objective calculation. Here, I would like to give you three methods of recognizing pieces of glasses from diamonds. I cannot provide examples since I have literally thousands of them, and choosing the typical ones requires a lot of time. Also, I don't have much space here.

1. Consider Laws of Probability: Fifty percent "tails" should not tickle you

When you are provided by a calculation labeled as "mathematical miracle" don't just accept the claim without investigation (17:36). Be extremely careful for the temptation to please the person who wants to give you a ride in his/her fantasy bandwagon. Especially, be more careful when the person is using God's name and praising him after his/her so-called discoveries. You may attribute the absurd and most stupid calculations and ideas to the Greatest Mathematician, the Most Wise. You should check whether the method of calculation is arbitrary, and the claimed relations are personal speculations.

Each differing method of calculation weakens the significance of calculations. Two methods of calculations increase the probability twice. Three methods of calculation increase thrice and four methods of calculations quadruple the probability. For instance, if you sometimes add a set of numbers and sometimes

55

put them next to each other and get multiple of 19, you need twice as many examples than usual in order for your arithmetic to go beyond probability and be considered interesting, or extraordinary.

Arbitrarily processing a bunch of arbitrarily selected chapter and/or verse numbers by an arbitrary process of adding or concatenating or both adding and concatenating, and using an arbitrary number of items in this combination and finding a mathematical relation has little significance.

Finding a multiple of 19 by selecting a combination of two, or three, or four, or five, or more verses out of more than 6000 Quranic verses and passing them through numerous different calculations does not support the purported claims. Unfortunately, I have received many letters and bulletins filled with thousands of examples of deceptive calculations ending with "Praise be to God" or "Subhanallah."

The numbers of combinations (sets) that can be created out of more than six thousand elements exceed the biggest number known to our dictionaries, except googolplex, of course. According to the law of probability, there should be trillions of combinations that will form multiple of 19, since one out every 19 numbers will be multiple of 19. Some (thousands) of them, may show some semantic relation, especially, if the book contains repetitions and the discovery of relation and interpretation is left entirely to our wish!

2. Pay attention to the mathematical properties. Learn basic principles of math before "discovering" math patterns

Not only mathematical illiterates, but also a college-educated person can be deceived by mathematical properties. They can see them as part of the divine mathematical design. For example, the total of all chapter numbers being multiple of 19, that is, 6555 (19 x 345), is a mathematical property. If the number of items in an addition is multiple of N, then, the total of those elements will be multiple of N too. Another example: The total of Chapter numbers from 9 to 27 (9+10+11+......+27) is 342, and it must be multiple of 19, since exactly 19 consecutive numbers are added. Therefore, 342 being a multiple of 19 is a mathematical property in that context, since, the total of every consecutive N whole numbers is a multiple of N.

For those who are curious about special properties of numbers and recreational mathematics, besides Martin Gardner's books, I recommend *Mathematical Amuzements and Surprises* by Alfred S Posamentier and Ingmar Lehmann.

3. Claims should be "falsifiable." The number of letters in this article may equal to the number of angels or daemons in your head; but how can someone else know that?

We can find or attach meaning or reference for any number. You can arbitrarily select some chapter or verse numbers and arbitrarily put them together by adding, dividing, multiplying, subtracting or concatenating, or by another method of your choice, and when finally you arrive to a number which you can relate to some Quranic or to a *special* number, you may think that you have discovered a great mathematical miracle.

We should pay special attention to the criterion of *falsifiability*, since everyone, including university professors, can fall victim in confusing expert manipulations with mathematical calculations. We should question those who arrive at a certain number after using unlimited ways of calculations and different methods and then try to assign a meaning or a relation to that product. Here is the question:

Is it logically possible to *falsify* your calculations and results? In other words, would it be wrong or a non-miracle if your result were Y instead of X? Or, in that case, couldn't you just find a relation or a meaning for Y in the Quran, or in the universe of numbers? If you had difficulty in putting that number in a context, wouldn't you add several more calculations and come up with a number that you could attach a meaning to? It is a great responsibility to label your speculative and manipulative calculations as a *miracle* by claiming that God has calculated all the numbers.

It is therefore significant that the beginning verses of Chapter 74 warn us: "Do not be greedy!" (74:6)

3
Contextual Facts on 19

1. *Code 19* provides a powerful evidence for God's existence, as expected, envisioned or demanded by some philosophers and scientists, such as Galileo Galilei, Isaac Newton, Gottfried Leibniz, David Hume, Paul Dirac, and Carl Sagan.

2. **Code 19 was hidden in the 74th chapter of the Quran *The Hidden*, for 19x74 (1406) lunar years, and was discovered in 1974 by my colleague, Dr. Rashad Khalifa, an Egyptian-American biochemist. The number 19 has been a major controversy since its discovery and the number has realized all its assigned functions according to the prophetic verses of Chapter 74.**

3. **Because of the implications of his discovery of the Secret, as well as his strong criticism of the sectarian teachings based on *Hadith* and *Sunna*, Rashad was declared a heretic/apostate by leading Sunni scholars from 38 countries who held an emergency conference in Saudi Arabia in 1989 to discuss the Salman Rushdie controversy. While Rushdi survived, Rashad was assassinated in this Masjid in January 31, 1990, by a terrorist group linked to al-Qaeda. The author of this book also received similar fatwa, yet he has escaped several assassination attempts, so far.**

4. *Code 19, which was also discovered in the original portions of the* Old Testament by Judah ben Samuel in 11th century, is simple to understand but impossible to imitate.

5. *Code 19* has little to do with numerology, since its literary-numerical (LitNu) pattern can be verified or falsified through scientific inquiry. It is radically different from the pattern demonstrated in *The Bible Code*, which has no statistical value.

6. Unlike regular metaphysical or paranormal claims, *Code 19* can be verified or falsified by virtually anyone, since the Arabic version of the Quran is available everywhere. Besides, for the most part, the reader does not need to know Arabic but only two eyes to see, an ability to count, a critical mind, and an open mind and heart to witness extraordinary signs as the fulfillment of a great prophecy.

7. **This discovery has created a paradigm change among those who witness it: —instead of joining a religious bandwagon by blindly believing a holy story or hearsay, we must be critical thinkers; we must question everything and seek truth through knowledge. The code suggests a "Copernican revolution" in theology of religions. Instead of Krishna-centered, or Jesus-centered, or Muhammad-centered religions we must turn to the original center, to the God-centered model. The message of rational monotheism has sparked an ongoing controversy in countries with Muslim-majority populations, e.g., Egypt, Pakistan, Malaysia, Saudi Arabia, Turkey, etc. Internet forums are filled with heated debates regarding this code.**

8. A majority of people, including many adherents of Sunni and Shiite religions so far have not been able to witness this extraordinary divine sign. *Code 19* proved one more time the Quranic maxim that the majority of people in fact do not seek nor follow the truth. It exposed billions of believers; contrary to what they pretend, they do not acknowledge the Truth but they are believers in their culture and are herds in various bandwagons (27:82).

9. *Code 19* is one the greatest discoveries in human history, since it provides an extraordinary and universal evidence for God's existence. The evidence is hundred percent objective, yet paradoxically it has a subjective pre-requisite, akin to the pre-requisite of keeping both eyes open in order to witness a 3-D picture made of random dots. This is exactly in accordance to the prophetic description in Chapter 74. As it is the case with computer generated random dot 3D stereograms, in order to witness one of the greatest miracles, some conditions need to be met.

10. By distinguishing those who regress and those who progress, *Code 19* has opened a new era. (74:37). Code 19 distinguishes rational monotheists from the blind followers of manmade hearsay collections and sectarian teachings, as well as from those who give lip service to the Quran to serve their political, financial, and personal agendas (27:82).

11. *Code 19* shed light on some multiple-meaning (*mutashabih*) and prophetic verses, in some cases enriching the meaning of many verses of the Quran. For instance, it explained the meaning and function of the combination of alphabet letters initializing 29 chapters.

12. *Code 19* showed that the Quran is not an ordinary book, but a very interesting and unique book.

13. *Code 19* showed us that there is indeed a force, negative energy called Satan, which exerts hypnotic power over its constituents. Those who decide not to use their God-given minds to their full potentials are blinded by their masters; they cannot see and appreciate the precise and obvious divine signs and their message (7:146).

14. The improbability and impossibility of the numerical structure of the Quran being produced by a medieval Arab genius becomes evident when we consider the following factors:

   • It includes simple elements of the Quran and goes deeper to an interlocking system of complex numerical patterns and relationships.
   • It involves not only frequencies of letters and words but also the numerical values of letters.
   • It involves not only an intricate numerical pattern but also a huge set of data consisting of units with multiple functions, such as letters that are also digits, words that are also numbers.
   • The nemeroliteral aspect of the Quran were not known by the adherents of the Quran until late 1960s and especially, 1974.
   • The literary aspect of the Quran has received praises from many literary giants throughout centuries.
   • The scientific accuracy of Quranic statements on various fields has been immaculate.
   • Muhammad was one of the busiest and greatest social and political reformists in human history.
   • The timing of the discovery of the code is precise and prophetic.
   • A series of prophetic events regarding the code has been fulfilled.

15. Whether we like it or not, the Creator of the universe decided to design creatures that could make independent decisions from their original program. For some reasons that we may not know now, the Omniscient and Omnipotent Creator, by bringing specific limitations to His powers, is testing some of these creatures with their given ability to choose. He will later discard those programs that are infected with viruses and will select those who made good and proper decisions. Put it in computer terminology,

Code 19 is one of those virus detectors. Its release year is 1974. It diagnoses the brains infected with the most destructive viruses called bigotry and polytheism, as well as the brains that are healthy. However, many people do not want to face the fact that the Hellfire and Smoke they see all over Chapter 74 is, in fact, the product of their own infected mind, their imagination.

16. You may understand nothing from the information shared in this book and you may wonder by saying "So what? What does this number mean?" You may choose to be duped by the ingrates or extreme skeptics who are good in hiding the truth and distorting the facts. Or you may witness one of the greatest miracles. If you become one of the few lucky witnesses with clear vision, nothing will be the same for you. You will experience the taste of eternal peace, love and happiness starting in this life on this lowly planet. You will respond to tragedies and miseries with resolve and dignity, and you will respond to blessings and gains with humility and appreciation. As an active peacemaker, you will stand against falsehood, superstitions, hedonism, corruption, oppression, racism, nationalism, misogyny, intolerance and injustices without fear and hesitation. You will no longer believe in religious stories, you will no longer accept others as holy power-brokers between you and your Creator, and you will no more compromise your brain with illogical and nonsensical dogmas and taboos. You will no longer be a sheep exploited and manipulated by kings and priests, sultans and mullahs, rajahs and maharajs, politicians and pundits, atheists and polytheists. You will no more accept the claims of any "holy book" without unequivocal evidence. You will no more turn yourself into a schizophrenic character by allocating prime regions of your brain for contradictory beliefs, bizarre stories and Trojan horses. You will be a perpetual seeker of truth and submit yourself to the Truth alone. And the Truth will set you free. You will be honest; whenever your errors are shown, you will correct yourself. You will *know* that there is God, a Loving God, and you will receive God's communication through countless yet consistent signs received from multiple channels: brain, heart, nature, scripture, and your personal experience. You will be as rational as a human can be, and you will have no doubt about the purpose of life, eternity and resurrection. Though you will not find all the answers to your philosophical questions, but you will find answers to the most important ones. Of course, as a natural consequence of your appreciation and responsibility as a witness, you will share this message of rational monotheism with your friends and relatives in the best possible manner and expecting no reward from them.

61

17. The author of this book challenges theologians, mathematicians, philosophers, thinkers, believers, agnostics and atheists, specifically, the living Pope, Evangelist Rick Warren, Atheist Richard Dawkins, Skeptic Michael Shermer, Mathematician Ian Stewart, professors at al-Azhar University, Ayatollahs in Qum, and other prominent scholars, including those who live in Ivory Towers, to respond to the function of *Code 19* that provides extraordinary evidence for the extraordinary claims made by the Quran. The author is ready for face-to-face public debates at the place of their choice. The author also claims right to the James Randy Educational Foundation's Million Dollar Paranormal Challenge and expects the JREF test his assertions regarding the fulfillment of the prophecies in Chapter 74 of the Quran.

18. The author invites all mathematicians and specifically the mathematicians who have signed a petition rejecting the assertions made by the so-called *Bible Code* to examine Code 19. They rightly rejected its claims in a petition published at CalTech.

> "We refer in particular to the paper *Equidistant Letter Sequences in the Book of Genesis*, published in Statistical Science in 1994. This experiment suffers from major problems concerning both its execution and the interpretation of its conclusions. Even without these concerns, we would not take such extraordinary claims seriously without a vastly more systematic and thorough investigation. No such investigation has been carried out, nor has the work so far established a prima facie case. In addition, word clusters such as mentioned in Witztum's and Drosnin's books and the so called messianic codes are an uncontrolled phenomenon and similar clusters will be found in any text of similar length. All claims of incredible probabilities for such clusters are bogus, since they are computed contrary to standard rules of probability and statistics."

http://math.caltech.edu/code/petition.html

Thus, I invite the same mathematicians who rejected the *Bible Code* to critically evaluate the merits of the Quranic assertion that the Quran itself is a numerically coded divine book, and the number 19 provides extraordinary evidence for its claims:

> Robert E. L. Aldred, Dror Bar-Natan, Jay H. Beder, Valentina M. Borok, Robert Brooks, Mark Burgin , George M. Butler, Gary A. Chase, E. B. Davies, Percy Deift, Persi Diaconis, Laurence S. Freedman, Fritz Gesztesy, Sheldon Goldstein, Lawrence F. Gray, Rami Grossberg, Lee O. Hagglund, A. Michael Hasofer, Tim Hesterberg, Peter Hines, Svetlana Jitomirskaya, Gil Kalai, Fima Klebaner, David Klein, Richard N. Lane, Joel Lebowitz, Nati Linial, Gary Lorden, Brendan McKay, Tom Metzger, Stephen D. Miller, Paul Nevai, Amos Nevo, Eli Passow, John Allen Paulos, Yehuda Pinchover, Alexander Pruss,

62

Maurice Rojas, Mary Beth Ruskai, Jeremy Schiff, Gideon Schwarz, Senya Shlosman, Barry Simon, Martha Simon, J. Laurie Snell, Terry Speed, Terence Tao, Ian Wanless, Thomas Ward, Herbert S. Wilf , Henry Wolkowicz, Abraham Wyner, Doron Zeilberger, Yakov I. Zhitomirskii...

Knowing that scientists and scholars too are not immune from popular prejudices, I do not expect much from mathematicians. Compared to religious communities, scientific communities provide more freedom to their members. Yet they too have their own conventions and taboos. There are red lines in each scientific community, and scientists who dare to challenge them may risk being penalized through excommunication, defamation, or loss of their jobs. For instance, questioning the atheistic spin on the theory of evolution could be considered an anathema in the biological field. Likewise, questioning the double standard in discriminating between the consumption of alcohol and other drugs might be costly for a scientist living in a world where the horrific harms of alcohol consumption to human health and morals are minimized, ignored, or even suppressed by interest groups —such as beer and wine industries, as well as the scientists and media personnel who financially benefit from them— under the mantra of beneficial effects of "drinking in moderation."

So I am not surprised if two scientist colleagues of mine, who had been supportive of this work for years, now decline to write positive reviews of my work. One of them is an accomplished mathematician and the other is a professor of statistics at a major US university. I cannot justify such fear, and I consider it a sign of weakness in the academic and business establishment. Phillip Davis of Brown University and Reuben Hersh of University of New Mexico support my impression:

> "It is the writer's impression that most contemporary mathematicians and scientists are agnostics, or if they profess to a religious belief, they keep their science and their religion in two separate boxes. What might be described as the 'conventional scientific' view considers mathematics the foremost example of a field where reason is supreme, and where emotion is does not enter; where we know with certainty, and know that we know; where truths of today are truths forever. This view considers religion, by contrast, a realm of pure belief unaffected by reason. In this view, all religions are equal because all are equally incapable of verification and justification."[2]

Many mathematicians and scientists apparently are constrained to consider spending even a few minutes on this work, let alone studying it carefully and objectively. They may confuse the teaching of the Quran with the backwards

---

[2]   Philip J. Davis & Reuben Hersh, *The Mathematical Experience*, Houghton Mifflin Company, Boston, 1981, 1983, 10th edition, p. 109.

religious teachings and practices of Sunni or Shiite sects. They may be suspicious of anything with religious connotations. They may be influenced by the political agendas and propaganda of Western imperialism against backwards countries with predominantly Muslim populations. They may have bad experience with similar claims such as the so-called *Bible Code* and confuse this work with esoteric numerology. They may also feel incompetent in assessing the value of the claims, since they assume it to be heavily dependent on the knowledge of Arabic, which is not.

As for Arabic-speaking scientists, or more generally, Muslim scientists, they may have extra impediments: the prejudices caused by Sunni and Shiite teachings, fear of violence from religious zealots who are extremely irritated when they see or hear about the number Nineteen, fear of condemnation by religious scholars who remain innumerate and ignorant, and/or fear of repressive governments that imprison and kill those who are declared as heretics...

On the other hand, I also know that critics have helped us to learn and appreciate more about this amazing evidence. Even if one of these mathematicians shows the courage and wisdom to study and then witness this prophetic sign, it might encourage their peers to get over their prejudices.

Though mathematicians are objective in their professions, when they are confronted with philosophical or theological arguments, most react skeptically, which is justifiable. However, there is a line between skepticism and dogmatism, between confidence and bigotry. As an example, I would like to share with you the response of a mathematician listed above to my letter titled: NINETEEN: God's Signature in Nature and Scripture:

Dear mathematician and skeptic, peace be on you.

This email is sent to the 55 mathematicians who signed the petition at Caltech rejecting the so-called Bible Code. I know, after hearing so many astrologists talking about stars, and alchemists about gold, religious people about numerology and God, you might not be interested at all to hear an astronomer talking about a star, a chemist talking about gold, and a philosopher about numbers and God.

I invite you to review my upcoming book, NINETEEN: God's Signature in Nature and Scripture.

If an ancient book has the frequency of the word MONTH exactly 12 times and the frequency of the word DAY exactly 365 times, wouldn't you give a little chance that the book might have some numerical structure embedded within its text? You do not need to believe that it is from God, but at least, wouldn't such an ancient book pull your attention? Please read the following 15 points and let me know whether you would like to receive the final draft of the book.

I challenge you to look at the following brief introduction and the link below, which leads you to a short article that summarizes the 19-based numerical structure in the Quran and original portions of the Old Testament. I would like to receive your feedback and criticism if you have any. The following introductory paragraphs are from the first pages of the book and your name is mentioned in the last one. ....

*****

Dear Edip Yuksel,

You cannot expect me (or anybody else, for that matter) to volunteer my time to study a topic which I am convinced is foolish. Time is valuable, as you know. I wish I had time to do all the things that I actually consider as worth doing.

Yet if you wish, my time is for sale. You can buy some hours of my time, and I promise to spend these hours reading your material and writing a sincere statement about my conclusions (which I believe will be that your material is worthless, yet I will be reading your material with an open mind). If you buy a short amount of time my opinion will be less considered than if you buy a large amount of time.

My starting fee will be $361 (US) per hour, to be paid in advance. This is less than what most lawyers charge, and my time is no less valuable. I will not negotiate and I will not read any of your material unless the fee is paid. I may increase my fee later on if I will find your material to be too boring to read for such a small fee.

Sincerely,

Dror Bar-Natan.

*****

Dear Mr. Bar-Natan:

I am glad that you did not charge me for reading my email to you. I was first inclined to accept your deal. I wanted to send you enough money to encourage you to read the manuscript for one or two hours, but later I gave up from doing that for the following reasons.

1. You assume that I am practicing law and charging an excessive hourly rate to my clients. You are wrong. I gave up practicing law after a brief experience in a small law firm where my moral values put me at odds with the major partner of the firm. (Not all lawyers are the same; in every profession you will find rotten apples.) You may not believe it, but there are still people out there that value truth and justice more than money. I have read many draft books and gave my opinion on them without asking for a penny. I receive several dozen emails daily and I try my best to respond them, regardless of the name and position of the sender. I do not expect you to be like me, but I want you to know that every person with law degree is not crazy about money.

2. Your prejudice and bigotry reminds me of Tertullian's dogmatism, the father of modern Nicene formula $1 + 1 + 1 = 1$, who asked his followers not read any book except the Bible. I respect a healthy doze of skepticism, but I have no respect for dogmatism and bigotry that lead an educated man like you to make general and unjustified statements containing "foolish" and "worthless" words.

3. With dollar signs in your brain, and with your tone of sarcastic arrogance and prejudice, you will never be able to appreciate the great mathematical miracle of the Quran, since it can be witnessed only by

66

those who honestly seek the truth and subscribe to the Socratic maxim of "unexamined life is not worth living." (7:146).

4. "On it is Nineteen" (74:30)

*****

Dear Edip,

A dogmatic prejudiced bigot I may be, but at least, a sincere one.

Best, Dror

*****

Nice to hear that, Dror.

Sincerity is one of the most important traits in the quest of truth, and I hope that one day you will open your heart and mind to a possibility of a numerically coded divine message. ... If you wish to receive the copy of the book as my gift when it is published, please give me a post address.

Peace.

4

Chronology of the Discovery

Below is the outline of the major revelations and discoveries regarding the mathematical design of the Quran in chronological order. It includes some of my own contribution and experience:

610: Prophet Muhammad starts receiving the first revelations of the Quran in month of Ramadan, including the verse "on it is nineteen" of Chapter 74.

630: Revelation and recording of the Quran is completed.

1200: Judah ben Samuel of Germany (b. 1140–d. 1217), the spiritual leader of the Ashkenazi Hasidic movement, discovers a numerical system in the original text of the Old Testament based on the number 19, thereby fulfilling the prophecy of the Quran in verse 46:10.

1938: Fuad Abdulbaqi of Egypt, publishes his landmark concordance/index of the Quran: al-Mujam ul-Mufahras li-Alfaz il-Quran il-Karim (The Indexed Concordance for the Words of the Holy Quran.)

1924: Mustafa Kemal Atatürk abolishes the Khilafa, the Ottoman dynasty that established monarchy and imperial power in the name of God. Ottoman sultans, like their predecessors Abbasid and Umayyad kings, continued exploiting religious teachings and dogmas. Like many heroes, Mustafa Kemal too would be idolized after his death in 1938. A fascinating 19-based pattern in Mustafa's life was first published in 1951 in an article by Kadircan Kaflı in Yeni Sabah newspaper.

1959-1968: Abdurazzaq Nawfal, an Egyptian scholar, using Fuad's concordance notices some examples of numerical symmetry between the frequencies of semantically related words and phrases.

1969: While first human landed on the Moon, another Egyptian, Dr. Rashad Khalifa, a biochemist working at St. Louis, decides to study the Quran via computer to solve the mystery of combinations of alphabet letters initializing 29 chapters of the Quran. He publishes his work under several titles, such as Muhammad's Perpetual Miracle. The connection between the two historic events (Landing on the Moon and Feeding the Quran into the Computer) is supported by interesting coincidences.

1973: Rashad publishes Miracle of the Quran: Significance of the Mysterious Alphabets in St. Louis. While the book lists the frequencies of each initial alphabet letters in each verse of their respective chapters, it never mentions their relationship with the number 19. This study is published by an Egyptian magazine called *Ahir Sa'a* (Last Hour).

1974: Rashad discovers the role of the number 19, prophetically promised in Chapter 74, The Hidden, exactly 1406 (19 x 74) lunar years after the revelation of the Quran. This "one of the greatest" that would distinguish those who wish to regress or progress, starts the modern Islamic Reform movement that rejuvenates the liberating and powerful message of *La ilahe illa Allah.*

1975: Abdurazzaq publishes his book *al-I'jaz al-Adadi fi al-Quran al-Karim* (Numerical Miracles in the Holy Quran) containing examples of numerical relationship between related words, synonyms and antonyms in the Quran.

1975: University of Medina publishes the book of top Saudi cleric Sheikh Ibnul Baz: *El-edilletün Naqliyyetu vel Hissiyatu Ala Cereyaniş şamsi ve Sukunil Ardi ve İmkanis Suudi ilal Kavakibi* (The Traditional and Empirical Evidences for the Motion of the Sun and Stillness of Earth, and Possibility of Ascending to Planets). The book claims that the earth is fixed and deems any Muslim claiming otherwise of becoming an apostate, thereby making him subject to death penalty.

1976: Rashad's discovery of Code 19 becomes popular in Muslim Countries and his work is translated into many languages. He becomes a celebrity and is invited to give lectures in major academic conferences and is welcomed by various universities, including al-Azhar. Ahmad Deedat, the founder of Islamic Propagation Centre in South Africa, using Rashad Khalifa's work, writes an eloquent book, *Al-Quran: the Ultimate Miracle*, and freely distributes it in tens of thousands from 1979 until 1986.

1977: Jerry Lucas and Del Washburn publish *Theomatics: God's Best Kept Secret Revealed*. As an extension of Ivan Panin's work (1890), the book contains numerous anecdotal, arbitrary, inconsistent, and some interesting observations on gematrical values of certain words in the Bible.

1978: Joseph Dan of the Association for Jewish Studies, unaware of Rashad's discoveries, publishes an article on the work of Judah ben Samuel, an eleventh century German scholar. Judah's discovery of the code 19 in the original text of the Old Testament is the fulfillment of a prophetic

Quranic statement in verse 46:10. For the article, See: Studies in Jewish Mysticism, Proceedings of Regional Conferences Held at the University of California, Los Angeles and McGill University in April, University of California press, 1978.

1980: As the Turkish representative of the Muslim youth, I meet Ahmad Deedat in an International Youth Conference at Çanakkale, Turkey. The conference, which lasted for two weeks in a seaside camping site with the participation of more than 200 youth leaders from 42 countries, was organized by WAMY (World Assembly of Muslim Youth) and Turkey's Ministry of Youth. Besides Ahmad Deedat, there I meet many scholars, mostly members or affiliates of Muslim Brotherhood, including the famous scholar, Yusuf al-Qaradawi. Ahmed Deedat's lecture on the number 19 pulls my attention for two reasons: it sounds like a philosophical and theological breakthrough and it is a number very familiar to me, the name of the youth group that I established in 1978: FT/19. (FT abbreviation of Fatih—a region in Istanbul, or Fetih—Victory. The Turkish police had made extensive research trying to find out the meaning of 19 in this illegal group, which drew a little attention from the Turkish media.) I get permission from Deedat to translate and publish the book. The same year, the famous mathematician Martin Gardner publishes an article on the code 19 in Scientific American. Rashad claims to have discovered the end of the world from the Quran, which would later prompt me to find exactly the same date independently. According to the coded prophecy, starting from the year in which Rashad discovered the prophecy, the end of the world: 300 solar or 309 lunar years later.

1981-1982: Rashad publishes *The Computer Speaks: God's Message to the World* and one year later, *Quran: Visual Presentation of the Miracle* (http://www.submission.org/miracle/visual.html)

1982: While I am in prison, after verifying the evidences presented, I translate Ahmed Deedat's book to Turkish. The following year, my research on the verses of the Quran with scientific implications and my additional discoveries on the mathematical structure of the Quran are published together with my translation of Deedat's book. *Kuran En Büyük Mucize* (Quran, The Greatest Miracle), which becomes a best-selling book in Turkey until 1987.

1982: Rashad, in his milestone work, Quran, Hadith, and Islam sheds the light of the Quran that had been clouded by the distortion and false doctrines of sectarian teachings that rely on hearsay sources called Hadith and Sunna. The book, which can be read in a few hours, exposes the falsehood of Hadith and Sunna through the light of the Quranic verses.

70

With is laser sharp focus and reasoning, the book brilliantly exposes the numerous distortion made by so-called Islamic scholars. The book debunks the two main tricks used by Sunni and Shiite clergymen: it refutes mistranslations of the Quranic verses by comparative analysis of the scripture, and puts the abused verses into their context. Thus, a paradigm-changing powerful purge of men-made religious teachings in the process of Islamic reform with the motto of "Quran alone" starts.

1982: Dr. Cesar Adib Majul, a retired professor of philosophy and logic, a Filipino-American, publishes his book, The Names of Allah in Relation to the Mathematical Structure of Quran. Adib finds an interlocking numerical system between the frequencies of four words of Bismillah and the numerical values of God's attributes. I had numerous phone and mail conversation with Adib, my namesake, before his ascension to God in 2004. I have some of his unpublished research on the numerical structure of the Quran.

1983: Rashad translates his book, *The Visual Presentation of the Miracle*, into Arabic under the title *Mugizat ul-Quran* and it is published by Dar el-Ilm Lil-malayin, Beirut, Lebanon, and it is still listed at their website for order: www.malayin.com/publ.asp?item=00468

1985: Rashad declares that the verses 9:128-129 in current manuscripts do not belong to the Quran, since they violate the 19-based mathematical structure of the Quran. It is remarkable that from 1974 until 1985 the table listing the frequency of the word ALLAH ended with the sum of 2698 (19x142). Many books published in various languages disseminate this information around the world. Though with the exception of the frequency of the letter Alif, I check every factual detail mentioned in Ahmad Deedat's book, like many researchers, I do not notice the error regarding one the extra word, because of Fuad Abdulbaqi's concordance. Later, I would consider this eleven-year delay as a divinely planned event to disseminate the information in the world before subjecting them to a test.

1986: After months of debate with Rashad via snail-mail, my faith in Hadith and Sunna weakens, and it reaches to a breaking point after receiving his book Quran, Hadith and Sunna. The night of July 1, 1986, when I was a drafted soldier in Samsun, I read the book from cover to cover turning that night into my personal *Laylat ul-Qadr* (the Night of Power). With tears in my eyes, I give up associating partners to God, and decide to devote my service to God alone by following the Quran alone. By this decision, I would risk my career, fame, my political future, my family, and life. In October 23, 1986, exactly 114 days after my conversion to monotheism in July 1, 1986, I experience a divine sign followed by a series of prophecies as God promises in 43:53. The same year, the Haley Comet that visits our planet in every 76 (19x6) years passed nearby. This was the 19th visit since Muhammad's time and it was year 1406 (19x74) according to Hijra Calendar. Rashad considered it as one of the fulfillments of 41:53.

1986-1987: I discover an asymmetry in Adib's remarkable studies, and I hypothesize a solution. Soon, I discover the missing divine attribute *ShaHeeD*, in Adib Majul's table on the numerical relationship of God's names, fulfilling my prediction based on the 19-based pattern. My discovery completes an interlocking table of attributes of God, and is published in my Turkish book, *İlginç Sorular-2* (Interesting Questions-2). Furthermore, I hypothesize that there are exactly 114 attributes of God mentioned in the Quran.

1989: My rejection of Hadith and Sunna through my books and articles receives a nationwide attention and my affiliation with Rashad makes me a target of threats and attacks. Receiving an invitation from Rashad, I immigrate to the United States. For a while I live in a room adjacent to Masjid Tucson.

1989: In February 1989, Ibn Baz leads top Sunni scholars to discuss the matter of Salman Rushdi, and the fatwa of the 38 Sunni scholars becomes headline news in the Muslim world: "Rashad and Rushdi, both are Apostates." Rushdie was represented and defended by a British law firm called Article 19. The following year, Rushdie publishes his second book, titled Haroun and the Sea of Stories, in which the protagonist, oddly named Rashid, tells political stories.

1989: Abdullah Arık, a Turkish-American engineer, by using a simple computer program tests the compatibility of long numbers with the code 19 by putting together the numbers of chapters, verses, sequence of letters and their gematrical values. He discovers an interlocking and cumulative system in the sequence of the 19 letters of Bismillah.

1990: In 30th of January, Rashad is assassinated in Masjid Tucson by a terrorist group, al-Fuqara or al-Fuqra, affiliated to the early stage of al-Qaida organization founded by Saudi militant Osama Bin Laden. The same year, Saddam Hussein declared Kuwait to be Iraq's 19th province. To discuss the matter, the Organization of the Islamic Conference immediately convened its 19th conference.

1992: Milan Sulc of Switzerland, founder of Faith-through-Science Foundation, discovers an impressive numerical structure in the table of frequencies of ĤM letters, thereby showing the intricacy of the numerical structure of the Quran that integrates unique numerical computations with code 19. Milan Sulc and Ali Fazely, professor of

72

physics at the University of Louisiana, have announced their discoveries of a different numerical phenomenon in the Quran based on prime numbers and composite numbers. Though some of their findings are interesting, in my opinion, they lack a system and thus appear to be anecdotal and speculative at this point.

1997: Michael Drosnin, in a bestselling book, the Bible Code, claimed that he discovered many prophecies in the Bible by searching its text for equidistant letter sequences (ELS). Soon, skeptics debunked the claim conclusively. www.nmsr.org/biblecod.htm

1998: To promote Islamic reform according to the message of the Quran alone, to share the paradigm-shifting mathematical miracle, I establish 19.org.

1999-2000: The practice of abbreviating the four-digit year to two digits in computer programs in order save memory space, created a problem came to known Y2K Problem or the Millennium Bug. "While no globally significant computer failures occurred when the clocks rolled over into 2000, preparation for the Y2K problem had a significant effect on the computer industry. There were plenty of Y2K problems, and that none of the glitches caused major incidents is seen as vindication of the Y2K preparation. However, some questioned whether the absence of computer failures was the result of the preparation undertaken or whether the significance of the problem had been overstated." (Wikipedia). Whether real or perceived, this major problem cost over 300 billion US dollars (BBC News, 6 January 2000), and it was caused by computer programmers ignoring the number 19.

2001: On September 11, 2001, the 19 hijackers of al-Qaeda attack Pentagon, World Trade Center and kill thousands of Americans. One of the main organizers of the attack, Wadih al-Hage, was also implicated for his role in Rashad's assassination. Prosecutors have also suggested that El-Hage and Jamaat ul-Fuqra were involved in the murder of Dr. Rashad Khalifa on January 31, 1990 in Tucson. They believe that El-Hage knows who killed Khalifa. And even if El-Hage was not himself involved, the prosecution asked why he had not told them what he knew. El-Hage's family says he was not in the country at the time of the murder.

2002: More than a decade after my public acknowledgment of numerous errors in the count of Alifs, one of the reasons for being excommunicated by submitters, Dr. Atef Khalifa, finally acknowledges the fact that his brother Rashad indeed made errors in the count of Alifs and thus undertakes the overdue study on the subject: www.submission.org/miracle/alif.html

73

2002: Dr. Caner Taslaman, a Turkish philosopher/theologian and founder of Kuran Araştırmalar Grubu (Quranic Research Group), publishes a comprehensive book on the scientific and numeric aspect of the Quran *Kuran Hiç Tükenmeyen Mucize* (Quran: a Perpetual Miracle) is available online in English at www.quranmiracles.org. The same year, a group of muslims establish Islamic Reform, www.islamicreform.org

2003: A group of programmers, working together with Atef Khalifa, designs the "Quran Reader", a computer program to count the letters of the Quran text on the internet. http://www.submission.org/quran/reader/

2005: After being challenged by Dr. Ayman through his article Idiot's Guide to Code 19, I start my second book-length debate with him in May 2005, and call it, Intelligent People's Guide to Code 19. It will be published in the next edition of Running Like Zebras, God willing.

2005: In August, Manifesto for Islamic Reform is issued.

2006: Quran: a Reformist Translation is published by Brainbow Press.

2008: Using a program developed by a Turkish colleague on a reformed Arabic text of the Quran, in which all later innovations such as *hamzas, maddas, shaddas* and all other diatrical marks are deleted, we prove that there are only 114 verses in the Quran containing ALL the 14 letters used in the Quranic initials (*huruf-u muqatta*). Several other significant discoveries are made.

2009: With the progress in DNA forensic technology, the evidence gathered in 1990 was re-examined and one of the murderers of Rashad Khalifa was arrested in Canada, 19 years after Rashad's assassination.

2011: Two decades after Rashad's landmark work was published in Visual Presentation of the Quran, this book, *NINETEEN: God's Signature in Nature and Scripture*, is published by Brainbow Press.

2014: Reformist muslims or rational monotheists are planning to hold their annual Critical Thinkers for Islamic Reform conference in Mecca, by God's will. Confrontation with one of the world's most corrupt and oppressive regimes is expected.

ا ي ل ب ه م ل م ح ل ن ح ه ل ه م ب
ت ن ا د ع ك ي ن د ا و ك م ي ر ا م ر
ه ل ت ع ا ي ل ط ص ي ت م ا ر ل ا د ا ي
ن ل ض ا ل م ي ع و غ ل ر غ
ل ن م ع ا ر ل د ح ا ي ر ا م ر ا ل ا س
ع س ك ي و ب ن ا ا ي ل م ي ل م ح ل ن ح
م ي ع م ن ن ذ ا ر م ق س ل ط ص ا ن ه ن
ي ل ا و ه ل ب ض م ا ي

74

5

Which One do you See: HELL OR MIRACLE?

(From 19 Questions for Muslim Scholars by Edip Yuksel)

From the smoke of hell-fire in their minds, hypocrites and unappreciative people are not able to witness one of the greatest miracles.

I know that the title and the subtitle of this article are quite challenging. If you have developed an attitude against witnessing the mathematical miracle of the Quran, you are justified to get upset with these words and perhaps get little angry. You might have already blinded your eyes and closed your ears with one or more false, contradictory, trivial, or irrelevant excuses, such as:

1. God uses literary styles to prove the authenticity of His word; but not mathematics. God is not a mathematician and His word has nothing to do with math.
2. If Muhammad was not aware of a mathematical miracle in the Quran, then it cannot be true. Muhammad knew everything in the Quran. The knowledge contained in the Quran is limited with Muhammad's knowledge and understanding more than 14 centuries ago.
3. If I accept a mathematical structure in the Quran based on number 19, then I will be denying two Quranic verses in the end of Chapter 9, and thus would contradict the majority of Muslims and their manuscripts. Besides, I would be contradicting the verses guaranteeing the perfect preservation of the Quran.
4. Some verses of the Quran are Mutashabih; none can understand their meaning except God. The proponents of 19-based mathematical system in the Quran are indulging in those Mutashabih verses.
5. The number 19 is heresy, since it is the holy number of Bahai sect.
6. Though I can balance my check and shop in the stores, I am innumerate when it comes to the 19.

7. We can find many mathematical phenomena if we spend enough time on any book. For instance, the Bible Code claimed that the distorted translations of the Bible contain impressive coded prophecies.
8. The Quran cannot contain secrets, mysteries or prophecies; it is a clear book.
9. What about the previous generations; will they all go to hell?
10. The claim of a 19-based mathematical code is a mythology.
11. The claim of a 19-based mathematical code is a magic.
12. There are many other numbers in the Quran and the number 19 has no special place.
13. There are problems in the count of letters or words. For instance, the first verse of the Quran does not have 19 letters; it has 18, 20, 21, or 22 letters, but not 19. For instance, the counts of A.L.M. letters have errors.
14. Part of the numerical claim is based on the Gematria system (ABJAD), a Jewish fabrication and a deceptive tool used by numerologists and astrologists.
15. The pronoun "it" in verse 74:30 is feminine and it refers to Hell not the Quran. Thus verse 74:30-31 is about guardians of hell, not a mathematical proof of authenticity.
16. Even if there is such a code in the Quran it is not important. We need to follow the instruction of the book. We need to focus on how to fight against infidels.
17. The discoverer of Code 19 claimed to be the messenger of God. He deserved divine retribution. Thus, he was killed in early 1990 by an Al-Qaida affiliate American cult, al-Fuqra, as Al-Qaida's first act in the USA.
18. If we believe in code 19, then we would end up rejecting holy teachings of *Hadith* and *Sunna*.
19. Code 19 is a Zionist trick.
20. I already believe in the Quran and I do not need miracles.
21. I believe in the Quran because its message is true.
22. I believe in the Quran because it does not contain contradictions.
23. I believe in the Quran on blind faith and I do not need any reason for my belief.
24. And, many more excuses, reasons or lack of them...

I will deal with all these excuses, and God willing, I will respond each of them one by one; but first thing first:

Whatever are your EXCUSES or REASONS to ignore, reject, or ridicule the number 19, this article will expose your true intentions, which you might be trying to hide even from yourself (11:5). If you are claiming to believe in the Quran, and after reading this article still continue ignoring or ridiculing the mathematical code of the Quran, you will be TORMENTED all your life by repeatedly witnessing an exciting fact and losing its sight afterwards, like a person who witnesses succeeding events of lightning in the darkness (2:17-20). You will neither appreciate nor comprehend the miracle, nor will you be satisfied with your denial. You will perpetually oscillate between momentary belief end prolonged disbelief, between private doubt and public denial. You will be doomed to SAQAR here and in the hereafter.

But, if you have acknowledge the truth and gained some goodness in your acknowledgement, then this article, by God's will, may wake you up from ignorance, and may change your paradigm (6:158). You will never be the same; you will be among the progressives (74:30-37). You will be one of the few who are blessed by God to witness one of the greatest miracles. You will be sure about your conviction, not based on self-deceptive claims, on conformity with a religious group, or on blind faith, or wishful thinking; but a faith based on knowledge gathered from empirical and rational evidence corroborated by spiritual experience (41:53; 74:31). You will experience God's presence in your life and attain happiness promised to believers and submitters (9:124; 10:64; 16:89). You will dedicate all your life to serve God alone without remorse or fear (2:62; 3:170; 10:62; 46:13). You will understand many Quranic verses that had not much meaning for you (2:1,108). Every time you see an unappreciative disbeliever denying or mocking the 19-based system, your knowledge-based faith will be justified. And the more people do not see what you and few others have witnessed will increase the importance, power and the wonder of the prophetic description of 74:31 in your mind. You will attain certainty (74:31; 27:1-3; 2:260; 13:2; 2:118; 45:20). However, witnessing this miracle will also put some responsibility on you (5:115; 47:25).

Let's start from the translation of verses 21-37 of chapter 74, The Hidden One:

74:21   He looked.
74:22   He frowned and scowled.
74:23   Then he turned away arrogantly.
        74:24   He said, "This is but an impressive illusion/magic!"
74:25   "This is human made."
74:26   I will cast him into SAQAR!
74:27   And what will explain to you what the SAQAR is?

74:28	It leaves nothing, it lets nothing (NOT MORE, NOT LESS; PRECISE; PERFECT);
74:29	VISIBLE (LAWAHATUN) to PEOPLE (BASHAR) (universal)!
74:30	On it is Nineteen.
74:31	As guardians of fire we appointed none except angels/controllers, and we assigned their number (1) to torment the unappreciative disbelievers, (2) to convince People of the Book, (3) to strengthen the acknowledgement of the acknowledger, (4) to remove all traces of doubt from the hearts of People of the Book, as well as the believers, and (5) to expose those who have disease in their hearts, and the unappreciative disbelievers; they will say, "What did God mean by this allegory?" God thus sends astray whomever He wills (or, whoever wills), and guides whomever He wills (or, whoever wills). None knows the soldiers of your Lord except He. IT (HIYA) is a reminder for the people.
74:32	Indeed, by the Moon.
74:33	By the night as it passes.
74:34	And the morning as it shines.
74:35	This (NUMBER) is one of the GREATEST (KUBRA).
74:36	A WARNING (NAZEER) for the PEOPLE (BASHAR).
74:37	For those who want to progress or regress.

Though the function of 19 is listed in Chapter 74, conventionally known *Muddathir* (The Hidden One), its implication and fulfillment was kept hidden according to God's will as the Quran's secret for 19x74 lunar years after its revelation to Muhammad. However, the All-wise God, unveiled this secret via a monotheist scientist in 1974, according to the most commonly used calendar, a calendar allegedly based on the birth of Jesus, which is considered a holy day and the sign of the End (19:15). As a result, the number 19, as the miraculous code of the Quran and the Bible strengthened and continue to strengthen the acknowledgement of believers, removed and continue to remove doubts in the heart of people of the book, and intellectually tormented and continue to torment hypocrites and unappreciative disbelievers.

Now let's first discuss the words written in CAPITAL letters in the translation of the verses above. Then, we will discuss each of the popular excuses used by the unappreciative disbelievers of the miracle 19.

We should understand the meaning of these words by first referring to their usage in the Quran. Their immediate context as well as their use in other verses

78

is usually sufficient to illuminate their meanings. For believers who have been lucky to witness this great miracle, the meaning of the verses above is good news; it gives them hope, enlightens them, informs them, and turns their position from wishful thinking or conjecture to knowledge-based paradigm. Thus, the rhetorical value of these verses is very high for believers:

> 74:27 → SAQAR (saqar, to be defined by the following verses)
> 74:28 → (NOT MORE, NOT LESS; PRECISE; PERFECT);
> 74:29 → LAWAHA (obvious; visible; tablet; screen)
> 74:29 → BASHAR (humans; people)
> 74:31 → HIYA (it; reference to the number 19)
> 74:31 → ZIKRA (reminder; message)
> 74:35 → HA (it; referring to the number 19)
> 74:35 → KUBRA (great miracles)
> 74:36 →NEZEER (warner to be embraced and supported)

On the other hand, those who have deprived themselves from witnessing the miracle 19 because of their ill intentions or their dogmatic rejection try hard to render these key words incompatible with the semantic context of the Quran. They conceive God of the Quran as an angry and despotic God who is not able provide any reasonable argument against those who question the Quran's authenticity, but only resorts to intimidation: "I will burn you in hell!!!" The God they depict has double standard: He asks the disbelievers to bring their evidence for their argument (2:111; 11:17; 21:24; 27:64; 28:75; 35:40) but for His argument He only wants to scare them! The opponents of 19-based miracle, by distorting the meaning of the words in these verses, manage to blind themselves to one of the most profound philosophical and theological arguments and evidences in history. Not only they divert themselves from the right path they try to divert others too (6:25-26; 22:3).

Thus, the understanding (more accurately, the misunderstanding) of those who cannot appreciate God as He should be (6:90-91), the argument for Quran's authenticity is scorching, burning, dark, hellish, misfortune, disastrous, and scary. Thus, in the minds of opponents of miracle 19, the rhetorical value of these verses is simply a threat to burn and torture:

> 74:27 → SAQAR (hell-fire)
> 74:28 → (NEITHER LEAVES THE FLESH NOR THE BONES; DESTROYS CONTINUALLY);
> 74:29 → LAWAHA (scorching; burning; shriveling)
> 74:29 → BASHAR (skin)
> 74:31 →HIYA (it; referring to hell-fire)

74:31 → ZIKRA (news of disaster)
74:35 → HA (it; referring to hell-fire)
74:35 → KUBRA (great punishment; gravest misfortune; dire scourge)
74:36 →NAZEER (warning to be escaped)

Now let's one by one discuss each of these words, which were widely and perhaps JUSTIFIABLY misunderstood by pre-1974 generations, and yet are intentionally distorted by the post-1974 opponents of one of the greatest miracles.

Does "Saqar" In 74:27 Mean Hell OR Something Else?

Though prominent Arabic dictionaries such as Lisan-ul Arab and specialized dictionaries such as Mufradat Fi Gharib-il Quran acknowledge that the word might be of foreign origin with no Arabic derivatives, these and other dictionaries and commentaries of the Quran do not hesitate defining it as hell or heat radiating from Sun. Lisan-ul Arab refers to a *Hadith* which uses a bizarre derivative of the word Saqar to mean "liars". The word SaQaR is mentioned four times in the Quran, three times in Chapter 74 and once in 54:48. In the later one, the word SaQaR is used in a statement warning that when criminals will be dragged to the fire they will be told: "taste the touch of Saqar". From this verse one might infer that Saqar is another word for fire; but a better inference is that Saqar is a negative feeling or state of mind one tastes after being committed to the divine punishment. In this case, there is no reason to think that this negative state of mind could not be obtained from experiences other than fire.

The Arabic word for hell is GaHyM or GaHaNnaM. The word NAR (Fire), though not a specific name for hell, is also frequently used to denote the same phenomenon. However, the word NAR (fire) is also used in its literal meaning, which is simply fire. For instance, verses 20:10-14 describe God's communication with Moses through fire. Obviously, this fire in the holy land where God spoke through cannot be the Hell. Similarly, the word SaQaR does not necessarily mean hell. In fact, the semantic connection of SaQaR with hell is allegorical, since SaQaR is a descriptive word derived from the verb SaQaRa rather than a noun like GaHYM, GaHaNnaM.

In any case, the verse 74:27 does not ask nor expect us to rush into defining the meaning of this word, which it appears to be its first usage in the Quran. We are warned against rushing to define Quranic words or attempting to preempt the Quranic definition by prematurely assigning a meaning to a verse before considering its immediate or Quranic context or fulfillment (20:114; 75:16-19). Doing so is a sign of pretension and arrogance. Since the word is rarely used in the Quran (total four and three of them is in this chapter) and this verse is most likely the first usage of this rare word, it is more appropriate to wait the Quran explain the word. In brief, rushing to limit/define the meaning of the word Saqar in the verse "Do you know what SaQaR is?" is a disrespectful act against the Quran. When the Quran asks "Do you know what X means?" it does not want us to try to understand the meaning of X in the very question about its meaning! It is a rhetorical question. God wants to pull your attention to its modified or new meaning. The Quran uses this question 13 times either to modify the meaning of an already used word or to add another nuance (See: 69:3; 74:27; 77:14; 82:17-18; 83:8, 19; 86:2; 90:12; 97:2; 101:3, 10; 104:5).

Nevertheless, we see that almost all commentators or translators of the Quran have rushed to translate SaQaR as Hell or Fire:

> **Yusuf Ali**: "Soon I will cast him into Hell-Fire!"
> **Marmaduke Pickthall**: "Him shall I fling unto the burning."
> **T.B. Irving**: "I'll roast him by scorching!"
> **M. H. Shakir**: "I will cast him into Hell."
> **MaulaHUM Muhammad Ali**: "I will cast him into hell."
> **N. J. Dawood**: "I will surely cast him to the fire of Hell. "
> **Muhammad Asad**: "[Hence,] I shall cause him to endure hell-fire [in the life to come]!"
> **Rashad Khalifa:** "I will commit him to retribution"

As you see, none leave the word SaQaR as it is. However, Rashad Khalifa, as the one who was chosen to fulfill this great prophecy, renders the word SaQar accurately by translating it with a general word, "retribution." Yasar Nuri Ozturk, a Turkish theology professor, in his post-1974 translation of the Quran, translated the word SaQaR under the light of the descriptive verses and related discoveries as the COMPUTER with succeeding screens manifesting the mathematical miracle of the Quran to all people.

So, we should not prime our minds or blind ourselves with prejudice by assuming SaQaR as HELL-FIRE before reading the following verses (75:16-19; 20:114).

Does "It Leaves Nothing; It Lets Nothing" in 74:28 Mean "Exact, Precise" OR "Destroys Flesh And Bones"?

Though the short verse is not generally mistranslated, but its meaning and implication is distorted. In the light of the context and post-1974 discoveries we should understand it is description of the 19-based mathematical structure that is exact. It does neither leave extra (BaQaYa) nor it let anything necessary go away (WaZaRa); in other words, it is perfect and precise; it is not one more not one less!

However, parroting the pre-1974 commentaries of the Quran, many translations and commentaries still convey the same misunderstanding. Since English translations usually do not comment on this particular verse and leave it to be understood within the context of CORPORAL PUNISHMENT, HELL and FIRE rather than the crucial role of the NUMBER NINETEEN, I will give you a sample of some popular Arabic commentaries of the Quran (which I have easy access in my personal library) in ascending chronological order. If the name of the commentary different than the name of the author his name will be indicated in the parenthesis. The number in parenthesis is the year of the authors' death:

> **Tabari (922):** Neither kills nor leaves alive; or devours them all and when they are recreated do not leaves them until eating them.
>
> **Tha'alibi (1035):** Does not leave the dweller of hell alone and burns them.
>
> **Bagawi (1122):** Neither kills nor leaves alive; eats everything thrown in; does not leave their flesh nor take their bones.
>
> **Nasafi (1143):** Does not leave the flesh nor does let the bones of its dwellers; destroys everything in it and then restores them back to their starting point.
>
> **Baydawi (1292):** It does not leave anything thrown in and does not leave it until destroys it.
>
> **Qurtubi (1272):** Does not leave for them neither flesh, nor bone, nor blood, but burns them. Or does not leave anything from them and then does not let them go when they are recreated.
>
> **Zad-ul Masir (Ibn Cawzi, 1200):** Does not leave but destroys their flesh and does not let them go when they are recreated.
>
> **Ibn Kathir (1372):** Eat their flesh, sweat, bones, and skin. They do not die in this condition or they live.
>
> **Jalalayn (Celaleddin, 1459):** Destroys flesh and bones, then starts it again.
>
> **Ruh-ul Maani (Alusi 1853):** Destroys flesh, bones of everything thrown in.

Interestingly, one can understand the verse as pre-1974 commentators understood. Though they all heavily relied on *Hadith* for understanding of this verse, and knowing that there were literally thousands competing to fabricate *hadith* for various reasons we cannot prove whether Muhammad and his believing friends too understood that way. In fact, since we know that

82

Muhammad and his friends had no clue about the mathematical structure of the Quran but had acknowledged its promise of QURANIC MIRACLES being manifested in the future (10:20; 25:4-6; 29:50-51; 41:53), it is highly conceivable that they understood the fact that these verses were about the importance of the number 19 and that their meaning would be fulfilled in the future. I further assert that Muhammad and those who dedicated their religion to God alone by upholding the Quran as the sole authority in their religion, stopped speculating on these verses as soon as they received the divine instruction in 75:16-19. As we know, after Muslims started following the fabricated *hadith*s, *sunna*, and man-made teachings of various sects (6:112-145; 6:159; 7:29; 9:31; 16:52; 18:57; 39:2,11;14; 39:29-37; 39:43-45; 40:14,65; 42:21; 98:5), they lost their capacity for understanding of the Quran (6:23-25; 17:46).[3]

However, it is important to remember that the verses 74:26-37 contain many words that can be understood both a description of Hell and a description of a great miracle, though the later is a much better fit. This linguistically marvelous aspect is well appreciated by those who witness the mathematical miracle and understand the original language of the Quran.

Does "Lawaha" In 74:29 Mean Scorching/Burning OR Obvious/Visible?

The derivatives of the word LWH are used in the Quran to mean a surface used for recording information, board, and flat wood; and nowhere is it used to mean scorch or burn. Before the fulfillment of the prophecy, translators and commentators of the Quran had difficulty in understanding the simple meaning of this word and thus, they resorted to external sources and often odd meanings, such as scorch, or burn. In fact, the drive to justify a particular meaning for some "difficult" Quranic words is one of the many reasons for fabricating Hadith.[4]

---

[3]   To see the examples of misunderstanding and distortion of Quranic verses by the followers of Hadith and Sunnah, please read the comparative section in the beginning of the Reformist Translation of the Quran.

[4]   Soon after Muhammad's death thousands of hadiths (words attributed to Muhammad) were fabricated and two centuries later compiled in the so-called "authentic" hadith books, to support the teaching of a particular sect against another (for instance, what nullifies ablution; which sea food is prohibited); to flatter or justify the authority and practice of a particular king against dissidents (such as, Mahdy and Dajjal); to promote the interest of a particular tribe or family (such as, favoring Quraysh tribe or Muhammad's family); to justify sexual abuse and misogyny (such as, Aisha's age; barring women from leading Sala prayers); to justify violence, oppression and tyranny (such as, torturing members of Urayna and Uqayla tribes, massacring Jewish population in Madina, assassinating a female poet for her critical poems); to exhort

83

Verse 74:29 is very interesting and crucial in understanding the rest of the chapter. Though it consists of only 2 words, this verse is translated in several different ways. Here are some examples from English Translations:

> **Yusuf Ali**: "darkening and changing the color of man"
> **Marmaduke Pickthall**: "It shrivelleth the man"
> **T.B. Irving**: "as it shrivels human (flesh)."
> **M. H. Shakir**: "It scorches the mortal."
> **MaulaHUM Muhammad Ali**: "It scorches the mortal"
> **N. J. Dawood**: "it burns the skins of men."
> **Muhammad Asad**: "making (all truth) visible to mortal man."
> **Rashad Khalifa**: "obvious to all the people."

Those who do not know Arabic might think that the words are really difficult to understand and translate. In fact, the meaning of these two words, LaWwaHa and BaSHaR is very clear in the Quranic context. The word LaWwaHa, which comes from the root LWH, is the sister of the word LaWH (85:22) and its plural aLWaH. The plural form aLWaH is used in verses 7:145, 150, 154 for the "tablets" given to Moses, and in verse 54:13 for broad planks used by Noah to build his ark. The medieval commentators, not knowing the mathematical implication of the verses, mostly chose an unusual meaning for the word: scorching, burning, shriveling, etc. Ironically, most of them did acknowledge the obvious meaning of the word as "open board, tablet" (See Baydawi, Fakhruddin Er-Razi, etc.) Few preferred the "obvious" to the obscure. For instance, Muhammad Asad, who had no idea of the mathematical code, preferred the most obvious meaning. Rashad Khalifa who fulfilled the prophecy and discovered the implication of the entire chapter reflected the same obvious meaning. That

---

more rituals and righteousness (such as, nawafil prayers); to validate superstitions (such as, magic; worshiping the black stone near Kaba); to prohibit certain things and actions (such as, prohibiting drawing animal and human figures, playing musical instruments, chess); to import Jewish and Christian beliefs and practices (such as, circumcision, head scarf, hermitism, using rosary); to resurrect pre-islamic beliefs and practices common among Meccans (such as, intercession; slavery;); to please crowds with stories (such as the story of Mirage and bargaining for prayers); to idolize Muhammad and claim his superiority to other messengers (such as, numerous miracles, including splitting the moon); to defend hadith fabrication against monotheists (such as, condemning those who find the Quran alone sufficient); and even to advertise a particular fruit or vegetables (such as, the benefits of date grown in Ajwa farm). In addition to the abovementioned reasons, many hadiths were fabricated to explain the meaning of the "difficult" Quranic words or phrases, or to distort the meaning of verses that contradicted to fabricated hadiths, or to provide trivial information not mentioned in the Quran (such as, Saqar, 2:187; 8:35...)

"obvious" meaning, however, was obscured by the smoke of "scorching fire" burning in the imaginations of generations before him.

In 7:145; 7:150; 7:154, the word alwah, the plural of lawha is used to depict the tablets on which the Ten Commandments were inscribed. In 54:13 it is used to describe the structure of the Noah's ship that made of wood panes. In 85:22 the same word is used for the mathematically protected record of the original version of the Quran. As for the lawaha of 74:29, it is the amplified noun-adjective derived from the root of verb LWH, meaning open tablets, succeeding screens, obvious, manifesto, or clearly and perpetually visible.

Ironically, the Quran uses different words to describe burning or scorching. For instance, for burning the derivatives of *haraqa* (2:266; 3:181; 7:5; 20:97; 21:68; 22:9; 22:22; 29:24; 75:10), or for scorching the derivatives of *salaya* (4:10; 4:30; 4:56; 4:115; 14:29; 17:18; 19:70; 27:7; 28:29; 29:31; 36:64; 38:56; 38:59; 38:163; 52:16; 56:94; 58:8; 69:31; 74: 26; 82:15; 83:16; 84:12; 87:12; 88:12; 92:15), or *nadaja* are used (4:56).

Again, we should note that the understanding of pre-1974 commentators was not without basis. Though their understanding did not rely on the Quranic usage of the words, and created some problems (such as explaining the verse 74:31), they had some justifiable excuses to understand the way they understood. The word *lawaha* also meant burn and *bashara* was another word for skin in Arabic language. As I mentioned above, the multiple meaning of these verses allowed the impatient pre-1974 generations to have an understanding, though a temporary and not primarily intended one. In fact, it was better for them to have patient and not rush to speculate on these verses without knowledge (20:114; 75:16-19). It was the computer generation destined to understand their real meaning (10:37-46).

Does "Bashar" in 74:29 Mean Skin OR Human?

The translation of the second word, BaSHaR is also among the distorted one. Many old commentaries translated it as "skin" rather then "human being" or "people" or "humans." For instance, N. J. Dawood parrots such a translation. The meaning of BaSHaR is also obvious in the context of the Quran. The word BASHAR occurs in the Quran 36 times. It is also mentioned as BASHARAYN (two bashars). If we exclude the BASHAR of 74:29 for the sake of the argument, we see that the word BASHAR is used to mean human beings in all 36 verses: 3:47; 3:79; 5:18; 6:91; 11:27; 12:31; 14:10; 14:11; 15:28; 15:33; 16:103; 17:93; 17:94; 18:110; 19:17; 19:20; 19:26; 21:3; 21:34; 23:24; 23:33;

23:34; 25:54; 26:104; 26:186; 30:20; 36:15; 38:71; 41:6; 42:51; 54:24; 64:6; 74:25; 74:29; 74:31; 74:36.

Then, why those who have allergy against the code 19 still insist to translate the word BaSHaR as skin while in its all 36 occurrences the word BaSHaR there is not a single instance where the word BaSHaR is used to mean skin; but always used to mean human beings? Especially, despite the fact that the word BASHAR occurs thrice in the same Chapter: after 74:29, in 74:31 and 74:36, and the fact that the verse 74:36 witnesses for the 36th times for "human beings," and that they were obliged to translate the last two occurrences as "human beings," how could they still insist on translating the BaSHaR of 74:29 as "skin." Since the prophetic verses did not make any sense for them before its fulfillment in 1974, translators and commentators of the Quran had an excuse to translate it differently before the fulfillment of the prophecy. However, after learning the discovery, none has an excuse to still continue parroting the overly stretched meaning.

Furthermore, the Quran uses "GeLD" for skin. This organic wrap that protects our body and provides tactile sense is referred in 13 verses with GLD and its derivatives: 4:56; 4:56; 16:80; 22:20; 24:2; 24:2; 24:4; 24:4; 39:23; 39:23; 41:20; 41:21; 41:22.

In short, while the Quran consistently uses GLD and its derivatives for skin and related words and while it consistently uses BShR for human beings, a convincing reason must be provided for ignoring all these examples and translating the BShR of 74:39 as "skin." The pre-1974 generations can be excused for trying to understand these verses even if it meant accepting some uncommon usages or dialects. As for post-1974 generation, if they are still stuck with hell and fire, then they share the same disease with the disbelievers of the past (2:75; 4:46; 5:13; 5:41).

After the fulfillment of the prophecy and divine clarification of its context, it is distortion to render this two-word verse 74:29 as "shrivels/scorches the skin" or "shrivels/scorches the man" rather than "it is clearly visible for human beings" or "it is a manifesto/successive-screens for human beings." However, the repetitive stretch and distortion committed by the post-1974 translators/commentaries on the many verses of Chapter 74 is bizarre and extraordinary. This pattern is prophetically described by verses 2:18; 3:7; 11:28; 41:44; 17:72; 25:73; 27:81.

Does The Pronoun "Hiya" (She) in the End of 74:31 Refer to the Hell OR to the Number 19?

We know that verse 74:30 does not qualify the reference of 19. It pulls our attention to a NUMBER ALONE. The verse does not say "on it nineteen angles" or "on it nineteen guards" or "on it nineteen this or that." The verse says "on it nineteen." Period. Verse 74:31, after informing us that the number of guardians of hell is 19, isolates this number from hell again describes its role.

Yet, those whose minds stuck in Hell cannot even notice this evident emphasis to the number. Thus, they violate the general grammar rules and universal linguistic logic and identify the reference of HIYA (SHE) in the end of 74:31 as HELL FIRE.

According to the grammar rules and common linguistic logic, the reference of a pronoun should be sought in proximity first. Sure, if there is no compelling empirical or rational reason for skipping a closer noun. For instance, "Yesterday, at Mary's home I saw Lisa. She was very thoughtful." In this sentence grammar and logic leads us to think that the thoughtful person was Lisa not Mary. But, "Yesterday, in Delhi's streets on the back of an elephant I saw Lisa. She was swinging her trunk/hose left and right and walking majestically." (I had given the original example in Turkish and in Turkish the same word is used for both elephant's trunk and hose. For didactic purposes please assume that it is the same in English.) Though grammatically the best candidate for reference of SHE is Lisa, we have an empirical reason to ignore the general rule since the one who would swing her trunk/hose is the elephant. But if these sentences are in a story and if we are told that Lisa was walking in the street with a hose in her hand, then the SHE in the sentence may equally refer to elephant OR Lisa. There will be ambiguity and if it is intentionally done then we wonder the purpose of the ambiguity. But, if the sentences continue, "When she hit the trunk/hose in her hand to a store sign...," then we become sure that it was a hose not trunk. "Your mother had given you a walnut and a book. Have you eat and finished it?" A sound mind can easily deduce that that "it" refers not to the book, but to the walnut. But, if the question were "Your mother had given you a walnut and a book. Have you finished it?" then, an ambiguity will arise. However, if the context of the question is known then we can deduce whether the query is about finishing the book or the walnut or both.

The pronoun HIYA (SHE), based on the rules of grammar, the emphasis of the verse in its context, and based on the rhetorical superiority of its meaning, must be referred to the grammatically feminine closest noun IDDATAHUM (their number, that is Nineteen). How can we explain the act of skipping

IDDATAHUM (their number, Nineteen) by ignoring the grammar, context, emphasis, and rhetoric, and insisting to reach the FIRE?

We will see in the miraculously elegant and prophetic expression of the Quran that hypocrites and unappreciative people have deserved the SaQaR penalty both in this world and hereafter. In this world as the 19-based mathematical code of the Quran; they will be intellectually tormented by its powerful evidences. And in the hereafter as the 19 guards of symbolic hell-fire; they will be convicted to eternally face the number 19 in the presence of guarding angles.[5]

Does The "Zikra" (Reminder) In 74:31 Refer To Hell OR The Evidence?

This brings us to the last phrase of verse 74:31 "it/this is a warning/reminder for mankind." Old commentaries and their parroting contemporaries refer "it/this" to the FIRE or HELL.

Those who have difficulty in accepting the possibility of an intellectual argument in The Hidden One (Muddathir), insert a HELL in the translation or the understanding of the last phrase of 74:31:

"This (HELL) is a warning for people."

This way, they transform the ZiKRa to a warning to be scared, to a penalty to avoid. However, all the derivatives of ZKR, including ZiKRa, are used in the Quran 273 times; and if we include DKR which is mentioned 7 times; this word is mentioned 280 times in the Quran and NOWHERE is it used to describe HELL or FIRE. You may check for yourself all the 21 verses where the word ZiKRa is mentioned: 6:68; 6:69; 6090; 7:2; 11:114; 11:120; 21:84; 26:209; 29:51; 38:43; 38:46; 39:21; 40:54; 44:13; 50:8; 50:37; 51:55; 74:31; 80:4; 87:9; 89:23. You may also check the two verses where the word is suffixed with pronouns ZiKRaha and ZiKRahum: 47:18; 79:43.

More interestingly, however, you will find another derivative of the same word "ZiKRa" (reminder) is used in verse 74:49 as "taZKiRa."

"Why do they turn away from this message/reminder (taZKiRa)?"

The disbelievers had ignored the reminder! Obviously, hell cannot be that reminder, since you are not supposed to embrace and face hell while you are alive on this earth. You are expected to turn away from Hell and turn to the

---

[5] For an article titled "Eternal Hell and Merciful God" visit www.yuksel.org OR www.19.org.

message (taZKiRa). Therefore, taZKiRa and ZiKRa cannot be both HELL and the MESSAGE of the Quran. All the derivatives of ZKR are consistently used in the Quran for the divine message that revives and bestows eternal happiness.

Because of the self-ignited and self-inflicted hell fire burning in their minds, the disbelievers and hypocrites ignore the PLEASANT meaning associated to the word 280 times, and fight with us to keep it as HELL. Well, this was exactly what they were promised and what they deserve. Their initial decision, their prejudice, ignorance and arrogance, have led them to bet their faith on hell, a questionable one chance against 280 chances!

How does the Number Nineteen Increase the Faith of Believers and also Remove Doubts From the Minds of People of the Book as Promised in 74:31?

With the exception of Rashad Khalifa, all the translations and commentaries listed above have had big problem with this question. Since they did not accept it as a prophecy left to be fulfilled in the future, and since they rushed to speculate on them without knowledge, they dug a hole for themselves. Surely, they always could deceive themselves and others by claiming that THEIR FAITH was increased because of the association of the number 19 with the guardians of hell. This claim obviously would be a non-falsifiable claim. One can say anything about his or her own faith. However, this claim could not explain why specifically 19 and not another number or another thing. Let's assume they could pretend that their faith was really increased by the nineteen guardians of hell, but how could they explain the other functions of the number nineteen? How could this number nineteen remove the doubts of people of the book? The proponents of HELL-19 just could not make up an answer for this, since the issue involved an OBJECTIVE fact, not a subjective claim.

Interestingly, those who saw the Hell-fire in these prophetic verses claimed that the people of the book would believe in the Quran because they would see that the number of guardians of hell was also nineteen in their own books! What? Please read it again. Yes, this is the explanation of all the prominent Sunni and Shiite scholars of the past and unfortunately it is exactly parroted by the contemporary ones!

First, there is no such a statement in the Bible. Second, even if there was such a statement there would not be anything special about nineteen, since there are many principles, stories of prophets, and instructions mentioned in the Quran that are also in the Bible. In fact, the very existence of similarity had an opposite effect on many people of the book; they claimed and still claim that the Quran

89

was plagiarized from the Bible. Third, none has been persuaded regarding the authenticity of the Quran because the number of guardians of hell is nineteen! I personally communicated and met dozens of Christians or Jews whose doubts about the authenticity of the Quran was removed because of their witnessing the mathematical miracle of the Quran based on the number nineteen. You can find some of these people on Internet forums. This is happening despite the aggressive misinformation and disinformation campaign carried out by numerous Sunni and Shiite groups. Compare the prophetic role of MIRACLE-19 to the role of HELL-19 in their powers of removing doubts from the minds of People of the Book. Two DECADES versus more than 14 CENTURIES, and several THOUSAND people versus more than a BILLION people.

The HELL-19 advocates cannot show a single Christian or Jew who converted to their version of Islam because their doubts were removed after they noticed that the number of guardians of hell is also 19 in their books! This failure alone should be sufficient to wake up those who are still eager to see Hell in these verses. Yet, they still shamelessly repeat this lie to defend their hellish version.

Some opponents of the CODE-19 even tried to use a phrase in verse 74:31 against understanding of the very verse. They quote the fifth function of the number nineteen, "to expose those who have disease in their hearts, and the disbelievers; they will say, 'What did God mean by this allegory?'" and then charge us: "You 19ers are asking this question; therefore, you have disease in your hearts." First, none of us have asked this question, including Rashad Khalifa. We happened to learn the meaning of number nineteen even before asking such a question. We know the meaning of this allegory and we do not challenge that meaning. Furthermore, this question is not a question of a curious person who sincerely wants to learn the meaning of the Quran. This question is the challenge of arrogance and ignorance. It is a question that many still millions of Sunni and Shiite people are asking themselves. They will keep asking this question as long as they follow their ancestors blindly and worship Prophet Muhammad and religious scholars.

Does the "Ha" In 74:35 Refer to Hell OR the Number 19? Does the Word "Kubra" in the Same Verse Refer to "Great Disasters and Calamities" OR to "Great Signs and Miracles"?

When we reflect on the context and the usage of the words of verse "This is one of the greatest" we can easily deduce that this refers to the number 19 and its prophetic function as a conclusive evidence for the authenticity of the Quran and a great test for people. Unfortunately, many of those who did not witness the incredible mathematical structure based on the number 19, related this verse to Hell, again.

Let's one by one see several different translations of this verse. (Again, the pre-1974 translations have an excuse in their mistranslations since they did not witness the fulfillment of the prophecy of the mathematical miracle):

> **Yusuf Ali**: "this is but one of the mighty (portents)"
> **Marmaduke Pickthall**: "Lo! This is one of the greatest (portents)"
> **T.B. Irving**: "Surely it is one of the gravest (misfortunes)."
> **M.H. Shakir**: "Surely it (hell) is one of the gravest (misfortunes)."
> **MaulaHUM Muhammad Ali**: "Surely it is one of the gravest (misfortunes)"
> **N. J. Dawood**: "it is a dire scourge."
> **Muhammad Asad**: "Verily, that [hell-fire) is Indeed one of the great [forewarnings]"
> **Rashad Khalifa**: "This is one of the greatest miracles."

I leave for you to assess the contextual reference of the pronoun in 74:35. Here, I will draw your attention to the descriptive word KuBeR, which is the plural of KaBeeR derived from KBR (great). The derivatives of KBR are used in the Quran to describe positive or negative events, people, or things. For instance, in verses 20:23; 44:16; 53:18; 79:20 it is used to describe God's ayaats or miracles. The derivatives of the same word are used for disaster in 79:34 and hell in 87:12.

In other words, if we do not consider the context of the group of verses, we may understand the reference of THIS IS ONE OF THE GREATEST as either HELL FIRE or one of God's greatest MIRACLES. Through this verse alone, one may see flames of fire or signs of a miracle. Nevertheless, considering the context of this verse it becomes clear that the first vision is a man-made hallucination and the later a divine gift.

Does the Word Nazer in 74:36 Refer to the Hell Fire OR to the Devine Evidences?

NAZER is an adjective derived from the root NZR and means "WARNER" Throughout the Quran the various derivatives of this word occur 130 times. Here are the verse numbers where the form NaZeR (warner) occurs 44 times: 2:119; 5:19; 5:19; 7:184; 7:188; 11:2; 11:12; 11:25; 15:89; 17:105; 25:1; 25:7; 25:51; 25:56; 26:115; 28:46; 29:50; 32:3; 33:45; 34:28; 34:34; 34:44; 34:46;

35:23; 35:24; 35:24; 35:37; 35:42; 35:42; 38:70; 41:4; 43:23; 46:9; 48:8; 51:50; 51:51; 53:56; 67:8; 67:9; 67:17; 67:26; 71:2; 74:36.

In NONE of these verses the word NaZeR is used for HELL or FIRE. In these verses NaZeR describe God's messengers, books, revelation, and signs.

Then Why Those Who Claim to be Muslims do not Understand or Appreciate the Prophetic Fulfillment of These Ayaat (Revelations/Signs/Miracles)? Why They See and Work Hard to See Hell-Fire, Instead of the Precise and Universal Great Miracle?

Surely, previous generations who were not aware the discovery of Code 19 via computer had an excuse to twist the meaning of obvious verses in order to make sense out of them. They could not imagine that the 19 was the code of an elaborate mathematical system. Nevertheless, some early commentators of the Quran who did not limit the understanding of the Quran with *Hadith* and *Sunnah* sensed something beyond HELL from these verses and discussed them as alternative understanding. For instance, Fakhruddin er-Razi in his famous Tafsir al Kabir, in his 23rd comment on Basmalah, speculates on the 19 letters of Basmalah and lists numerous implications, such as the difference between the number of 5 prayers and the number of hours in a day; the number of physical, intellectual, and emotional faculties that humans are blessed with, etc. Among the modern commentators who lived before 1974, Muhammad Asad is the only one that subscribes to Razi's interpretation of the number nineteen of 74:30.[6] However, Asad too could not avoid but see Hell everywhere.

---

[6] After translating the verse 74:30 as "Over it are nineteen (powers)" Muhammad Asad explains his parenthetical comment with the following footnote:
"Whereas most of the classical commentators are of the opinion that the "nineteen"' are the controllers that act as keepers or guardians of hell, Razi advances the view that we may have here a reference to the physical, intellectual and emotional powers within man himself: powers which raise man potentially far above any other creature, but which, if used wrongly, bring about a deterioration of his whole personality and, hence, intense suffering in the life to come. According to Razi, the philosophers (arbab al-hikmah) identify these powers or faculties with, firstly, the seven organic functions of the animal - and therefore also human - body (gravitation, cohesion, repulsion of noxious foreign matter, absorption of beneficent external matter, assimilation of nutrients, growth, and reproduction); secondly, the five "external" or physical senses (sight, hearing, touch, smell and taste); thirdly, the five "internal" or intellectual senses., defined by Ibn Sina - on whom Razi apparently relies - as (1) perception of isolated sense-images, (2) conscious apperception of ideas, (3) memory of sense-images, (4) memory of conscious apperceptions, and (5) the ability to

92

The opponents of the miracle 19 are no different than those who rejected the entire Quran in the past. The description of these verses fits both groups, which is another miraculous aspect of the Quran:

74:21 He looked.
74:22 He frowned and scowled.
74:23 Then he turned away arrogantly.
74:24 He said, "This is but an impressive illusion/magic!"[7]
74:25 "This is human made."

In Conclusion:

Those who reject to witness the miracle 19 are insulting the wisdom of God. According to them God's only answer to someone who challenges the authenticity of the Quran is: "Get lost! I will burn you in the hellfire" Depicting God as someone who cannot engage in an intellectual argument with His opponent, but can only employ the cheapest method of persuasion (threat) cannot be the path of believers who are warned by the incredible prophecies of this wonderful chapter, The Secret.

Turning back to the question: Then why those who claim to be Muslims do not understand or appreciate the prophetic fulfillment of these ayaat (revelations/signs/miracles)? Why they see and work hard to see hell-fire, instead of the precise and universal great miracle?

The Quran provides prophetic answers to this question in numerous verses. Please read and reflect on them:

---

correlate sense-images and higher apperceptions; and, lastly, the emotions of desire or aversion (resp. fear or anger), which have their roots in both the "external" and "internal" sense-categories - thus bringing the total of the powers and faculties which preside over man's spiritual fate to nineteen. In their aggregate, it is these powers that confer upon man the ability to think conceptually, and place him, in this respect, even above the controllers (cf. 2:30 ff. and the corresponding notes; see also the following note)."

[7] Magic is not considered an extraordinary, paranormal event by the Quran. Its influence on people is described by its two aspects: illusion and suggestion/bluffs/hypnosis (7:103-120). Magic is an art of influencing gullible people via illusion and hypnosis (7:11; 20:66; 2:102). So, when the unappreciative disbeliever describes the Quran OR the mathematical miracle of the Quran as "influencing magic," what should be understood is that it is just a manipulation and hyperbole.

29:1   A1L30M40

29:2   Did the people think that they will be left to say, "We acknowledge" without being put to the test?

29:3   While We had tested those before them, so that **God** would know those who are truthful and so that He would know the liars.

29:4   Or did those who sinned think that they would be ahead of Us? Miserable indeed is their judgment!

2:17   Their example is like one who lights a fire, so when it illuminates what is around him, **God** takes away his light and leaves him in the darkness not seeing.

2:18   Deaf, dumb, and blind, they will not revert.

7:146   I will divert from My signs those who are arrogant on earth unjustly, and if they see every sign they do not acknowledge it, and if they see the path of guidance they do not take it as a path; and if they see the path of straying, they take it as a path. That is because they have denied Our signs and were heedless of them.[8]

29:54   They hasten you for the retribution; while hell surrounds the ingrates. (Please read from 29:48)

3:7   He is the One who sent down to you the book, from which there are definite signs; they are the essence of the book; and others, which are multiple-meaning. As for those who have disease in their hearts, eager to cause confusion and eager to derive their interpretation, they will follow what is multiple-meaning from it. But none knows their meaning except God and those who are well founded in knowledge; they say, "We acknowledge it, all is from our Lord." None will remember except the people of intellect.[9]

25:73   Those who when they are reminded of their Lord's signs, they do not fall on them deaf and blind.

---

[8]   The plural word AYAAT means revelation, sign, evidence, or miracle. However, its singular form AYAT is mentioned in the Quran 84 times and in all its occurrences it means sings, evidence or miracles, not literary revelation of the Quran.

[9]   This verse is a crucial verse in understanding the Quran and ironically this very verse is one of the most commonly misunderstood verses in the Quran. For a detailed argument on this verse please visit: www.yuksel.org or www.19.org.

6:158   Do they wait until the controllers will come to them, or your Lord comes, or some signs from your Lord? The day some signs come from your Lord, it will do no good for any person to acknowledge if s/he did not acknowledge before, or s/he gained good through his/her acknowledgement. Say, "Wait, for we too are waiting."

6:25   Among them are those who listen to you; and We have made covers over their hearts to prevent them from understanding it, and deafness in their ears; and if they see every sign they will not acknowledge; even when they come to you they argue, those who reject say, "This is nothing but the tales from the past!"

6:26   They are deterring others from it, and keeping away themselves; but they will only destroy themselves, yet they do not notice.

27:84   Until they have come, He will say, "Have you denied My signs while you had no explicit knowledge of them? What were you doing?"

78:27   They did not expect a reckoning/computation.

78:28   They denied Our signs greatly.

78:29   Everything We have counted in a record.

74:56   None will take heed except if **God** wills. He is the source of righteousness and the source of forgiveness.

6

A Letter to Signs Magazine

This letter was written 1995, when I was still with the Submitter's International, before an idol-carving gang infected the group and became its loud voice.

". . . Indeed, we believe in the message he brought" (7.75)

Editor of Signs, peace be on you.

I was very happy to see the Signs magazine promoting the message of the Quran alone. However, the third issue was a surprise. You have dedicated most of the pages of the magazine to attacking Dr. Rashad Khalifa, a monotheist believer. Though I do appreciate some points you raise, I find your evaluation of Rashad's work biased.

Personally, I consider myself lucky to have met him and assisted him in Masjid Tucson. I was a Sunni political activist and author advocating Sunni Islam in books published in Turkey before 1986. I was tortured and imprisoned for four years because of my articles in magazines calling for an Islamic revolution in Turkey. After a year-long correspondence and debate with Rashad I came to the conclusion that the Quran is the only source of guidance. It was no surprise that my acceptance of the Quran alone brought excommunication, attacks from religious media and death threats from my previous readers.

I believe that Rashad Khalifa was a messenger of God. If you ask why do I believe, I can list quite a few reasons. However, I prefer to repeat the answer of the believers who supported Salih: "indeed, we believe in the message he brought." Rashad is now with his Lord, and he is dead as far as we are concerned. He will be judged by God Almighty on the day of judgment, when every one of us, including all the messengers will worry about their own neck (21:28). Rashad IS no longer a messenger. The Quran is a living and talking messenger until the Last Hour. Therefore, I'll stick with the Quran alone, inshaallah. I hope you will receive this letter with empathy, not with paranoid theories of "conspiracies."

To make it convenient for the readers to follow the argument, each point is numbered. May God guide us to the truth, since The Truth is one of the attributes of God. I'm willing to dismiss any part of my argument if you enlighten me with sufficient reason and evidence from the Quran, inshaallah. May God give us the patience to understand and tolerate each other. We may not be able to agree on every issue; however, we should be extremely careful about labeling or condemning each other.

Rashad was a fallible human messenger

1. Rashad never claimed to be infallible, nor did he claim that his translation, with parentheses, footnotes and appendices is a revelation, since he never doubted that the Quran is the last book revealed to the last prophet. However, I agree with him that his translation is authorized by God for its clear emphasis on worshipping God alone and not adding any other sources (including Rashad's) to God's word, which is perfect and fully detailed. Rashad was a student of the Quran, trying to purify his mind from the atmospherics of his traditional past that was preventing him from getting the clear broadcasting of the divine message. He was usually humble in acknowledging his errors. For instance, he encouraged us to edit and discuss his second revision of his translation verse by verse. During this intense consultation period we had numerous discussions. We continuously learned from each other according to God's will. During that period, he was persuaded to correct some of the mistakes of the first edition. For instance: 2.106; 3:97*; 7.75; 11.87; 11.88; 12.88; 18.83; 21.96; 21.112; 24.35; 27.42; 37.63; 38.44; 39.6; 43.61; 56.79; 72.7; 72.18; 74.31; 96.2.

Later, when he re-revised his translation, he continued correcting his errors. For instance, 4.176!; 6.74*; 12.88; 18.83; 30.3; 38.59; 95.5!, etc.

Briefly stated, he never claimed to be infallible. The three revised editions of his translations are blatant witnesses to the fact that he was in a continual learning process and open-minded to reasonable criticism. If he were alive, he would surely make many corrections in his third revision. In fact, it is the experience and fate of all translators. Every time I edit my Turkish translation of the Quran which has been in my computer since 1990, I find errors caused by insufficient information, imperfect attention, shortcomings, linguistic problems and unintentional mistakes. Nevertheless, I still believe that my translation will deliver the message, inshaallah. Thank God we have the original Quran that we can refer to anytime we have a question. It would be an unfortunate repetition of history if one day some of those who responded to Rashad's call to worship God alone claim that Rashad's translation, Quran The Final Testament, is "the English Version of the Quran," or is "the ultimate English translation of the Quran," or "a revelation from God." I do see signs pointing to this constant human tendency: hero-worship.

All messengers were fallible humans for a good reason

2. All previous messengers were humans, not angels. During their mission they lived like their contemporary fellows. They made mistakes, sometimes grave ones. Those mistakes, paradoxically, functioned as a blessing for believers, and an excuse for disbelievers and a test for hypocrite idol worshipers: believers would focus on God alone and stop idolizing messengers, while disbelievers would use those weaknesses and mistakes as an excuse for their disbelief, and idol worshipers would claim the infallibility of the messenger and try to defend the evident human errors and attribute them to The Most Wise.

Please imagine that you are dwelling in Madina during Muhammad's time. He has some friends and many enemies. You hear conflicting news from people about his personality and his message. Meanwhile, you witness some of his weaknesses and shortcomings. For instance, you see him trying to hide his intention or revelation from people regarding the estranged wife of his adopted son (33.37), you hear him favoring a rich person and ignoring a poor blind man (80.1-11), you see him rushing into speculating on some verses without sufficient knowledge (20.114; 75.16-19), you experience the "devil's interference" in his wishes followed by a great communal chaos and feud (22.52-55), you witness his tendency of trying to compromise with his categorical enemies (17.74), etc. What would you do? Obviously, you could react in three ways. You could either dismiss his claim of messengership, or accept him as a human messenger, or idolize him as an infallible messenger and interpret those errors and shortcomings as virtue.

I believe that if you use the same standard of criticism you will not dwell on some human errors and weaknesses on the part of Rashad Khalifa. You can appreciate many positive things if you can overcome your prejudices. Please compare Rashad's translation with other translations regarding the following verses: 2.1; 2.26; 2.54; 2.106; 2.171; 2.222; 2.224; 3.7; 7.157; 8.35; 10.1; 15.87; 16.44; 17.46; 20.15; 21.87; 22.15; 24.31; 25.5; 33.56; 38.44; 43.61; 56.79; 61.6; 74.29-35.

Rashad did not receive a book; he advocated Quran alone

3. The footnotes, subtitles and appendices of Rashad's translation, "Quran the Final Testament," are not revelations besides the Quran, but his personal understanding that is subject to human filtration, misunderstanding, interpretation and speculations. He advocated the Quran alone; the Arabic and mathematically coded book revealed to Muhammad, containing 114 chapter and 6346 verses including all *Basmalah*s.

Unfortunately, there are now some who consider his translation infallible, and accept the footnotes, subtitles, and appendices as the second source besides the Quran. To justify their new source, they frequently refer to the mistranslated verse 49:1. For instance, they prefer "consensus" of Appendix 36 to "consultation" of 42:38. To me, those people are no different than the pioneer Muhammadans. They are not helping Rashad by sanctifying the obvious mistakes in his translation. Just two simple examples: To the criticism regarding the subtitle of verse 2:172, which reads "Only Four Meats Prohibited," one of his companions tried to defend the error by claiming that "blood is meat." Rashad's unintentional error was transformed to a bizarre assertion by his gullible defender.

Another example--which is prevalent in all translations--is the result of translating two different words by a single word thinking that they are synonymous. These kinds of errors are difficult to avoid, since it is sometimes impossible to find one-to-one correspondence in lexicon of languages. For instance, by translating the two slightly different words, "Qaryah" and "Ummah," with a single word, "community," Rashad created a contradiction between verse 25.51 and 35.24 unintentionally. Rashad's translation, I believe, will guide the sincere seekers of the truth, and simultaneously it will misguide those who consider or expect him to be infallible. Indeed, it is the nature of the infallible original Quran to guide and misguide people depending on their innermost intentions (17.82; 3.7).

For those who claim the infallibility of Dr. Khalifa, I would like to give a sample of verses that I think carry some minor or important translational problems: 2.114; 2.233; 2.275*; 2.282*; 4.34&*; 4.79*; 7.193; 7.157; 8.64; 10.34!; 11.54; 11.87; 12.37; 14.4; 16.75*; 18.16*; 19.26!; 20.96&*; 20.114; 21.96*; 21.90x21.73; 22.78x10.72; 25.30*; 29.12x29.13; 32.5!; 34:41; 35.24x25.51; 43.11x41.12*; 43:36*; 47.11x42.15; 49:1x38.26&7.3; 56.83-85; 65.12*x42.29; 73.15!; 75.27; 75.31?; 87.6 (Asterisks are for footnotes and/or subtitles, exclamation marks for missing phrases, and "x" for contradictions.) No one can represent God

4. The language of the article titled 'Khalifa' Satan, a temporary god on Earth? was distasteful, at best. Nevertheless, I agree that the idea "Satan: A Temporary 'god'" is problematic. It may contradict 43.45; 8.17; 3.189; 5.120; 36.83, if it is taken literally. Knowing Rashad and being familiar with his work, I believe that he did not intend the literal meaning of "god." He tried to convey his intention by putting it in quotation marks. Personally, I think that Adam is more suitable to be "Khalifa" (a ruler) in the context of verse 2.30, and the entire Quran, since the word "Khalifa" is consistently used for humans. But, it is still possible to understand the "Khalifa" of 2.30 as Satan, meaning "a ruler." To support this understanding one can refer to 7.15, 143; 15.8,38, 42; 36.60; 16.99; 82.19, etc.

I agree with the author of the article that the traditional translation of the word "Khalifa" in 2.30, that is, "representative," is wrong. I applaud his statement that "the concept of 'representative' is the invention of self-interested individuals who wanted to exploit people in the garb of 'God's vicegerency.'" If we want to be fair to Rashad's translation, however, then there is a problem with the criticism. If by the word "representative," the translator (Rashad), meant "the one who acts on behalf of God" then the criticism is right to the point. However, Rashad could not have meant that, since he used the word "representative" of 2.30 to refer to Satan, the enemy of God. Obviously, Satan cannot be both God's enemy and His representative! It seems that Rashad's translation has a major language problem in this particular verse.

The Mathematical Miracle

5. I am really surprised about your apathetic attitude regarding the assessment of the 19, as the code of the mathematical system of the Quran. I don't think that you really need the confirmation of any "recognized authority" to understand and appreciate the mathematical system of the Quran, which is very simple to understand and easy to examine. All you need is a basic knowledge or intuition of mathematics and the will to dedicate some time to verify the count by yourself. You can reduce the amount of time needed for sufficient examination by dividing your friends into groups and comparing their results. Also you can use Fuad Abdulbaqi's Index of the Quran to verify the count of words. I did this when I was in prison. With the help of 15 inmates I confirmed most of the counts of letters. This manual examination was more than enough to convince me that the Quran had a superhuman mathematical structure embedded in its natural language. You can do a similar examination, since you have the time and resources to publish an international magazine.

Here, I want to add that I do not agree with all of the calculations presented in Rashad's translation and Muslim Perspective. Some of them have no mathematical significance. It is easy to dismiss them with the laws of probability. I often witness manipulations of numbers and speculations made by the members of my community. For instance, some of them claimed that a huge asteroid would hit Saudi Arabia on May 19, 1990 to fulfill the prophecy of Smoke in 44.10. With an odd blend of sincerity, arrogance and ignorance they attributed this doomsday "prophecy" to God Almighty.

Their lack of basic knowledge and intuition of mathematics led them to this embarrassing and dangerous claim. They produced many pieces of so-called mathematical evidence collectively. Unfortunately, some of them were published in the Muslim Perspective despite a strong opposition from some of us. The same group who produced or esteemed all those manipulations are still peddling their new discoveries of "miracles." Furthermore, some of the members of this doomsday gang are working constantly to carve an idol of Rashad! Nevertheless, this should not be an excuse for intelligent people to undermine the great mathematical miracle of the Quran. You can easily distinguish diamonds from pieces of glass, if you don't close your mind.

How long does it take to examine one word?

6. You ask "Another conspicuous discrepancy worthy of note is that in the early translation in appendix 1 under the title of 'simple facts,' the word 'God' is noted as having a count of 2,698 (19x142). In the later translation the count is also given as 2,698. But the two verses of Sura 9, one with the word 'God,' have been eliminated. How can the total remain the same?" You continue, ". . . It can be assumed that other calculations may also be wrong, but it is not within the scope of Signature Publications to carry out the necessary research." It will take less than 45 hours if you spend 60 seconds to check each word God. Let's be more generous and make it 90 hours. If you have 9 people (that will make three groups of three individuals) on your editorial board, you can independently examine each word three times in 30 hours. This is the slowest way of examining this crucial part of the "mathematical code" of the Quran, which according to your magazine is part of an impressive conspiracy against the Quran. You could reduce the needed time to less than one hour by comparing Rashad's results with Abdulbaqi's Index. Therefore, I believe that this job was well within the "scope" of your magazine before publishing this article. It is embarrassing for the students of the Quran not to know the frequency of the word "Allah" in the Quran during the last two decades! Please forget Rashad's errors (which are more even than you have noticed), just consult the Quran in your hand. I bet you will find its frequency is 2,699 in numbered verses, if your Quran has the "two extra verses" at the end of chapter Nine.

7. Regarding the mathematical structure of the Quran, I urge you to be open minded. By your initial bias, you may disqualify yourself from seeing the great miracle of the Quran that is mentioned in Chapter 74. If you close one of your eyes you cannot see the 3-D pictures that others can see. I have discussed this issue briefly in the last argument of 19 Questions For Muslim Scholars. I would like to hear your answers to my questions at the end of that argument.

Age of forty

8. Our "eccentric" belief regarding the age of 40 is an inference based on verse 46.15. Does age 40 bring an extra responsibility? Is it different from age 39 or 41 when guidance and righteousness are involved? Obviously, God distinguishes age 40 and indicates its importance as a crucial time for repentance. If repentance has any role in determining a person's situation in hereafter, then, why should it be "wishful thinking" to believe that The Most Merciful, The Forgiver will not punish a person who did not reach the age when repentance is strongly advised?

You refer to 4.48 "God does not forgive idolatry, . . ." and ask the question, "What about the those who idolize others beside God? Would all their sins also be absolved? Or would they be subject to certain qualifications?" After quoting the verse you conclude, "The Quran says nothing about all sins being forgiven for those under forty." If you take the Quran as a whole, you will see exceptions to the rule stated in other verses. For instance, if a person after practicing idolatry repents and practices righteousness, then God will forgive that person's previous sins. Obviously, this exception, that is repentance, is not stated in verse 4.48. Therefore, why do you not consider 46.15 bringing another exception, that is, age?

9. Do you really think that an 11 year-old person who goes to Church with his parents and worships Jesus willingly will go to hell forever if he dies before his 12th birthday? What about a child of 12, or 13, or 14, or 15, and so on and so forth? If you tell me that age is not important, the person's capacity and qualifications are important, then you should define those qualifications. You may say that you don't need to know at what age a person is supposed to be eternally responsible. Well, we will respect this position and expect you to respect others who happened to know that age by inference from a Quranic verse. Your lack of desire to understand something should not lead you to blame those who inquire.

A Muslim Jew, A Muslim Christian, A Muslim X

10. I agree with you that a Hindu cannot be considered righteous according to the Quran if he or she continues to believe in a cast system. However, I see there are some problems if we claim that a person cannot be considered a Muslim if he doesn't follow the Quranic laws. (I'm not sure whether you meant this or not.)

Verse 5.44 describes the Jewish prophets who ruled according to Torah as Muslims. The verse ends with the phrase "Those who do not rule in accordance with God's revelations are the disbelievers." Then, verse 47 of the same chapter reads, "The people of the Gospel shall rule in accordance with God's revelations therein. Those who do not rule in accordance with God's revelations are the wicked." Verse 48, after ordering prophet Muhammad to rule among them according to the Quran, states a divine fact, "For each of you, we have decreed laws and different rites. Had God willed, He could have made you one congregation. But He thus puts you to the test through the revelations. You shall compete in righteousness."

From these five verses we can derive the following points:

11. Ruling in accordance with the Torah or the Bible was a righteous act. 2. Muhammad's contemporary Jews and Christians were ordered to rule according to their own scripture. 3. There are differences among the laws and rites of those scriptures. 4. Each community should follow their own book consistently, without abusing or distorting the law therein. 5. Muhammad and his followers should rule according to the Quran alone. If the above conclusions are correct then the following or similar questions will be worthy of research: If a Jew does not eat the fat of any animal on the basis of the Torah, can we claim that he is an idol worshiper by referring to 6:145?

Therefore, considering the verses mentioned above, a person can be considered a Muslim Jew or a Muslim Christian, or a Muslim X, since the word Muslim means Submitter. However, I see the difficulty in details.

"Community of Rashad"?

12. The articles published in Submitters Perspective (SP) don't reflect the view of all the members, and does not publish all the submitted articles. None of the members is taking it as a source of guidance. (That is one of the reasons I'm still writing there!). We don't have Pope, nor do we have his official bulletin. I believe (and hope) that the majority of this little community is still alert against the idolization of Rashad. Therefore, please don't accuse all members of this community based on a particular article written by an individual.

I read the article that you were referring to labeling us as the "Community of Rashad." I found the expression exactly as you have quoted. But, the author was not using the "community of Rashad" to describe her community. Obviously, you had taken the odd expression out of context while rushing to find a label for us. If you read it again you will see that she was using that expression in the same way as the "community of Noah, community of Lot, etc," is used in the Quran. Briefly, she was referring to the disbelievers, not believers with the "community of Rashad."

I would like to remind you that we are trying our best not to be the "Community of Rashad." Inshallah we will always be able to say, "We are God's supporters." (61.14)

The two false verses

13. As far as 9.128-129. . . From 1974 (when the miraculous function of number 19 of chapter 74 was discovered), until 1985 (when the two false verses were exposed), for eleven years, no one knew that Rashad's computer data was one short regarding the frequency of the word God. Both the proponents and opponents of the miracle did not notice this crucial error for more than a decade. Why? Because, the most popular and accurate index of the Quran, Al-Mujamul Mufahras Lielfazil Quranil Kariym, had the same count 2698, when you add the one missing. (The index accepts the Fatiha's *Basmalah* as the first verse of the Quran, but fails to mention it in the list of the word "Allah")

If I had noticed the error in the beginning, I would, most likely, never have had the courage, or interest in studying the mathematical structure of the Quran. I would have refuted it outright and labeled it as a blasphemy of numerology. The miracle of the Quran would have been stillborn in the first days of its discovery. No one would have considered it worthy of examination. Even Rashad himself, then a traditional Muslim, most likely would not have accepted it under those circumstances. We strongly believe that God Almighty deliberately delayed this important problem for a simple reason: to publicize the miracle of the Quran, and give the Muslim world a chance and time to study it. After everyone got a good idea what 19 was all about, God let us discover the error. Everyone was put to the test: either follow the testimony of the Quran, or the testimony of your parents and the majority of people. Those who strengthened their faith according to the prophecy of 74.31 chose the testimony and signs of God, those who followed their parents and peers rejected the testimony and signs of God. This process, I believe, was a fulfillment of 6.158 and 3.179.

14. Please note the emphasize on "We" in verse 15.9: "We, indeed We, yes We have revealed the Zikr (reminder), and We will preserve it." I don't remember such an emphasis anywhere else in the Quran. The short verse refers to God Almighty four times in regard of preserving His book. What does that mean? To me, the message is obvious: it is not you (people) who will preserve the Zikr, it is God who will preserve it. Please note, the relation between the revelation and preservation in the verse. What is more appropriate than looking for a relation between the revelation and its protection? Is it possible that the revelation contains an automatic protection? How does God preserve the Zikr? Obviously, with its own revelation. How do we know, or what is the proof? We know this through the Zikra (74.31), which is "Nineteen" (74.30) and "One of the greatest" (74.37). Zikra (74.31) is the virus protection of the Zikr and it is encoded in it in a miraculous way. Don't you feel the allusion between Zikr and Zikra? "Sad, and the Quran that contains the Zikr. Those who disbelieve have plunged into arrogance and defiance" (38:1).

15. You can see many ancient documents, tablets and books in museums and special sections of libraries. They are preserved by human effort and technology for hundreds and even thousands of years. If the preservation of the Quran was similar to this "normal" and "ordinary" human affair, then, why should God emphasize Himself and give the impression that the preservation of the Quran is a "divine" and "unique" feature? It is not necessarily a divine merit for a book to be preserved by its zealot followers. For instance, Bukhari is a document which has been well-preserved for approximately 12 centuries.

16. After we have seen the Satanic nature of hadith and sectarian jurisprudence produced by our great ancestors and respected ulama, how can you expect us to trust them as the preservers of the Quran? According to your argument we have to rely on the history of Muslims to be convinced that the Quran is well preserved. Well, that history itself is written by those people who you claim to be the biggest liars on earth. When we look at their history regarding the Quran, it creates more doubt than it erases. According to that history, the argument regarding extra verses started just after prophet's death, not 19 or so years later.

What is your answer if one asks you the following question: You refer to the history (books) written by Muslims and claim that thousands of believers memorized the Quran that we have today. According to the history that you refer to, the Quran was not compiled in a book during the time of prophet Muhammad. Any claim by a Muslim (?) would be accepted if they could produce the second witness. In fact, some of the verses were accepted even without the second witness. There are hundreds of arguments regarding the Quran during the reign of the first four Caliphs. It is a historical fact (accepted by both Sunnis and Shiites) that Marwan burned the original Manuscript in order to stop the accelerating arguments over the Quran, which provides the best explanation for the disappearance of Muhammad's manuscript. What if then, the corrupt leaders of the Umayyad Caliphate officially accepted and added verses 9.128-129 to the manuscripts since they were accepted by many people? What if the opponents were oppressed and forced to hide their opposition? What if Shiite's underground claim regarding the distortion of the Quran is an exaggerated continuation of the early reaction against the original distortion?

If x million of them could agree on the authenticity of Bukhari, then it is possible that 2x million of them could agree on the authenticity of the Quran. If millions of Shiite could insert Ali's name in the Azhan, then it is possible that millions of Muslims could insert Muhammad's name in the Azhan. (Ironically, hadith books confirm this claim by narrating Azhan containing 19 words.) If millions of Christians can believe that their Bible is preserved by ignoring Nicene Conference and important textual problems, then it is possible that millions of Muslims could believe that their Quran is preserved by ignoring the early arguments, Marwan, and narrated textual conflicts. Obviously, there are differences in the degree, but not in the nature of the thing.

17. Now there are thousands of copies of the original Quran that do not contain 9.128-129 are circulated and read all around the world. Does this version of the Quran refute the divine promise in verse 15.9? If your answer will start, "No, the distortion made by a community of diverted people does not refute the verse, because. . . . ", then our answer will also start, "No, the distortion made by idol worshipers after prophet Muhammad centuries ago does not refute the verse, because. . . . " What is the difference between these two cases? If you resort to the number of years or the number of followers, then we resort to the number 19, which is a Zikra, a divine warning for human race. Additionally, we remind you of verse 5.100.

18. The Quran is in the heart of those who are blessed with knowledge (29.49). It is a numerically structured book (Kitabun Marqum) which is witnessed by those close to God (83.20,21), and not recognized by the rejecters (83.9,10). Even if idol worshipers add "verses" to the Quran, those who "have received knowledge" will be able to recognize God's revelation.

If you add or amalgamate a cheap element, say copper, to a golden ring, an expert who knows the property of the prime element gold, will be able to expose the counterfeit. The expert does not need to rely on the testimony of sellers who bring the gold to him. He will rely on the infallible testimony of the gold itself. He will test and examine the physical properties of the matter and make his decision. Similarly, a believer who is blessed by God to know the mathematical structure of the Quran, which is based on the prime number nineteen, can easily distinguish the false from the genuine. In this sense, both the golden ring and the Quran are preserved from any falsehood. This is why The Most High emphasizes Himself in preserving the Quran. The preservation of the Quran is unique and extraordinary! It shares this feature with God's signs *ayaat* in the nature! Praise be to God, the Possessor of Infinite Bounties.

19. The validity of a negative statement or challenge (such as in verse 41.42) cannot be proved unless we witness the failure of attempts against it. It is a circular argument to say that it is preserved because it claims so, since the claim of preservation also is under question. Similarly we cannot say that it is preserved because it is preserved. For instance, if I claim that no one can climb the walls of my castle, my challenge will not make sense until some people try and fail to do so. If the trials and failures of some people demonstrate that my castle's walls have been protected by an impenetrable surface with virtually zero friction supported by high voltage electricity and an automatic alarm system, then my challenge is proven.

20. An example of such proof is the word "*BasTata*" in 7.69. The code 19, demonstrated that falsehood can never enter the Quran (please remember the parable of the gold ring that is also preserved from falsehood). If you look at the Quran in your hand you will see that the word "*BasTata*" in verse 7.69 is misspelled with "Sad." We detected this minor printing error through the Quran's mathematical code. When we studied the oldest available versions, for instance, the Tashkent Copy, we found that it is exactly the way that we had predicted according to the mathematical structure. We took the photograph of the verse and published it in Appendix 1 of Rashad's translation.

21. You ask about previous generations, wondering how is it possible that they possessed a Quran containing two false verses for centuries. As an answer I'll repeat Moses' answer given to similar question centuries ago in 20.52: "The knowledge thereof rests with my Lord in a record; my Lord never errs, nor does He forget." Those people were responsible according to the knowledge they were given. Ironically, most of them had no problem in accepting volumes of fabrications and preferred them to the Quran. Adding two apparently harmonious sentences unknowingly is not the same as adding thousands of contradicting paragraphs knowingly. Therefore, please my brother, worry about your own test when you are surrounded by overwhelming signs of God (10.39; 6.158).

22. You may ask: "What about subtraction? How can you know that any verses of the Quran are not missing?" Here, I would like to summarize my answer that took several pages in "Notlar" (Notes) published in Turkish. I have at least three reasons for believing that today's Quran is complete: 1) The mathematical structure of the Quran based on number 19 was not known before Rashad's discovery. Any subtraction would have a detrimental effect on the integral parameters of the interwoven mathematical design 2) The extraordinary examples of a harmonious mathematical design and their practical functions increase our faith regarding God's promise for protecting His word. 3) God will not hold us responsible for the things that are beyond our means. Please reflect on these points if you entertain such a question.

23. You are confusing the scribes who are criticized by Rashad Khalifa with the scribes praised in 80.15-16. These verses are not praising all scribes who wrote the Bible or the Quran. They refer to the believers. In fact, they refer to messengers who wrote the revelation. Please note that "Suhuf (Scriptures)" is plural and refers to all divine books. "Safarah" is the plural of "Safeer" that means ambassador, or messenger.

24. Besides, how can you answer the question regarding the authenticity of the very verse that claims the preservation. What if, after the dangerous arguments started soon after the death of prophet Muhammad, some zealots constructed or fabricated the verses 15.9 and 41.42, and it was well accepted by pragmatist leaders to heal the social discord in their land? After all, there were no shortage of people who could produce Arabic statements in the name of God or the prophet.

108

25. Here are several examples of the Quranic testimony that 9:128&129 are not authentic. This testimony is made by the code of its "great" mathematical structure mentioned in Chapter 74.

- The frequency of the word God (Allah) is 2698 (19x142), without those two.
- The sum of the numbers of all verses containing the word God is 118123 (19x6217), without those two.
- The frequency of the word God, from the beginning of the Quran until the end of Chapter 9 is 1273 (19x67), without those two.
- The frequency of the word God, from the first initialed chapter until the last initialed chapter is 2641 (19x139), without those two.
- From the missing *Basmalah* of Chapter 9 until the extra *Basmalah* of Chapter 27, the word God is mentioned in 513 (19x27) verses, without those two.
- The frequency of the word "Elah (god)" is 95 (19x5), without those two.
- The frequency of the word "Arsh" referring t
- God's domain is 19, without those two.
- The number of all verses, including 112 unnumbered *Basmalah*s is 6346 (19x334), without those two. The absolute value of this number is 19, without those two.
- The frequency of *Rahim* (Merciful) is 114 (19x6). Verse 9.129 creates a curious single exception by using it for prophet Muhammad. The other names mentioned in *Basmalah*, that is, Allah and *Rahman* are never attributed tothers than God.

26. In the initial announcement made in March 1985 issue of Muslim Perspective, Rashad presented 9 reasons to reject 9.128-129. His reasons were 9 words. According to his count, the frequencies of each word were one extra with those two verses, that is, they all would be multiple of 19 without them. When I checked his evidence I found that some of the counts (of Anfus, Tawallu, Tawakkaltu, Rabb) had nothing to do with his claim. He was obviously wrong in his hasty count. These and similar errors are caused by our weakness and desire to increase the number of evidence (74:6). (Later, he discarded them after a face to face discussion.) However, there were two outstanding words that whose would be one extra with those two verses: Allah and Raheem. I could even eliminate the one extra Raheem for a simple reason: It is used for Muhammad not God, as many other names of God used for humans as in 11.24; 76.2; 9.114, etc. However, I could not explain and reconcile the extra word "God."

After I rejected his claim by labeling him a "disbeliever" in a jotted letter, I tried ways to avoid this conflict. For instance, I tried to accept the first *Basmalah* as unnumbered verse, like the others in the beginning of chapters. Already there were some Muslim scholars who were claiming that Fatiha's first verse starts with the letter Alif, not Ba, that is, "Alhamdulillahi Rabbil Alemeen." In this case, I had to include "Raheem" of 9.127, ignoring its meaning and reference. But, this time I would have a problem with *Rahman*: it would be one short. On one hand, being absolutely sure about the mathematical code of the Quran, and on the other hand, knowing the verses 15.9 and 41.42 that guarantee the preservation of the Quran, put me in the most difficult dilemma that I had ever experienced.

However, my Lord, The Gracious, saved me from that dilemma by a miraculous experience. He showed me his clear signs as He promised in 41.43. He referred me to 3.41 and erased all my doubts regarding those two false verses which were exposed by the miraculous mathematical code of the Quran.

> 74:54-56 Verily, this is a reminder. For those who wish to take heed. They cannot take heed against God's will. He is the source of righteousness; He is the source of forgiveness.

***

A personal experience

My personal experience regarding this issue may not seem appropriate in the context of an objective discussion. However, we learn from the Quran that God shows his signs in the horizons and in ourselves to convince us (41.53). I will narrate my story since I believe that God encourages me to do so (93.11), and hoping that it may cause you to examine your motivations.

My personal experiences are obviously nonfalsifiable subjective cases. But, they can be supported by witnesses and physical evidence. Here, I'll tell you the most fascinating one. The one that dramatically changed my entire life. This paper will not be enough to put it in its context. Thus, consider this as a snapshot picture from the middle of a continuing story.

In 1 July 1986, I made the greatest decision in my life. I came to the conclusion that the religion that I inherited from my parents was abysmally corrupted. The introduction of my ninth book, The "Sakincali Yazilar" (Dangerous Articles, 1988) starts by mentioning the importance of that day in my life. I had to criticize and reject most of my previous religious position published in my previous bestseller books. I rejected the conventional traditional religion. My inquiry brought me to a startling conviction: Traditional Islam had nothing to do with Muhammad's original teaching. It could not be God's religion.

Several months after that crucial decision, I encountered a big intellectual and spiritual problem. I found myself in a dilemma. The mathematical structure of the Quran was blinking at the two last verses of Chapter 9. This was a very serious issue, since the Quran claims that it is perfectly preserved.

I was confused, I was scared. I could not solve the problem. The mathematical code of the Quran, which I had no doubt about, was exposing those two verses as man-made insertions. Indeed, there was some historical evidence about controversial arguments over those two verses. However, the consensus of Muslims was clear.

The problem needed a crucial "Yes" or "No" from me. But, it would determine my fate, both in this world and in the hereafter. It was a very important issue. I could be killed by fanatics if my answer was "Yes". But, I was more concerned about finding the truth.

For approximately two weeks I was lost. I was persistently praying to God, asking for a "sign" to save me from that dilemma. "God, give me a sign" was my repeated prayer. One day, on October 23, 1986, at around 1:30, I was sitting alone in my office trying to finish the second volume of "Interesting Questions." I could not concentrate; the terrible paradox was eating away at my soul. I prayed again in Turkish: "Please give me a sign." Suddenly, an unusual thing happened. My heart started beating vigorously as if I had run five miles.

It was the first time in my life, that I had that kind of heart beat for no apparent reason. Shortly, I heard a very clear voice from my HEART, repeating in Turkish: "Uc Kirkbir! Uc Kirkbir! Uc Kirkbir!", that is, "Three Forty One, Three Forty One, Three Forty One." I don't remember exactly how many times it repeated. My excitement was at a peak. I was shocked. The only thing that came to my mind at that moment was to look at the Quran, 3.41 (Chapter Three, Verse Forty One). I cannot describe my excitement and joy. Verse 3.41 was exactly repeating my Turkish prayer in Arabic with its Quranic answer:

> 3.41   He said, "My Lord, give me a sign." He said, "Your sign is that you do not speak to the people for three days, except by signals. You shall commemorate your Lord frequently, and meditate night and day.

111

This extraordinary event not only saved me from the worst situation I have ever had, but it also taught me a great lesson: Don't worry about what people think about you. Seek the truth without any personal agenda.

Later, somehow, I wanted to see whether there was any relation between this incredible experience and my accepting the Quran alone as the source of my religion. I was assured by an astounding mathematical relation. The number of days between 1 July 1986 (the most important day in my life), and 23 October 1986 was exactly 114 (19x6) days, which is the total number of the chapters of the Quran.

I have studied philosophy and some engineering and psychology. I'm perceived as a skeptic by my friends; but, I cannot doubt that event. I cannot ignore or depreciate its factual existence in my history. I am aware of paranormal problems. Here I will list some of the possible objections by skeptics:

The narrator's subconscious, under strong stress, may have remembered the verse number where his prayer is mentioned.

1. It may be a schizophrenic event. The verse number and its matching text is coincidence.
2. The narrator is lying.

I would not argue against any of these, since I'm not trying to prove anything by telling you this experience here. As far as I'm concerned, I'm as sure about my experience as you are sure that you are reading or hearing these words.

After my crucial decision in 1 July 1986, as an ex-convicted political activist, I started to fight the government to get a passport. Though I had two uncles in the National Congress, it took me two years to receive a passport. Interesting enough, the date of issue on my passport was 1 July 1988.

I was single until my early thirties. Verse 3:41 mentioned above was related to Zechariah and his son Yahya (John). Thus, I sympathized with them. Just after I experienced the incredible paranormal phenomenon, I gave a silly promise to God: "If I marry, and if I have a son, I will name him Yahya." This promise remained a secret between God and me, until my wife surprised me with another "coincidence":

In 1989 I married here, in US, with an Iranian-American lady. When she got pregnant, I started to wonder: how can I convince her about the name Yahya if the baby is a boy? I was waiting for a good day and mood to talk about this

112

issue. An incredible thing happened. One night, two or three weeks after learning about the pregnancy, she came to me and for first time talked about the name of the baby. She suggested only one name: Yahya. (This name is a rare name in Turkey and even rarer in Iran.) I thanked God Almighty, and told her my story regarding my silly promise to God.

In the meantime, we received two interesting letters. One was from a close friend from Turkey, who had just heard about the pregnancy. He did not have any idea about my promise regarding the name. In his letter he wrote a prayer: "May God raise your child like Yahya." Why like Yahya? We had numerous heroes in our history. Another coincidence!? My mother in-law's letter (again, within several weeks of the pregnancy), contained a poem about our coming baby. The name of the poet was Yahya.

Similar signs continued. Therefore, I was convinced by these signs that our child was a boy and he would be born on the 1st of July, as God's reward for my decision to follow- the Quran alone. I announced my prediction regarding the gender and birth-date of our child to more than thirty people, in a Quranic study, in Masjid Tucson. The baby failed the predictions of doctors and came to the world on the predicted day, at 10:53, morning of July 1st, 1990. We both hugged him by saying: "Welcome Yahya." Indeed, he was a boy.

> 10:53   And some people ask you, 'Is all this true?' Say, 'Yes indeed, by my Lord, this is certainly true. . ."

Praise be to God.

7

The Story of
a Precious Crown

Assume that you have a precious crown that has been inherited within your family for centuries. This priceless golden crown with numerous diamonds has a strange inscription on it: "This is a unique crown and it is protected by its maker from tampering, from alteration, addition or subtraction. Fraudsters cannot touch this crown!" None of your parents, uncles, aunts, brothers or sisters had any question regarding this allegedly hoax-proof crown. As far as you know, the same was true about the previous generations. They all believed the claim on faith. You have learned that any of those who dared to question the authenticity of the crown and the veracity of its promise were either excommunicated or subjected by gangs of ignorant adherents to insults, injury or even death.

One day, you show the courage and wisdom to question the long-held dogma, the family's faith in the authenticity of that crown and the truth-value of its promise for perfect protection. You take the crown to a jeweler nearby. You ask him to assess the authenticity of the crown. The jeweler finds the crown very interesting. The design of the little diamond pieces and their unique arrangement pulls his attention. He starts studying it by using his tools, including a microscope. After some time, he comes up with the news: "This is a very interesting crown. I have never seen something like this before. It is priceless. Of course, its gold and diamonds have great value. But its intricate design distinguishes it from all other crowns!" You are very excited to learn that indeed you have overall a priceless crown. You call your family inform them about the good news, that you have verified the claim through an expert testimony.

However, while the expert is still studying the crown with awe, he notices an anomaly. He asks you to give him more time to study further a cluster of diamonds. You take advantage of that moment and share the news with some of your friends who had similar doubts about the authenticity of the crown.

A few minutes later, however, the jeweler comes to you with a bad news. "I found a fake 'diamond' on the crown!" You cannot believe your ears. You make him repeat the news. "Indeed, this golden crown contains 361 real diamonds that are arranged according to an intricate geometric design. However, I am sure that this is not a diamond!" You are disappointed. Initially, you get angry with the expert. You feel like punching him in the mouth. But, you compose yourself and decide to act rationally. You ask for the evidence for his claim. Ironically, until recently neither you nor any of your family members had demanded for evidence for their affirmative claim regarding the authenticity of the crown and its inscription. You have accepted both claims on faith.

The expert jeweler does not ignore your demand for evidence by pointing at your family-size traditional inconsistency! He does not question your sincerity in respecting this antique crown, when he learns that some of your family members who lived centuries ago made up many crowns from fool's gold and then they considered it equal or complementary to this unique and original crown. The jeweler provides you with plenty of evidence. For instance, he pulls your attention to one of the 19 hexagons surrounding the crown, each containing 19 little diamonds. A hexagonal diamond in the center is surrounded by six hexagonal diamonds, and those six are surrounded by twelve hexagonal diamonds, thereby creating a bigger hexagon with 19 diamonds. The expert then shows you a diamond on the corner of the 9th hexagon. "You see this one? This is not a real diamond. This is not made by nature; this is a piece of manmade glass! It is virtually worthless anywhere, and absolutely worthless on this crown! Come let me show you under the microscope and see the difference." After showing the fake diamond to you, the jeweler makes an offer: "If you allow me I can easily remove this fake diamond from here. Even the glue used to attach it has poor quality. You do not need to pay me for this simple task"

You are shocked. You cannot believe that the inscription was a false promise. You feel confused and contradicted. On one hand, you know now, not on faith, but through knowledge and evidence, that the crown is unique and priceless; on the other hand you also know that it contains a fake diamond and a false inscription! What to do with this?

When your extended family members learn about the bad news about the existence of a fake diamond, most of them call the jeweler a liar, and a heretic. Some fanatic zealots among them even suggest killing him for insulting the crown and the honor of the family members who always believed in the perfection of the crown and the authenticity of the inscription. You go to the jeweler and beg him for changing his verdict. But, the jeweler is a wise man. He tells you this:

"You must be an idiot. In fact, by allowing me to discover the fake diamond which was attached to it later by fraudsters, the maker of the crown proved both his creativity and his promise, beyond shadow of doubt. Indeed, this crown is unique and priceless. The inscription on it is now verified: it is protected by its peculiar design. It is protected not by your family or ancestors, but in spite of the fraudsters among your ancestors. It is protected by the ingenuity of the maker of the crown. You have now a reason, an incontrovertible reason to respect and trust the inscription on this crown." And now you have equally disturbing reason to be suspicious of the immaculate reputation of your ancestors.

The news about the fake diamond in the crown becomes widespread in your town. People do not talk much about the ingenious design of the crown neither do they reflect on the jeweler's words. The leaders of your extended family hold an emergency meeting. They decide that the jeweler confused many and he became an infidel by defaming their crown; thus, he should be killed. A few days later your cousins carry out the decision. The jeweler is stabbed to death in his jewelry store.

Prophetic verses in the end of Chapter Nine

What about the Quran's prophetic description of the conspiracy leading to the addition of two verses to the Manuscript —not the Quran, which is perfectly protected!— such as the highest frequency of the word *Zada* (increased) in the Quran, three times in 9:124-126? What about the reference to the revelation of a chapter? Increase in their foulness? What about the depiction of hypocrites who secretly sneak and ask each other about the revelation of a chapter? Here are the last four verses of the Chapter 9 with an intriguing prophetic language:

> *Conspiracy of Hypocrites Increases Only Impurity*
>
> 9:124   When a **chapter** is revealed, some of them say, "Whose acknowledgement has this **increased**?" For **those who acknowledge**, it **increased** their **acknowledgment**, and they rejoice.
>
> 9:125   As for those who have a **disease in their hearts**, it only **increased foulness** to their foulness, and they died as **ingrates**.
>
> 9:126   Do they not see they are **tested** every calendar year once or twice? Yet, they do not repent, nor do they take heed.
>
> 9:127   When a **chapter** is revealed, **they looked at one another: "Does anyone see you?",** then **they turn away.** God turns away their hearts, for **they are a people who do not comprehend.**

In sum, by demonstrating the perfect protection of the Quran through code 19, this **Chapter** has **increased acknowledgement** of **those who acknowledge** the truth, and **increased foulness** of the **ingrates** by becoming a **test** for them as well as their *salaf* who **conspired against this chapter** and **turned away** from truth. These are **people who do not comprehend..**

Have you also noted that the key words above —increase, acknowledgement, acknowledge, disease in their hearts, ingrates, test— are common key words with those in verse 74:31? Are all these coincidence?

In sum, besides proving the divine origin of the Quran, the Code 19 leaves no doubt about the divine promise of the perfect preservation of the Quran, since it functions as an error-correcting code (ECC) on various manuscripts.

Idolizing an Iconoclast: The Story of a Cult after Dr. Rashad Khalifa

The discoverer of the Code 19, Dr. Rashad Khalifa, was assassinated in Tucson, Arizona, by an international terrorist group al-Fuqra or al-Fuqara, which was affiliated to Usama Ben Laden, in 1990. However, the power of this message is promising a new era of reformation in religions, particularly Islam and Christianity. Ironically, some of Khalifa's "followers" reverted back to the days of ignorance by trying to turn the community of Submitters into a cult after the departure of that brave iconoclastic scientist. Though there are still many monotheists in the group, as it seems, a gang is actively busy in repeating the history of ignorant people by carving a new idol for the group. The gang claims infallibility of Rashad Khalifa and considers all his writings, appendices, footnotes, fingernotes, and articles in the newsletter and his speeches in video recordings to be divine revelation. The gang members consider Rashad's re-re-revised translation to be error-free, and even justify the obvious errors such as the spelling ones. They have replaced their former idol with a new one. The Submitters community has been fatally infected by the idol-carvers and with the rate of infection; the group may excommunicate all the monotheists within a decade or so. I have exposed the distortion of this gang in a booklet titled, *United but Disoriented.* In 1990s many considered some of my predictions to be "unreal" or "paranoid", but my recent debate with the gang members proved my predictions. They are now making claims that none expected to hear, including themselves. Like most polytheists, they have mastered their arguments for their version of polytheism while pretending to be monotheists. The book's first edition is out of print, but it is available on the net. God willing, an updated edition of the book tracking the group's devolution since departure of Rashad will be published by Brainbow Press, most likely under the title: *Idolizing an Iconoclast: The Story of a Cult after Dr. Rashad Khalifa.*

117

8
Running Like Zebras
Abu Ameenah Bilal Philips is:

a) **A Salafi scholar**
b) **An accomplice for murder**
c) **A hoaxer**
d) **A geocentric antique**
e) **A slanderer enemy of the Prophet Muhammed**
f) **A number-phobic innumerate**
g) **A father of 20 children from four concubines**
h) **Some of the above**
i) **All of the above**

Who is Abu Ameenah Bilal Philips? For sure, he puts a Dr. before his name. In fact, he loves that title and uses it as a prefix for his name everywhere. So what? What is wrong with someone with a doctorate degree to put the word Dr. before his name? I have no problem with that, but here there is more to that title than meets the eye. Before you are impressed by his academic credentials, you should first know about where he got most of his education from and his role in the assassination of my friend, Dr. Rashad Khalifa. According to his website, Abu Ameenah:

> "Shortly after his reversion to Islam, he embarked on a spiritual academic journey to the other side of the world seeking Islamic knowledge. This journey took him to Saudi Arabia where he completed a BA in Islamic studies in Madeenah, and an MA in Islamic Theology in Riyadh, then to the University of Wales, UK, where he completed a PhD in Islamic Theology in the early nineties."

In other words, he was the student at a university when the notorious Bin Baz was the top cleric in Saudi Arabia and the most revered and influential professor at the universities there. Now, let's learn a bit about the deeds of the top cleric who approved Abu Ameenah's BA and MA diplomas and trained him as a model Salafi. I will be quoting from the introduction of the 2ⁿᵈ edition of Abu Ameenah's Hoax book:

> "Doubts about the correctness of his initial premise of 19 being the mathematical key to the Qur'aan were then raised and a number of emotional articles were written against both him and his theory. In fact, one of the leading scholars of Saudi Arabia, Shaikh 'Abdullaah ibn 'Abdul 'Aziz ibn Baz, wrote a *fatwaa* (religious ruling) concerning Khalifa's heretical claims in which he declared him an apostate.[10] "

Note the words HERETICAL and APOSTATE above, which are justified by Abu Ameenah, a Sunni scholar, a student of ibn Baz. So what? He has the freedom to consider Rashad Heretical and Apostate as Rashad considered him and his sheikh *mushriks* and ignorant gang. But there is a great difference between the labels because of their implications in their context.

"Heretic" and "Apostate" are code words for DEATH penalty according to the Sunni jurisprudence. Ameenah's father has the audacity to refer the reader to his book for the translation of his guru's death fatwa. Ironically, in a book titled *In Defense of the Qur'an and Sunnah*, he brags about "defending the Qur'an" by justifying and encouraging the murdering of people who dare to disagree with his ignorance and idolworship! His relation with ibn Baz was not mere formality. Here is some information about his personal life (emphasis is mine):

> "Son of Jamaican parents born on 7th, July, 1947 in Jamaica, he grew up in Canada and lived and studied in Malaysia. **He is the father of 20 Muslim children** and has a black belt in Tai Kwon Do and currently studies Tai Chi. Among the **more notable aspects of his academic life** is that he sat in the circles of **famous Hadith scholar**, Shaikh Naaseer udeen Albani for 4 years in Madinah (1976-1980), as well as the circles of the famous Mufti of Saudi Arabia, **Shaikh Ibn Baz in both Madinah and Riyadh, Saudi Arabia. He also translated for both Shaikh Ibn Baz and Shaikh Ibn Jibreen on numerous occasions."

---

[10] An English translation of the *Fatwaa* can be found in *In Defence of the Qur'an and Sunnah*.

It is important to understand the real reason why Abu Ameenah is full of bigoted hatred against the number 19 and its discoverer. He has studied too much *hadith*, the volumes of contradictory and silly stories fabricated by ignorant people and he sat too long in the circles of the world's most ignorant and bigoted shaikhs.[11]

It is now evident that he was an accomplice in inciting and perhaps planning the killing of Rashad Khalifa, like the monotheist mentioned in 40:23-35. So, before refuting their criticism of 19, it is important to understand the backwardness, the bigotry and the venomous religious views of Abu Ameenah and his late guru Abd al-Aziz ibn Baz.

Abu Ameenah's Shaikh: "Earth is Fixed; or You are Dead"

Abd al-Aziz bin Baz was the chief cleric of Saudi Arabia, the head of the Council of Senior Religious Scholars for three decades. He had great impact in regressive and oppressive laws in Saudi Arabia, including the ban on women's driving. A book authored by Bin Baz was published in 1975 carried the following title: *Scientific and Narrative Evidence for that the Earth is Fixed, the Sun is Moving and it is Possible to Go to the Planets.*

The book was not published by any publishing house; it was published by no other than the Islamic University of Medina. In that book, Bin Baz complains about a new heresy; he is saddened to see, well more accurately, hear, people talking about motion of the earth and he wants to put a stop to that heresy. In page 23, after listing some hadiths issues a fatwa, **Bin Baz asserts that those who believe that the earth is rotating are *kafirs* (disbelievers), and if they were Muslims before they become apostate.**

The Saudi Sunni leader does not stop there and explains the ramification of fatwa: any Muslim believing in the rotation of earth loses his or her right to life and property; they should be killed! This top cleric, İbn Baz, was the head of an international conference of Sunni scholars representing 38 countries discussing the Salman Rushdi affair. Then, Saudi Arabia was competing with Iran regarding leadership in the Islamic world and this was the hot issue then. The conference issued a unanimous fatwa in 19 March 1989, condemning Rushdi and Rashad, as apostates. The Western world by then knew Rushdi, but not many westerners were familiar with the second name in the fatwa. Dr. Rashad

---

[11] For those readers who are not familiar with our resons for rejecting hadith and sunna, I recommend them to read one of the two booklets: 19 Questions for Muslim Scholars, or Manifesto for Islamic Reform. Bot hare available as boks or freely available on Internet, such as www.19.org or www.islamicreform.org

Khalifa, the late leader of modern reformist movement and the discoverer of the mathematical code of the Quran, in less than a year after this fatwa would be assassinated by a Sunni group affiliated with the Saudi terrorist Osama bin Laden in 31 January 1990, in his Masjid in Tucson, Arizona.

In his book, Bin Baz quoted some verses and many hadith to support his position that the earth is fixed. After expressing his religious verdict of death penalty for the apostates who believe in a moving earth, he included the following reasoning as his scientific evidence:

> "If the world was rotating as they assert, countries, mountains, trees, rivers, seas, nothing would be stable; humans would see the countries in the west in the east, the ones in the east in the west. The position of the *qibla* would change continuously. In sum, as you see, this claim is false in many respects. But, I do not wish to prolong my words."

Towards the end of the 20th century, a "university" of a Sunni country publishes such a nonsense authored by the highest ranking religious leader of that country! Considering that the Ottoman chief clerics banned the import and use of printing machine for about 300 years, it becomes clear the reason why the so-called Muslim countries are so lagging behind civilization, socially, politically, and in science and technology.

Ibn Kathir is a popular commentary of the Quran, which is respected because of its reliance on hadith to "explain" verses of the Quran. Ibn Kathir (d. 1372), in the classic commentary carrying his name, makes the following remarks on verses 2:29 and 68:1. For this commentary, he relies mainly on a hadith from *Abud Dawud*, one of the so-called authentic Sunni holy hadith books:

> "Ibn Abbas told all of you by Wasil b. Abd al-Ala al-Asadi-Muhammad b. Fudayl- al-Amash- abu Zabyan- ibn Abbas: the first thing god created is the pen. God then said to it: write!, Whereupon the pen asked: what shall I write, my lord! God replied: write what is predestined! He continued: and the pen proceeded to (write) whatever is predestined and going to be to the coming of the hour. God then lifted up the water vapor and split the heavens off from it. Then god created the fish (*nun*), and the earth was spread out upon its back. The fish became agitated, with the result that the earth was shaken up. It was steadied by means of the mountains, for they indeed proudly (tower) over the earth."

After learning the intellectual level of the believers, collectors, narrators, and commentators of the above hadith, such as Abu Dawud (d. 888), al-Tabari (d. 1516), and Ibn al-Baz (1995), it becomes clear why Muhammed would utter the words in 25:30.

As a Sunni scholar, as the student of the geocentric death-fatwa issuer Ibn Baz, Abu Ameenah is not a regular academician as you expect. Besides his connection with the assassination, he promotes pedophilia and polygamy. You can find his articles on the Internet and his speeches at YouTube defending marriage with girls at age 9, stating that it was "the norm of the time". Like other enemies of the Prophet Muhammed, he follows the fabricated hadith that falsely accuses the Prophet Muhammed marrying Aisha at age 9 while he was 55![12] Salafis, like all other followers of fabricated hadith and Sunna, take the ethical role-modeling of the Prophet Muhammed outside the Quranic context and reduce it to trivial local cultural aspects such as growing beards and entering the bathroom with left feet, etc.

No wonder Abu also promotes the distorted practice of polygamy with virgins. Now, does it take a rocket scientist to guess about how many wives he has and their ages? He brags about the number of his children (20 so far), but he leaves his audience in darkness regarding the mothers of his children. They are invisible ghosts in his *haram*; they are women with no identities, with no voice of their own. In fact, they have no faces either. When they are allowed to go out they have to walk in black sacks their faces veiled according to the teaching of his *mazhab*!

Are these *ad hominem* attacks? No! They are all connected. They explain why Abu hates code 19; because he hates the truth, and mathematics is the ultimate truth! Without learning the horrible mindset and backward religious dogmas of this guy you will not be able to understand his modus operandi. Only after you learn his religious views, connections and environment, then you will understand why he and his ilk hate the truth so much. Only after that you will be able to comprehend the reason behind their contempt against the greatest miracles and their contempt to its discoverer.

---

[12] I recommend an excellent article written by TO Shanavas on this subject. "Was Aisha a Six-Year-Old Bride?" has been published in the first volume of anthology after our first conference: *Critical Thinkers for Islamic Reform*, editors Edip Yuksel, et al, Brainbow Press, p. 39-46, 2009.

"According to Holy Hearsay Reports the Quran is indeed a Miracle!"

- "Since the Quran claims that it is a miracle, then it is a miracle"
- "Muhammad is the messenger of Allah; and the book delivered by him must be a miracle"
- "I believe in the Quran; therefore, it must be a miracle"
- "I find the information in the Quran perfectly conforming to my culture, knowledge and expectations. Therefore, the Quran must be a miracle."
- "I am not an expert on every scientific issue, history or archeology, and I am not a logician; but since I cannot find any contradiction in the Holy Book that I was raised with its beautiful recitation, therefore it must be a miracle."
- "I do not really know how to prove it; but I believe that the Quran is a literary miracle!"
- "Neil Armstrong, the astronaut who first landed on the Moon, and Captain Custo, the renowned French explorer and many other scientists became Muslims. Therefore, the Quran must be a miracle."
- "Our great scholars and saints all have asserted the miraculous nature of the Quran. How could all be wrong! Thus, it is a miracle."
- "If more than a billion people believe in it, then it must be a miracle!"
- "I do not need a miracle to believe that a book is from God. Faith needs no justification. If I believe then it becomes a miracle!"

No wonder, millions who resort to silly and fallacious "reasons" like the ones I listed above have no problem in following silly, backward, harmful and contradictory teachings of Hadith and Sunna; no wonder they act blind, deaf and dumb when they are informed about the verses of the Quran or invited to engage in critical thinking.

Those who have no clue about the meaning of miracle are wrong even if the object of their claim is indeed a miracle. Compare these two: the assertion "67 is a prime number" to the assertion "My glorious ancestors believed that 67 is a prime number and I am following their verdict without questioning". While both of these agree on the numerical nature of 67, the difference between the two is akin to the difference between graphite and diamond. Though both are made of pure carbon, the diamond is one of the hardest matters in the world, while graphite is one of the softest.

You should not be a surprised if a person declaring that "According to reliable reports, $2 + 2 = 4$", also declares "According to reliable sources $1 + 1 + 1 = 1$".

Why? Because such a person is not using his or her own mind; they have put their brains in the blind basket of religious adherence. Has become sheep in a herd... Has placed her eggs in in a nest that no bird has hatched! The blind bandwagoners do not witness miracles through critical and analytical thinking. All they do is copying, memorizing, parroting, imitating the religions, sects and cultures of their ancestors whom they have idolized!

Those who believe in the Quran blindly betray the Quran blindly, and unfortunately, they also lack the awareness and capacity to notice their contradictions, their ignorance and arrogance!

Signs/miracles (*aya*): As a divine response to the epistemological interrogation and objection to messengers, they are endowed in accordance to God's law (*Sunnatullah*) with divine signs/evidences consisting of extraordinary and superhuman special events/phenomena that cannot be explained by our collective experience and knowledge yet can be witnessed by our senses and/or by our reasoning faculties.

Back to Abu Ameenah

I concede that Abu Ameenah is not as stupid as his late sheikh and he is clever to hide his backward and bloody religious views. It is now evident that he was the one who informed his blind sheikh about Rashad Khalifa and used him to issue the death fatwa against Rashad after he wrote his book against him (1987) and his historical discovery. You may expect such a scenario from Hollywood movies, but it is real.

Though, unlike his sheikh he was raised in the Western world and studied in secular schools, he got his share of backwardness from the same "university" where Sunni students are trained in ignorance. In his Hoax book, Abu Ameenah briefly mentions my name:

> "Since his death in 1990, no new leader has emerged to take his place in the cult. However, Edip Yuksel, a well-educated Turk provided religious leadership during the 1990s (His father is a Muslim scholar who has condemned his son's writings.) Followers of Rashad Khalifa are currently located in such places as Phoenix, Arizona; Riverside and

124

the Bay area in California; Vancouver, British Columbia; and other parts of the world. (*Mission to America,* pp. 137-165)"[13]

Apparently, Abu Ameenah, that is, Ameenah's Father, knows about me a little more than I expected him to. However, he is again sneaky and ill-intentioned. His description of my father as "Muslim scholar" and referring to his "condemnation" of me is another insidious message to his Sunni followers. A short but well-understood message:

"Remember, how some brave Sunni brothers followed the fatwa of my teacher, the Muslims Scholar ibn Baz, and assassinated Rashad Khalifa, the chief heretic! This guy named Edip too is a heretic and has received death fatwa from his Sunni father. Thus, if any idiots among you wish to do some jihad, you may as well plan to put that fatwa into action."

No, I am not reading too much into Abu's statements about me and Rashad Khalifa. I know these enemies of freedom very well. The meaning of his statements is obvious to anyone who is familiar with their diabolic teaching. According to all Sunni sects, an apostate deserves death. As you will notice, this so-called Doctor of theology does not use a single word condemning the barbaric assassination of Rashad Khalifa. To the contrary, he presents it as well-deserved job! Here how he introduces his book:

"However, there still remains the question of whether the basis on which Rashad Khalifa's theory is built is valid or not, as it continues to attract the unsuspecting and the ignorant. Hence, it is not sufficient to merely discredit Rashad Khalifa based on his many heretical statements, any one of which is sufficient to remove him from the fold of Islam. This controversy has to be laid to rest by a factual dismantling of its foundations."[14]

"The following chapters of this book systematically and methodically challenge and disprove the vast majority of the so-called fundamental 'facts' of Rashad Khalifa's theory. Furthermore, they clearly expose his deliberate falsification of data and alteration of the Qur'aanic text in order to bolster his theory of 19 as the miraculous numerical code of the Qur'aan. This work will demonstrate, without a shadow of a doubt – God willing, that the Theory of 19 is a shoddily concocted hoax

---

[13] Dr. Abu Ameenah Bilal Philips, THE QUR'AN'S NUMERICAL MIRACLE: Hoax and Heresy, 2nd Edition, 2002, p. 9

[14] Ibid, p. 16

unable to withstand serious scientific scrutiny. However, before proceeding with the refutation of Dr. Rashad Khalifa's claims, the following summary of his theory has been prepared for readers unfamiliar with its core issues."[15]

I will not spend much space here responding Abu Ameenah Philip's book, **since his criticism was fully utilized by two critics: Lomax and then by Ayman.** To avoid too much redundancy, I will here briefly respond to the "conclusion" of his criticism, which he published in the end of his Hoax Book in pages 99-101. He introduces the summary or the conclusion as: "The following is a brief summary of the main points by which Dr. Khalifa's theory has been proven false in the preceding chapters of this book."

> **1. Misinterpretation of Quranic Texts:** One of the foundations of Dr. Khalifa's argument is that verse 30 (over it are nineteen) or Surah al-Muddaththir refers to the miracle of 19 being over anyone who claims that the Qur'an is false. However, it has been clearly shown that the number 19 refers to the guardian angels over the Hellfire as understood by the classical commentators of the Qur'an.

As you will read the section titled *"Which One do You See: Hell or Miracle"* and the rest of this book, *Running Like Zebras*, you will learn that Abu Ameenah is among those who are exposed by the prophetic verses of the Quran in 74:31-37, as an ingrate and one of those who regress. Here, I will quote a few paragraphs.

The punishment issued for the opponent is very interesting: nineteen. Almost all numbers mentioned in the Quran is an adjective for a noun. Forty nights, seven heavens, four months, twelve leaders… But here the numerical function of nineteen is emphasized. Nineteen does not define or describe anything. The disbeliever will be subjected to the number nineteen itself. Then, what is the mission or function of this nineteen? Those who tended to understand the meaning of *Saqar* as "hell" naturally understood it as the number of guardians of hell. However, the punishment that is described with phrases such as, difficult task, precise, and universal manifestations, was an intellectual punishment; a mathematical challenge. Indeed, the following verse isolates the number nineteen from the number of controllers and lists five goals for it.

> 74:31 We have made the guardians of the fire to be angels/controllers; and We did not make their number except as a test for those who have rejected, to convince those who

---

[15] Ibid, p. 17

126

were given the book, to strengthen the acknowledgment of those who have acknowledged, so that those who have been given the book and those who acknowledge do not have doubt, and so that those who have a sickness in their hearts and the ingrates would say, "What did God mean by this example?" Thus God misguides whoever/whomever He wishes, and He guides whoever/whomever He wishes. None knows your Lord's soldiers except Him. It is but a reminder for people.

Traditional commentators of the Quran had justifiably grappled with understanding this verse. They thought that disbelievers would be punished by 19 guardians of hell. That was fine. But they could not explain how the number of guardians of hell would increase the appreciation of believers and convince the skeptical Christians and Jews regarding the divine nature of the Quran. Finding no answer to this question, they tried some explanations: the Christians and Jews would believe in the Quran since they would see that the number of guardians of hell is also nineteen in their scripture. Witnessing the conversion of Christians and Jews, the appreciation of Muslims would increase.

This orthodox commentary has three major problems. First, neither the Old, nor the New Testament mentions number nineteen as the guardians of hell. Second, even if there was such a similar statement, this would not remove their doubts but to the contrary; it would increase their doubts since they would consider it one of the many evidences supporting their claim that the Quran plagiarized many stories from the Bible. Indeed, there are many Biblical events are told by the Quran, though occasionally with some differences. Third, none so far converted to Islam because of guardians of hell.

Some scholars noticed this flaw in traditional commentaries. For instance, Fahraddin el-Razi, in his classic commentary, *Tafsyr al-Kabyr*, offered many speculations, including that the number nineteen indicates the nineteen intellectual faculty of human being. Tough it is a clever interpretation, it fails to explain the emphasis on the number nineteen itself, and it also fails to substantiate the speculation....

On the other hand, those who have deprived themselves from witnessing the miracle 19 because of their ill intentions or their dogmatic rejection try hard to render these key words incompatible with the semantic context of the Quran. They conceive God of the Quran as an angry and despotic God who is not able to provide any reasonable argument against those who question the Quran's authenticity, but only resorts to intimidation: "I will burn you in hell!!!" The God they depict has double standard: He asks the disbelievers to bring their evidence for their argument (2:111; 11:17; 21:24; 27:64; 28:75; 35:40) but for His argument He only wants to scare them! The opponents of 19-based miracle, by distorting the meaning of the words in these verses, manage to blind themselves to one of the most profound philosophical and theological arguments and evidences in history. Not only they divert themselves from the right path they try to divert others too (6:25-26; 22:3).

Thus is the level of the understanding (more accurately, the misunderstanding) of those who cannot appreciate God as He should be (6:90-91). The argument of the god of their imagination for Quran's authenticity is no more than scorching, burning, dark, hellish, misfortune, disastrous, and scary punishment! Thus, in the minds of opponents of miracle 19, the rhetorical value of these verses is simply a threat to burn and torture.[16]

> **2. Incorrect Letter Count Totals:** Another foundation principle of Dr. Khalifa's theory is that the *Basmalah* (i.e.*Bismillahi-Rahmaanir-Raheem),* some verses and some chapters consist of 19 or a multiple of 19 letters. Since the Qur'an was not revealed in the written form, this argument becomes meaningless. In fact, strictly speaking, the actual total of the letters composing the *Basmalah* is 22 and not 19.

This very criticism is another miracle of the Quran, since in order to blind themselves to its numerical structure, ingrates and bigots have denied the most obvious and simple fact: the numbers of letters in Bismillah or Basmalah. Lomax, who was inspired by bin Baz and his student Abu Ameenah, parroted the same claim.

I knew that some well-known Sunni critics of the code 19 counted the letters in *Basmalah* ending up with 18, 20, and even 21. For instance, according to Dr. Süleyman Ateş, an Al-Azhar educated scholar, Basmalah has 21 letters. After losing a live TV debate with me on a popular Turkish TV program, he wrote the following:

> "The number 19 is the number of Bahai cult and has nothing to do with the Quran. It is modern cabbalism and nonsense. Quran's first verse, Bismillah, does not have 19 letters as they claim, but it has 21 letters. I have written a book on this deviation... Recently numerous engineers and doctors started following this path. One of those who follow his imaginations is the guy who was once imprisoned for participating in terror activities in Turkey and now

---

16 For a comprehensive refutation of Abu Ameenah and other Sunni or Shiite 19-phobic critics, see the chapter titled, *Which one Do you See: Hell or Miracle?*

living in America. I met that guy first time during the Ceviz Kabuğu TV program and then I immediately understood that he had idolized his ego. He was rude and did not recognize any rules of etiquette."[17]

But Abu Ameenah Bilal Philips is the only Sunni scholar who came up with 22! A whopping addition of 3 letters to the actual 19, which is 16% inflation! If Abu Ameenah counts the number of his children the same way he is counting the Arabic letters then he must have 17.27 children rather than 20. Of course, the number of the fingers in his two hands, according to the "invisible fingers" theory, which is an adaptation of his "invisible letter" theory, would be 11.58 fingers! Joke aside, this very claim of "22 letters in Basmalah" alone is sufficient to expose their delusional state of mind, their desperate attempt to turn their followers blind like themselves.

This example is enough to demonstrate how Muhammedans are twisting the simplest facts in order to cast doubt on the mathematical miracle of the Quran. It is noteworthy that not a single Muslim scholar had a different count for the letters of *Basmalah* **before** the discovery of the code. Whoever mentioned the numbers of its letters acknowledged the simple fact: *Basmalah* consists of 19 letters. For instance, Molla Jami starts his Persian Divan by referring to the 19 letters of *Basmalah*. Fakhraddin Al-Razi, in his 30 volume commentary, et-*Tafsir al-Kabir*, links the 19 letters of *Basmalah* with 19 guardians of Hell. Furthermore, the Abjad (Gematrical) value of *Basmalah* is well known as 786 for centuries, which is the numerical value of its 19 letters. Many a Muslim still use this number on top of their letters, instead of *Bismillah*...

What about this? Copy and paste a Bismillah (not its picture, but typed in Arabic characters) in MS Word. Then select it and ask the program to count the number of letters in it and check the "characters (with no spaces)"!

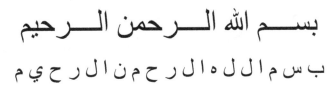

---

[17] Süleyman Ateş (76), former head of the Department of Religious Affairs in Turkey (1976-1978), theology professor and author, Turkey. At his personal website: http://www.suleyman-ates.com/

You should be careful against the Hoaxers who use this very word to cover up their own character. Our Sunni "alims" who are well-trained in promoting ignorance, might try to add "invisible" letters or try to confuse Arabic letters with sounds. Don't fall for their deceits! You should remember that there are 28 letters in Arabic alphabet and that all Arabic dictionaries follow this simple fact. You should also remember that no letter can be connected after Alif, since it would be confused with another letter, that is, Lam. So, if they tell you that there is an invisible Alif after B of *Bismillah* or a "short Alif" after Mim of *RaHMaN*, you should refuse to be tricked by reminding the following three facts: There is no dwarf or dumpy Alif in Arabic alphabet. If there were an Alif after B or Mim, there would be a space before the letter S or letter Nun. Besides, we are not counting invisible or ghost characters; we are counting written, visible letters of the Quran!

> **3. Letter Count Inconsistencies:** A major pillar of Dr. Khalifa's claim depends on his statement that "All Quranic initials, **without exceptions,** exist in their surahs in multiples of 19." However, this is only the case in three of the 29 Surahs having prefixed Arabic letters, namely Surah Qaf, Surah YaSin and Surah Maryam.

This criticism is partially true and mostly false. Indeed, Dr. Khalifa made some errors in the count of Alif and there are a few issues with other letters; but MOST of the initial letters are multiple of 19 and the initials that do not fit the system needs a comparative study on the oldest available manuscripts. I have extensive debate on this issue with Lomax and Ayman in this book.

> **4. Manipulated Letter Counts:** Dr. Khalifa achieved multiples of 19 in the letter counts for 13 chapters having Alif in the beginning of their "Quranic Initials" by counting the Hamzah as an Alif in some instances and not in others.

This is exactly the same allegation as in number 3, above. The example of the allegation in number 3 is perhaps listed as a different category in order to inflate the numbers.

> **5. Falsified Letter Count Data:** In order to artificially create multiples of 19 in some of his letter count totals, Dr. Khalifa has doctored his date in the following ways: (a) Some non-existing letters have been counted; (b) Some existing letters have not been counted; (c) The text of the Qur'an has been changed in order to either add letters to the text, or delete letters from the text.

Again, this is exactly the same allegation as in number 3 and I have dealt with it extensively in NINETEEN: God's Signature in Nature and Scripture. Besides, I responded to them in my debate with Lomax who used Abu Ameenah's book.

> **6. Word Count Inconsistencies:** A number of proofs used by Dr. Khalifa are based on the total number of words in verses and chapters being 19 or one of its multiples. This has been achieved by following an inconsistent system of letter counting whereby three or four words are sometimes counted as one word.

This is another highly exaggerated allegation, which we discussed extensively in this book.

> **7. Falsified Word Count Data:** Dr. Khalifa's claim that every word of the *Basmalah* occurs in the Qur'an either 19 times or one of its multiples is only true in the case of one of the main four words (i.e. *ar-Rahman*) and even in this case it is only achieved by *excluding* the 112 occurences of the word found in the 112 *Basmalahs* preceding the Surahs, yet he includes all the *Basmalahs* in his letter counts.

I have extensively discussed this issue with Lomax as my response to his "Claim number 13". Let me briefly summarize it here.

The mathematical system puts an end to the chronic arguments among various sects whether *Basmalah* is the first verse of Chapter *Al-Fatiha* or not. Now, it is clear that the *Basmalah* in the beginning of Chapter *Al-Fatiha* (The Opener) is the first verse while other 112 *Bismillahs* crowning the other chapters are un-numbered verses.

Our counting excludes the 112 unnumbered *Basmalah*s. Had we include them, Abu Ameenah would object again by labeling it as an "arbitrary inclusion of 112 unnumbered initial invocations which are merely repetition." Our exclusion is not arbitrary, since the other 112 *Basmalah*s are not numbered. Abu can see this fact in his own version of the Quran. Our method follows and justifies the well-known difference between the other two *Basmalah*s (1:1 and 27:30) and the 112 *Basmalah*s repeating in the beginning of chapters. It is a discovery which brings an explanation for this curious distinction. Though in the early manuscripts the verses were not numbered, they were ordered and separated from each other by dots which I believe justifies our numbering them. In fact, if there is a beginning and order of items of the same category, there is an implicit numbering in the structure. Therefore, early scholars were not wrong when they decided to number the verses as we know and use today.

131

8. False Claims for Surah Qaf: According to Dr. Khalifa, the term *"Qawm قوم"* is used to refer to Prophet Lot's people everywhere in the Qur'an except in Surah Qaf in order to keep the total number of *Qafs* in Surah Qaf a multiple of 19. However, this claim is totally false because there are not one but four other places in the Qur'an wherein the term *"Qawm"* is not used in references to Prophet Lot's people.

First, Rashad's point about the usage of the *Qawmu Lot* (Lot's People) had no substantial connection with the code; it was merely an observation regarding the usage of a word. Even if his comment was wrong, still the number of letter Qaf in Chapters starting with the letter Qaf would be 114 (19x6). Second, it is not Rashad but it is Abu Ameenah who is making false claim. Abu Ameenah is practicing his hoaxing skills as he did in the number of letters in Basmalah: He deliberately confuses the People of Lot, who are condemned, with the Family of Lot, who were saved.

Let me list all the verses where the **People** and **Family** of Prophet Lot are referred to:

Qawm (People):	11:70; 11:74; 11:89; 11:78; 22:43; 26:160; 27:56; 38:13; 54:33; 7:80; 27:54; 29:28.
Aal (Family):	15:59; 15:61; 27:56; 54:34
Ikhwan (Bretherns):	50:13

Contrary to what Abu Ameenah wants us to believe, verse 27:56 makes clear distinction between *Qawm* Lot (Lot's **People**) and *Aal-i* Lot (Lot's **Family**); they are not the same but opposing groups. The word *Ikwan* (Brethren) in verse 50:13 is referring to the **People** of the Lot not his Family.

In sum, to describe the people of Lot, the Quran uses the word *Qawm* (group, people, nation) 12 times. However, the 13th verse of the chapter that starts with the letter Q is an exception. Why then instead of the word *Qawm*, the word *Ikhwan* (brethrens) is used? For the answer, check the beginning of the chapter that contains that verse. As you see, the chapter starts with the letter Q.

Thus, Rashad Khalifa was right in his observation and Abu Ameenah is again proven to be confused.

9. False Claims for 19: The doctor claims that 19 was divinely chosen as the numerical code of the Qur'an because it translates into Arabic letters as *"WaahidQil"* (lit.one) and as such means "God is One", which he proposes is the message of the Qur'an. This claim is also incorrect as it is based on a system of numerology which has absolutely no place in Islam and is clearly rejected by Islamic law.

This is another evidence of blatant ignorance of Quran and the history of the number system used during the revelation of the Quran. Before and during the life of the Prophet Muhammed Arabs were using their alphabet as numerals. When Arabs decided to adopt the Hindu numerical system two centuries after the revelation of the Quran, they abandoned ABJAD and they changed the order of their alphabet to the current one about. I have presented conclusive archeological and Quranic evidence for the ABJAD system in the following chapters.

Abu Ameenah concludes his book with the following assertions:

> "From the preceding thorough refutation of the "facts" of Dr. Rashad Khalifa's Theory, it may be concluded that the theory of 19 as a miraculous numerical code for the Qur'an has no basis in the Qur'an itself and the few instances where 19 and its multiples do occur are merely coincidences which have been blown out of proportion by Dr. Khalifa. It may be further concluded that the Doctor's record of data falsification, textual changes and figure manipulation clearly indicate his dishonesty as a researcher and expose the low levels to which he stooped to invent support for his hoax."

> "Hence, 19 and its multiples may not be used to interpret anything of the Qur'an or Islam and all those sincere Muslims who have publicly propagated this theory in ignorance are Islamically obliged to publicly disown and discredit it, and immediately cease the publication, distribution and sale of books and tapes which support it."

> "With that I hope that all the doubts and queries surrounding the authenticity of Dr.Rashad Khalifa's "findings" have been finally laid to rest."

The rest of this book exposes the deception and ignorance of Abu Ameenah and his ilk. They are described by the numerous verses of the Quran as ingrates, those who have disease in their hearts, arrogant, ignorant and backward people. After reading the following verses you will learn the real source of Abu Ameenah's inspiration:

> 1:1   In the name of God, the Gracious, the Compassionate

2:17 Their example is like one who lights a fire, so when it illuminates what is around him, God takes away his light and leaves him in the darkness not seeing.

2:18 Deaf, dumb, and blind, they will not revert.

2:118 Those who do not know said, "If only God would speak to us, or a sign would come to us!" The people before them have said similar things; their hearts are so similar! We have clarified the signs for a people who have conviction.*

3:7 He is the One who sent down to you the book, from which there are definite signs; they are the essence of the book; and others, which are multiple-meaning. As for those who have disease in their hearts, eager to cause confusion and eager to derive their interpretation, they will follow what is multiple-meaning from it. But none knows their meaning except God and those who are well founded in knowledge; they say, "We acknowledge it, all is from our Lord." None will remember except the people of intellect.

6:4 Whenever a sign came to them from their Lord, they turned away from it.

6:5 They have denied the truth when it came to them. The news will ultimately come to them of what they were mocking.

6:25 Among them are those who listen to you; and We have made covers over their hearts to prevent them from understanding it, and deafness in their ears; and if they see every sign they will not acknowledge; even when they come to you they argue, those who reject say, "This is nothing but the tales from the past!"

6:26 They are deterring others from it, and keeping away themselves; but they will only destroy themselves, yet they do not notice.

6:104 "Visible proofs have come to you from your Lord; so whoever can see does so for himself, and whoever is blinded, does the same. I am not a guardian over you."

6:158 Do they wait until the controllers will come to them, or your Lord comes, or some signs from your Lord? The day some signs come from your Lord, it will do no good for any person to acknowledge if s/he did not acknowledge before, or s/he gained good through his/her acknowledgement. Say, "Wait, for we too are waiting."

7:132 They said, "No matter what you bring us of a sign to bewitch us with, we will never acknowledge you."

7:146 I will divert from My signs those who are arrogant on earth unjustly, and if they see every sign they do not acknowledge it, and if they see the path of guidance they do not take it as a path; and if they see the path of straying, they take it as a path. That is because they have denied Our signs and were heedless of them.

14:25 It bears its fruit every so often by its Lord's leave. God cites the examples for the people, perhaps they will remember.

14:26 The example of a bad word is like a tree which has been uprooted from the surface of the earth, it has nowhere to settle.

20:133 They said, "If only he would bring us a sign from his Lord!" Did not proof come to them from what is in the previous book?*

20:134 If We had destroyed them with retribution before this, they would have said, "Our Lord, if only You had sent us a messenger so we could follow Your signs before we are humiliated and shamed!"

20:135 Say, "All are waiting, so wait, and you will come to know who the people upon the balanced path are and who are guided."

25:4 Those who rejected said, "This is but a falsehood that he invented and other people have helped him with it; for they have come with what is wrong and fabricated."

25:5 They said, "Mythologies of the ancient people; he wrote them down while they were being dictated to him morning and evening."*

25:73 Those who when they are reminded of their Lord's signs, they do not fall on them deaf and blind.

27:82 When the punishment has been deserved by them, We will bring out for them a creature made of earthly material, it will speak to them that the people have been unaware regarding Our signs.*

27:83 The day We gather from every nation a party that denied Our signs, then they will be driven.

27:84 Until they have come, He will say, "Have you denied My signs while you had no explicit knowledge of them? What were you doing?"*

27:85 The punishment was deserved by them for what they transgressed, for they did not speak.

27:92 "That I recite the Quran." He who is guided is guided for himself, and to he who is misguided, say, "I am but one of the warners."

27:93 Say, "Praise be to God, He will show you His signs and you will know them. Your Lord is not unaware of what you do."*

29:1 A1L30M40

29:2 Did the people think that they will be left to say, "We acknowledge" without being put to the test?

135

29:3 While We had tested those before them, so that God would know those who are truthful and so that He would know the liars.

29:4 Or did those who sinned think that they would be ahead of Us? Miserable indeed is their judgment!

29:54 They hasten you for the retribution; while hell surrounds the ingrates. (Please read from 29:48)

38:1 S90, and the Quran that contains the Reminder.*

38:2 Indeed, those who have rejected are in false pride and defiance.

38:3 How many a generation have We destroyed before them. They called out when it was far too late.

38:4 They were surprised that a warner has come to them from among themselves. The ingrates said, "This is a magician, a liar."

38:5 "Has he made the gods into One god? This is indeed a strange thing!"

38:6 The leaders among them went out: "Walk away, and remain patient to your gods. This thing can be turned back."

38:7 "We never heard of this from the people before us. This is but an innovation."

38:8 "Has the remembrance been sent down to him, from between all of us!" Indeed, they are doubtful of My reminder. They have not yet tasted My retribution.*

38:29 A book that We have sent down to you, that is blessed, so that they may reflect upon its signs, and so that those with intelligence will take heed.

41:53 We will show them Our signs in the horizons, and within themselves, until it becomes clear to them that this is the truth. Is it not enough that your Lord is witness over all things?*

46:10 Say, "Do you see that if it were from God, and you rejected it, and a witness from the Children of Israel testified to its similarity, and he has acknowledged, while you have turned arrogant? Surely, God does not guide the wicked people."*

54:1 The moment drew near, and the moon was split.*

54:2 If they see a sign, they turn away and say, "Continuous magic!"

74:25 "This is nothing but the words of a human."

74:26 I will cast him in the Saqar.

74:27 Do you know what Saqar is?

74:28 It does not spare nor leave anything.*

74:29 Manifest to all the people.*

74:30 On it is nineteen.*

74:31 We have made the guardians of the fire to be angels/controllers; and We did not make their number except as a test for those who have rejected, to convince those who were given the book, to strengthen the acknowledgment of those who have acknowledged, so that those who have been given the book and those who acknowledge do not have doubt, and so that those who have a sickness in their hearts and the ingrates would say, "What did God mean by this example?" Thus God misguides whoever/whomever He wishes, and He guides whoever/whomever He wishes. None knows your Lord's soldiers except Him. It is but a reminder for people.

74:32 No, by the moon.*

74:33 By the night when it passes.

74:34 By the morning when it shines.

74:35 It is one of the great ones.

74:36 A warning to people.

74:37 For any among you who wishes to progress or regress.

74:38 Every person is held by what it earned;

74:39 Except for the people of the right.

74:40 In paradises, they will be asking

74:41 About the criminals.

74:42 "What has caused you to be in Saqar?"

74:43 They said, "We were not of those who offered support (or observed contact prayer)."*

74:44 "We did not feed the poor."

74:45 "We used to participate with those who spoke falsehood."

74:46 "We used to deny the day of Judgment."*

74:47 "Until the certainty came to us."

74:48 Thus, no intercession of intercessors could help them.

74:49 Why did they turn away from this reminder?*

74:50 Like fleeing zebras,

74:51 Running from the lion?

74:52 Alas, every one of them wants to be given separate manuscripts.

74:53 No, they do not fear the Hereafter.

74:54 No, it is a reminder.

74:55 Whosoever wishes will take heed.

74:56 None will take heed except if God wills. He is the source of righteousness and the source of forgiveness.

78:27 They did not expect a reckoning/computation.*

78:28 They denied Our signs greatly.

78:29 Everything We have counted in a record.

78:30 So taste it, for no increase will come to you from Us except in retribution.

83:18 No, the record of the pious is in Elliyeen.

83:19 Do you know what Elliyeen is?

83:20 A numerical book.

83:21 To be witnessed by those brought near.

9

Running Like Zebras

Edip Yuksel vs. Daniel (Abdul*Rahman*) Lomax

PREFACE

This book contains a debate between Mr. Edip Yuksel, a proponent of the Quran's numerical structure based on the number 19, and its detractor and skeptic "par-excellence" Mr. Daniel (Abdul*Rahman*) Lomax. The actual polemic took place on an Internet computer bulletin board, where Mr. Lomax has for the past several years posted his refutations of the Quran's mathematical structure.

The original work of decoding the Quran's mathematical pattern based on the number 19 (found in sura 74 verse 30, a chapter entitled "The Hidden") was done by the late Dr. Rashad Khalifa in the seventies and eighties. His work was initially welcomed in the Muslim world, until he arrived at the unavoidable conclusion that Muslims should follow the Quran, and the Quran alone, as a source of religious guidance. This was too hard for the power that be in traditional Islam (the religious scholars) to swallow. Dr. Khalifa's works were eventually banned in the Muslim countries (he was murdered in 1990), and ideas similar to his were either suppressed or hotly refuted.

In a sense, then, the debate contained in this book is an ongoing struggle between people who follow the Quran alone and those who follow hadith, sunna and consensus of the religious scholars and clerics. A debate between those who are satisfied with God's revelation and those who uphold religious traditions based on man-made dogmas. A polemic between the supporters of the true message of Muhammad and the supporters of the man-made sayings attributed to the man. In the end, the tone of the argument is not unlike the classic exchange of arguments between ahl al-Quran and ahl al-Hadith in early Islam.

In spite of the advances in the methodology of scientific investigation that our generation is blessed with, the heart of the main argument remains the same. True, one needs to be skeptical and to verify everything. At the same time, one must also be open minded, listen to all views and follow the best idea. There is a limit to skepticism beyond which lays a danger of closing of the mind from accepting any truth because of the contentiousness nature in us (Quran 41:45, 4:155; 18:54)

When it comes to understanding God's signs in the scripture, in the universe around us or within ourselves, we need to shift our paradigm, our way of looking at things, accordingly. In this case, the paradigm is given by this rhetorical question from the Quran: "Is God not sufficient for His servant?"(39:36). For those who answer affirmatively, God will manifest His signs accordingly. "We will show them our signs in the horizons, and within themselves, until they realize that this is the truth. Is your Lord not sufficient as a witness of all things?" (41:53)

Gatut S. Adisoma, Ph.D.

Tucson, Arizona, August 14, 1995

CLAIM 1

There is a whole Sura about the secret numerical code entitled "The Hidden Secret," namely Sura 74. In this Sura, God informs us that if anyone claims that the Quran is man-made (verse 25), God will prove him wrong by the number 19 (verse 30). God says that this number serves five functions: a) to disturb the disbelievers, b) to convince the Christians and Jews that this is a divine scripture, c) to strengthen the faith of the faithful, d) to remove all traces of doubts from the hearts of Christians, Jews, as well as the believers, e) to expose those who harbor doubt in their hearts and the disbelievers. They will say, "What did God mean by this allegory?"

LOMAX: The message of the passage in question has been distorted by this description, which is perhaps 90% accurate. What is not at all clear from that passage is that the Nineteen is an argument against the human creation of the Quran. It is clear that the number has these five purposes, but the last purpose could certainly be read as applying to those who attempt to find the hidden meaning of the Nineteen. This interpretation is very much in harmony with 3:7:

"He it is who has sent down to thee the book; in it are verses basic or fundamental; they are the foundation of the book: others are not of well-established meaning. But those in whose hearts is perversity follow the part thereof that is not of well-established meaning, seeking discord, and searching for its hidden meanings, but no one knows its true meanings except Allah. And those who are firmly grounded in knowledge say 'We believe in the book; the whole of it is from our Lord:' and none will grasp the message except [those] of understanding." (Translation following Yusuf Ali.)

YUKSEL: The description does not distort the message of the verse 74:31. Claiming that "perhaps 90% accurate" without clarifying which part of the description is inaccurate is a dubious tactic, intended to put doubts on the entire description. I expect Lomax to unveil the 10% inaccuracy in our translation or understanding of the verse. If I try to be nit-picking, I can see only one problem: the translation of the Chapter's title which is "The Hidden Secret." It is a redundant title. A more accurate translation should have been "The Hidden One" which still carries an implication to the nature of the mathematical code. The title can be also translated as "The Enfolded One."

A statement in his objection puts doubt in my mind about either Lomax's literal capability or his honesty: "What is not all clear from the passage is that the Nineteen is an argument against the human creation of the Quran." This is an incredible statement made by a person who is highly educated and reads the

141

translation of the Quran in his mother tongue. How can a person doubt about the context of Nineteen after reading until the 31st verse of Chapter 74? How can a person blind himself to the obvious connection between verse 25 and 30? I cannot believe that his attention span is less than five short verses. I bet that most of the secondary school students would be able to see the purpose of Nineteen CLEARLY after reading the translation of those verses. Obviously, I am not suggesting the archaic language of Yusuf Ali, who mimics the style of King James Version and is very good in complicating simple statements. Probably, it would be a good advice for Lomax not to "follow Yusuf Ali" alone. He can find many other translations written with much simpler English. However, a clear relation between the claim in verse 25 and the following verses is clear for an objective reader even from Yusuf Ali's translation Here are the translation of verses which according to Lomax are NOT CLEAR to indicate that "the Nineteen is an argument against the human creation of the Quran."

In the Name of God, Gracious, Merciful

1. O you (who are) hidden:
2. Come out and warn.
3. Extol your Lord.
4. Purify your garment.
5. Forsake what is wrong.
6. Be content with your lot.
7. Be steadfast for your Lord.
8. When the trumpet is sounded.
9. That will be a difficult day.
10. For the disbelievers, not easy.
11. Let Me deal with one I created as an individual.
12. I provided him with lots of wealth.
13. And children to behold.
14. I made everything easy for him.
15. Yet, he is greedy for more.
16. He stubbornly refused to accept our revelations and/or miracles.
17. I will punish him increasingly.
18. For he reflected, then decided.
19. Miserable is what he decided.
20. Miserable indeed is what he decided.
21. He looked.
22. He frowned and whined.
23. Then he turned away arrogantly.
24. He said, "This is but impressive (or old) magic!

25. "This is nothing but word of human."
26. I will commit him to *Saqar*.
27. Do you know what *Saqar* is?
28. It is no more, no less (precise).
29. Succeeding screens (obvious) for people.
30. Over it is nineteen.
31. We appointed angels to be guardians of Hell, and we assigned their number (1) to disturb the disbelievers, (2) to convince the Christians and Jews, (3) to strengthen the faith of the faithful, (4) to remove all traces of doubt from the hearts of Christians, Jews, as well as the believers, and (5) to expose those who harbor doubt in their hearts, and the disbelievers; they will say, "What did God mean by this allegory?" God thus sends astray whomever He wills (or, whoever wills), and guides whomever He wills (or, whoever wills). None knows the soldiers of your Lord except He. It is a reminder for the people.
32. Absolutely, by the moon.
33. And the night as it passes.
34. And the morning as it shines.
35. This is one of the greatest (miracles).

Now, just ask yourself these two questions: "Who is subjected to *Saqar* retribution and number Nineteen?" Obviously, the answer is the disbeliever. Then, why? Obviously, he did not believe that the Quran was authored by God; he claimed that the Quran was man-made. Therefore, he is immediately challenged by *Saqar* and Nineteen. I don't understand why this simple and obvious fact became difficult and ambiguous for Lomax. Furthermore, the following verse lists the objectives of the number Nineteen. According to that verse (74:31), it will increase the faith of believers, and remove doubts in the heart of Christians and Jews, etc. How can you disconnect this objective from previous argument regarding the doubt or rejection the divine nature of the Quran?

Finally, Lomax uses a mistranslation of 3:7 to make his erroneous point, that the verses about the number 19 are *mutashabih* and only hypocrites try to understand their meaning. First, I want to present my translation, and then explain why I think this is the correct translation, and why Yusuf Ali's translation is both wrong and dangerous. Please compare the important difference in punctuation of the last two statements:

"He revealed to you this scripture, consisting of straightforward verses, which are the essence of the scripture, as well as multiple-meaning verses (*mutashabihat*). Those who harbor doubt in their hearts will pursue the multiple-meaning verses to create confusion, and to seek their meanings. No one knows their meaning except God and those who are deeply rooted in knowledge. They say, 'We believe in this; it all comes from our Lord.' Only those who possess intelligence will take heed." (3:7)

First, we should try to understand the meaning of *mutashabih* The word *mutashabih* comes from *shabaha* (became similar), and its usage in other verses clarifies its meaning as "similar." It describes the things or words that can be confusing for a novice because of the similarities. (see: 2:118; 2:70; 4:157; 6:99; 6:141; 2:25) Verse 39:23 uses *mutashabih*at for the entire Quran, referring to the overall similarity, in other words, its consistency. In its narrow meaning, *mutashabih* refers to all verses which can be understood in more than one ways. This includes all kind of allegories. The various meaning or implications of the same words require some qualities in the audience: attentive mind, positive attitude, contextual perspective, and patience for research.

Now, let's come to the crucial part. It is one of the intriguing features of the Quran that the verse about *mutashabih* verses of the Quran is itself *mutashabih*, that is, has multiple meanings. Indeed, the verses about understanding the Quran themselves are not understood by those who betray the Quran by trading it with volumes of hadith books which are full of fabricated stories falsely attributed to the Prophet Muhammad (See our book "19 Questions For Muslim Scholars," and other literature published by ICS/Monotheist Publication, P.O. Box 43476 Tucson, AZ 85733-3476, Telephone/Fax (520) 323-7636) Thus, those verses become an empirical proof for their own claim as it is expressed in the following verses:

17:46   We place shields over their hearts, that they should not understand it, and deafness in their ears. If you mention your Lord in the Quran alone, they run away turning their backs in aversion.

41:44   Had We made it a non-Arabic compilation, they would have said, "If only its signs were made clear!" Non-Arabic and Arabic, say, "For those who acknowledge, it is a guide and healing. As for those who reject, there is deafness in their ears, and they are blind to it. These will be called from a place far away."

56:77   It is an honorable Quran.

144

56:78   In a protected record.
56:79   None can grasp it except those pure.

The last part of the verse 3:7 can be understood in two different ways by merely a punctuation change. If you stop after the word God, then you will (mis)understand like many Sunni scholars did. According to your punctuation, even those who possess knowledge will not be able to understand *mutashabih* verses. However, if you don't stop there its meaning will change to the opposite: Those who possess knowledge will be able to understand their meaning. Let's list our reason for preferring this later understanding:

1. The intention of those who try to understand the multiple-meaning verses is important. With the disease in their harts, they try to confuse others by focusing on multiple meaning verses. Since only sincere believers possess the quality to understand the Quran (17:46; 18:57; 54:17), hypocrites will miss the true meaning of multiple-meaning words. They either try to take them out of context or disregard other verses that bring explanation to them. For instance:

16:44   ". . . And we sent down to you this message, to proclaim (*litubayyena*) for the people everything that is sent down to them, perhaps they will reflect."

The word *lituBaYyeNa* is a derivative of "BYN", which is a multiple-meaning word. It means: 1) To reveal what is concealed. 2) To explain what is vague. The first meaning is the antonym of "hide", the second is the antonym of "make vague". God orders Muhammad to proclaim the revelation which is revealed to him personally. Indeed, this is the whole mission of the messengers." (16:35)

Prophets sometimes experience difficulty in proclaiming the revelation (33:37; 20:25). If the Quran is a profound Arabic book, if it is explained by God, and if it is simple to understand (5:15; 26:195; 11:1; 54:17; 55:1-2), then the prophet does not have an extra mission to explain it. Furthermore, the verse 75:19 does not leave any room for an extra human explanation. Thus, the word *litubayyena* of 16:44 is similar to the one in 3:187. Verse 3:187 tells us that the people who received the revelation should "proclaim the scripture to the people, and never conceal it."

The Quran broadcasts a very clear message. However, the problem is with our receivers. If our receiver does not hear the broadcast or cannot understand it well, then something is wrong with our receiver and we have to check it. If the signal is weak, we need to recharge our batteries, or reset our antennas. If we do not receive a clear message, we need to tune into the station, to the station of Quran alone, in order to get rid of the noises and interference from other sources. We may ask some help from knowledgeable people or experts for this task. If the receiver does not work at all, then we have to make a sincere effort to fix the broken parts. However, if we believe that the problem is in the broadcast, then nobody can help us. The divine broadcast can be heard in detail only by those who sincerely tune in, i.e., those who take it seriously and act accordingly. The condition of our receiver and the antenna, the power of the battery, and the precision level of our tuning are very important in getting the divine message properly.

      56:79   "None can grasp it except the sincere."

The Quran is simple to understand (54:11). Whoever opens his/her mind and heart as a monotheist and takes the time to study it, will understand it. This understanding will be enough for salvation. Beyond this, to understand the multi-meaning verses or allegorical descriptions you do not need to be a messenger of God. If you have a good mind and have studied the Quran as a believer, that is, if you have a deep knowledge, then you will be able to understand the true meanings of multiple-meaning verses. The verse 3:7, which is about the multiple-meaning verses, points this fact in a multiple-meaning way: ". . . No one knows their true meaning except God and those who possess knowledge..." Obviously, those who are confused by contradictory teaching of ignorant clergymen (falsely called *ulama,* the knowledgeable ones) do not possess knowledge and unable to understand this verse.

2. In order to believe in all the verses of the Quran, one does not need to be deeply rooted in knowledge. To be a "believer" is a sufficient condition to beleive all the verses. However, one needs to have deep knowledge of the Quran in order to understand *mutashabih* (multiple meaning) verses accurately. Therefore, 3:7 mentions a narrow category (those who deeply rooted in knowledge) in relation with those multiple meaning verses.

3. God, the teacher of the Quran, encourages believers to study the Quran with patience. A portion of the Quran requires intensive analytical study. It advises us not to rush into understandin without sufficient knowledge (20:114). Nevertheless, the Quran, in general, is easy to understand (54:17...). By mistranslating 3:7, you try to discourage and scare believers, including those who are deeply rooted in knowledge, from understanding the *mutashabih* verses, without a clear definition and identification of *mutashabih.*

146

4. If we follow your translation of 3:7, then, we must have a clear definition of *mutashabih* verses in order to avoid trying to understand them. According to your translation, if one tries to understand a verse and that verse happens to be a *mutashabih*, then that person is in danger to become a hypocrite. Therefore, you must be able to identify ALL the *mutashabih* verses. Can you provide us with a list of *mutashabih* verses? Can you justify your criteria for your selection? How can your lack of understanding be criteria for others? Someone's lack of understanding of one verse cannot make a verse "taboo" for all other people. Otherwise, the lowest degree of understanding will be the common denominator in understanding of the Quran. The more verse you don't understand the more you will be safe, and the more verse you understand the more you will be a subject for accusation to be a hypocrite.

5. There are some Sunni commentators who support our understanding. For instance, the classic commentary of al-Baydawi prefers this understanding. Please note that Yusuf Ali also acknowledges this fact in the footnote of 3:7: "One reading, rejected by the majority of Commentators, but accepted by Mujahid and others, would not make a break at the point marked Waqfa Lazim, but would run the two sentences together. In that case the construction would run: 'No one knows its hidden meanings except God and those who are firm in knowledge. They say', etc." If you are not determining the truth by the number of votes (majority), then, I suggest you reflect on the reasons I have listed here.

6. It appears that you agree with us regarding the first four function or objectives of the number Nineteen. (1) to disturb the disbelievers, (2) to convince the Christians and Jews, (3) to strengthen the faith of the faithful, (4) to remove all traces of doubt from the hearts of Christians, Jews, as well as the believers.

How can you explain these statements? How can this Quranic description or prophecy occur without understanding the meaning or implication of the number Nineteen? How many "believers" increased their faith without understanding the meaning of the number? How many Christians or Jews accepted the Quran because of this number, without understanding its meaning? How many people's doubt were removed by your version of Nineteen, that is, a Nineteen devoid of meaning for humans?

The way of understanding the implication of number Nineteen, on the other hand, entirely fits these Quranic description. It is an incredible prophecy of the Quran that the message of number 19 mentioned in Chapter 74 was unveiled in

1974, exactly 1406 (19x74) lunar years after the revelation of the Quran. God Almighty has increased the faith of believers (thank God I am one of them), some Christians and Jews have accepted the message of the Quran and the fanatic disbelievers and hypocrites who do not really care about divine revelation have been disturbed by this number; they get furious and nervous when they hear or see the number 19. This reaction is well described in the last verses of the same chapter, especially 74:49-51.

7. The question asked by disbelievers and hypocrites in 74:31 "What does God mean with this?" is not a question that expresses their curiosity for the meaning of God's word, but it is a question of ignorance and avoidance. They cannot or will not understand allegories, in other words, *mutashabih* verses. Please read verses 2:26 and 47:15-16, and note that after allegorical description of "heaven" disbelievers ask similar question. In the context, it is obvious that they don't want to understand the implied meanings of those allegories. Their arrogance and ignorance lead them to read those verses literally which brings nothing but ignorance, disbelief and confusion for them (17:60 and 37:62-66; 17:82).

Lomax' response is quoted after LOMAX. I have interrupted his response in order to make our argument easier to follow. You should consider Lomax' part as a continuous objection interspersed with my defense.

LOMAX: Mr Yuksel objects to my statement, which was not intended to be precise, that the presentation of the meaning of Sura 74 is "perhaps 90% accurate," and he questions my motives in not stating exactly what was 10% inaccurate. However, he goes on to acknowledge that the title "The Hidden Secret" is redundant (I would say misleading: Hidden is reasonably accurate, Secret is not), and he also objects to my specific statement that "What is not at all clear from that passage is that the Nineteen is an argument against the human creation of the Quran." This was the point. The sura, it must be acknowledged, threatens those who reject the divine source of the Quran with Hell. Then it says that "over it (Hell, presumably) are nineteen." This is a description of Hell, not necessarily an argument against human creation.

YUKSEL: For a person who is not a nitpicker there is little difference between the meaning of "The Hidden One" and "The Secret." They mean the same. Random House Webster's College Dictionary defines "secret" as "hidden from sight; concealed." Accepting the "The Hidden One" as the accurate translation of *Al-Muddathir* won't enable Muhammedans to re-hide the code which was unveiled after 14 centuries through a monotheist biochemist, Dr. Rashad Khalifa. The same dictionary describes the attitude of Muhammedans regarding

148

the mathematical code of the Quran under the entry "hide": (1) to conceal from sight; prevent from being seen or discovered. (2) to obstruct the view of; cover up. (3) to conceal from knowledge or exposure; keep secret, etc.

The Quran repeatedly uses three words for hell: *Jahannam, Jaheem,* and *Naar* (fire). The word *Saqar* is the only time used here (74:26)[18]. It is obvious that the word *Saqar* did not have any conventional meaning related to hell, since the following verse defines or describes its meaning (74:27-29). Previous commentators who did not know about the mathematical structure of the Quran had the excuse to understand it as Hell. They even forced the meaning of *Lawwahatun lil-bashar* (Obvious, successive screens for humans) (74:29). They translated the verse as "scorching the skin," despite the fact that the usage of the Quran did not justify such a meaning. For instance the word bashar is always used for "human being." Quran also uses the word *jild* for "skin."

It is noteworthy that Muhammad Asad in his translation "The Message of The Quran" does not follow the traditional bandwagon. Though he died before the discovery of the code, he demonstrates a good intuition. His translation of 74:29: "making (all truth) visible to mortal man." In the footnote he defends this translation: "Most of the commentators interpret the above elliptic phrase in the sense of 'changing the appearance of man' or 'scorching the skin of man'. The rendering adopted by me, on the other hand, is based on the primary significance of the verb laha- 'it appeared', 'it shone forth' or 'it became visible'. Hence, the primary meaning of the intensive participial noun lawwah is "that which makes (something) visible'. . . "

Finally, even if those verses are understood as the description of Hell, still the Quran treats Nineteen as a number. The number Nineteen is isolated from Hell both in verse 74:30 and 74:31. **No matter how hard you try, you cannot hide the number Nineteen in your imaginary Hell.**

It is a short but good step for Lomax that finally he acknowledges the theme of the chapter as a "threat" to those who claim human origin for the Quran. However, he is not able to see the "intellectual challenge." What he can see is a "threat with Hell." Well, some prefer Hell.

LOMAX: Then the Sura goes on to state the reasons for the setting of this number. This part of the Sura is consistent with there being some kind of phenomenon involving 19 in the Quran. However, it is quite clear that this

---

[18]   This was an erroneous statemet. The word Saqar is mentioned four times in the Quran, three times in Chapter 74 and once in 54:48.

phenomenon is a trial for the disbelievers, and it also consistent with the passage to understand that those who seek the meaning of the 19 are the people "in whose hearts is a disease and the disbelievers." (74:31)

YUKSEL: Yes, this phenomenon is a trial for disbelievers and hypocrites. But Lomax is craftily trying to hide the rest of the verse. He knows well that the rest of the verse cannot be explained with his "meaningless, incoherent, anecdotal and coincidental" nineteen. Again, he repeats himself without answering my criticism regarding his interpretation of the question asked by disbelievers and hypocrites. I gave him verses 2:26 and 47:15-16 as a reference for similar questions. Lomax should go back and read the last paragraph of my first round defense for Claim 1.

LOMAX: Mr. Yuksel wonders what my problem is that I cannot see what he sees in this verse. The "problem" is that I saw what he describes years ago, and now I see more, not less.

YUKSEL: Evidently, he never saw the miracle, since this miracle requires independent research and good state of mind, not blind acceptance. He probably pretended to see the miracle by his "faith." When he saw some problems with Rashad's work he became a disbeliever of it. He became paranoid and an extreme skeptic, resorting to every possible excuse to reject or cast doubt on this mathematical pattern. I understand this psychology. There is a proverb: "a scalded man fears cold water." He has volumes of stories and scholastic confusion at his disposal to take his revenge.

He is upset since he feels fooled years ago. Instead of blaming Rashad, I think he should blame his early ignorance. I agree that he can "see more" now. Samaritan too saw more, centuries ago (20:96).

LOMAX: It should also be made clear that I read the Quran in Arabic. I do not depend on Yusuf Ali. The translation which I use, personally, more than any other, is Muhammad Ali; I use Yusuf Ali when I want to present a standard, widely accepted translation. But there are obvious problems with Yusuf Ali, as there are with any translation. The translation given by Yuksel is highly interpretive; it reflects the conclusions of the translator, not the literal Arabic. I do not have time to exhaustively describe this. If Mr. Yuksel wants to start a topic on this translation, he is welcome. In the meantime, just look in a number of Qurans and see how far Yuksel's translation is from the rest. What is seen by everyone else as a description of Hell (which is multiply confirmed in the passage) is transformed into a description of the "miracle" of the 19 by a very loose translation.

150

YUKSEL: The claim that "everyone else" saw the verses as a description of Hell is false. This shows that Lomax still needs to see more! I have just quoted from Muhammad Asad's translation. If he wants to see more, I urge him to go to a library and do some research. For instance, he can see the commentary of famous scholar Fahreddin ar-Razi.

LOMAX: For example, 74:28. Yuksel has "It is exact and precise." M. Ali has "It leaves naught, and spares naught." (this is quite literal). Y. Ali has "Naught does it permit to endure, and naught does it leave alone." Arberry has "It spares not, neither leaves alone." Sale has "It leaveth not (anything unconsumed), neither doth it suffer (anything) to escape)."

YUKSEL: Well, I am ready to accept M. Ali's translation. After the discovery of the code it makes more sense. The literal translation of 74:28 is a description of exact and precise. We can translate the verse as "Neither does it leave, nor does it bear (no less, no more)."

Mr. Lomax has entirely ignored my lengthy criticism to his interpretation of verse 3:7 on *mutashabih* (multiple meaning) verses. When I questioned him by e-mail, he said that his answer would have been redundant. "Actually, the whole discussion is somewhat beside the point. The central issues are raised in the Draft FAQ: 19, Study Problems." In our previous e-mail conversation he invited me to answer his paper titled "On Dr. Khalifa's Theory of the Nineteen in the Quran." This paper is being posted over and over on Internet, Compuserve, America Online, etc. When he is pressed on concrete and specific examples, he is showing me another paper, which is an abstract argument expressing his confusion and acute skepticism. I will deal with the Draft FAQ later, God willing.

CLAIM 1a

The feminine pronoun "ha" (it or her) in the verse "Over it is Nineteen" (74:30) refers to *Lawwaha* (screen, plate, visually obvious) or to *Saqar* (challenge, difficult task, retribution). As for the feminine *heya* (it or she) in the last phrase of verse 74:31, "It is a reminder (*zikra*) for people" refers to the number Nineteen.

LOMAX: On the pronoun in verse 30. I noted that it is in the feminine (and therefore could refer to the blaze but not to the man). The "it" (feminine) at the end of 31 is unclear in reference to me. 'Ad+ (number) is masculine, as are "*tis'a+ ashar*" (19), "*kitaab*" (book), and "*masal*" (symbol). Feminine are *Naar* (fire) and "*Saqar*"(blaze), or perhaps just the general topic.

151

YUKSEL: It is obvious that the pronoun "ha" (it or her) cannot refer to the man, since man is masculine. It refers either to "lawwaha" or *Saqar*. These two words, however, are being mistranslated by many as "the thing that scorches" and "blaze," respectively. Many linguists claim that the word *Saqar* is a foreign word. Here, we really do not need to find out the conventional meaning of the "word" *Saqar*. Whenever the Quran uses a word followed by the question "do you know what that means?" then, the Quran adds or attaches a new meaning to that word in question. Please see 69:3; 77:14; 82:17-19; 83:8, 19; 86:2; 90:12; 101:3,10; 104:5. Therefore, it is better not to rush into the translation of *Saqar* since God Almighty is going to describe or define it. If you notice, in my translation I left the word *Saqar* as it is. You can derive its meaning from its context and description. Besides, there can be another candidate (sure, a remote one) for the reference of verse 30: *ayaat* (our signs and/or revelation) mentioned in verse 16. None of the possible references refute the mathematical implication of the number Nineteen.

Unlike Lomax, I am very clear regarding the reference of feminine "it" at the end of verse 74:31. Verse 31, in its beginning phrase, switches the subject from *Naar* (fire) to the number 19. Obviously, some people misunderstood the implication of 74:30. Later, verse 31 clarifies such misunderstanding. The number *tis'ata ashar* (Nineteen) in verse 30 can be also considered feminine. The word *eddatahum* (their number) in verse 31 is feminine too, and both words can be references for the pronoun *heya* (it or she) in the last phrase: "it is a *zikra* (reminder) for the people." Referring to *eddatahum* (their number) is more meaningful because of the context and proximity of the word.

We can easily disqualify the feminine words *Naar* (fire) and *Saqar*, since they are not *zikra* (reminder) in themselves. You can scan all the verses where various forms of this word is mentioned. It is noteworthy that in verse 49 of the same chapter we see a slightly different form of the same word, *tazkira* (reminder). It is used for something good, not bad: "Why are they so averse to this *tazkira* (reminder)?" (74:49). Obviously, hell is something to be disliked. Therefore, the word *zikra* (reminder) in the end of verse 31 cannot refer to hell.

LOMAX: This "claim" was written AFTER the "objection" which follows. My original writing was investigative, not argumentative. Thus Yuksel can make my comments look dim-witted. Here is my actual response, written now:

From the context, it is clear that the "it" in 74:30 refers to *Saqar* (Hell-fire), or, less likely, to *lawwAHa* (a darkening of color). This word *Saqar*, according to Penrice, is "supposed to be of foreign origin," though he gives the meaning of the root as "to injure by heat." *lawwAHa* is a verbal adjective of intensity, used with *li* of. The root *lAH* means to appear (a star); Yuksel is apparently following modern meanings or possibly speculating to come up with his translation.

YUKSEL: My purpose was not to make your comments look dim-witted, but clarify the issue. I apologize if I caused such an impression.

A person who is a little bit familiar with the Quran does not really need to learn the meaning of *Saqar* from Penrice, or someone else who is merely endorsing one of the many speculations. There is no need to speculate on the meaning of this "supposedly foreign" word, since the Quranic verses defines or re-defines the word *Saqar* in 74:26-29. As for *lawwaHa*, Lomax has finally discovered one of its meanings, "to appear with intensity or frequency." It is not clear why Lomax tries to undermine my translation, since I agree with this meaning. Is there a contradiction or a big difference between "to appear intensely" and "visually obvious?"

LOMAX: The "it" at the end of 74:31, it appears, would refer either to the "number" at the beginning of the verse (which, by the way, establishes that nineteen is the number of the companions of the Fire [!], which confirms that *Saqar* is the Fire), or to the verse as a whole.

YUKSEL: Another good step! Now you disagree with the commentary of the majority of orthodox scholars who try to skip the number 19 and refer the pronoun "it" at the end of 74:31 to Hell. I will not argue further on this issue, since you are determined not to see the clear statements in 74: 30 and 74:31 that isolate the number *nineteen* from the *fire*, a fire that you are so eager to see everywhere. Indeed, the number 19 is an intellectual punishment for disbelievers and hypocrites in this world, and it is also the eternal punishment as the number of the angels guarding the Hell. Disbelievers and hypocrites can never escape from this number!

> 74:49-51 Why are they so averse to this reminder? Running like zebras, who are fleeing from the lion!"

LOMAX: I have never denied that 74:30-31 may be read as referring to the topic of "nineteen" and the reactions of mankind to this "miracle." Rather, I point out aspects of this reference which are overlooked by Khalifites. One of the great dangers in reading the Quran is to assume that references to *kafiruwn* (unbelievers) are to "those people." It is essential, to be among those reminded, that we apply these words to ourselves, that we may be purified from *kufr*. To do otherwise is arrogance and pride, the opposite of *taqwa*. This passage is a

153

warning to all of us, not just to people who do not accept the "miracle of the nineteen."

YUKSEL: Trying to confuse the issue among the so called other aspects, and fighting against one of the greatest divine evidence is not righteousness. Lomax should follow his own advice since he prefers the so-called consensus of his scholars to the Quran.

LOMAX: By the way, Yuksel has edited my material somewhat, changing the transliteration scheme I had used. In a few places, this may be significant.

YUKSEL: I did indeed modify your "transliteration scheme" for two reasons. First, I wanted to keep our reference of the same words consistent. Second, I found Lomax' transliteration bizarre and meaningless, since it contained some symbols which did not correspond to any sound nor letter. I found it confusing and difficult to follow. I wanted the audience to follow the argument without getting tired from ostentatious transliteration schemes. I believe that the changes were not significant. Besides, Lomax should not complain about these "itsy-bitsy" things. He has answered my "19 Questions for Muslim Scholars" without presenting my argument. His answer, which is downloaded on several network libraries, is an unfair presentation.

CLAIM 2

The first verse in the Quran, known as the *Basmalah* consists of 19 Arabic letters.

LOMAX: Arguably true, but anecdotal. Of course, there are other ways to count the letters, but the method used by Dr. Khalifa here is reasonable.

YUKSEL: The number of letters in *Basmalah* is certainly true and significant. Though Lomax accepts this fact, he is trying to do his best in order to undermine this obvious physical fact. First, there is no reasonable argument about the number of letters in *Basmalah*. If he is referring to the absurd arguments raised by modern Sunni scholars who hallucinate by counting non-existing letters in order to blind people to the mathematical miracle of the Quran, then, he must know that everything can be arguable. With the same logic we can claim that the existence of the external world, including the Quran, is "arguably true", since some skeptical philosophers entertain doubts about it. I challenge him to show me "other ways of counting the letters" of *Basmalah*. Our counting is not merely reasonable, but the ONLY reasonable way of counting the Arabic letters in *Basmalah*. It is the simplest thing to do. Nineteen letters of *Basmalah* is a well known fact, since the well known formula of 786 which is used for centuries to represent *Basmalah* is the gematrical value of exactly 19 letters. In fact, there was no argument on the number of its letters until the discovery of the 19-based mathematical system. Nevertheless, I congratulate him for not following those Sunni scholars who pontificate that the number of letters in *Basmalah* is either 18, or 20, or 21; but never 19.

Second, how did he decide that this fact is anecdotal? Isn't *Basmalah* the FIRST, THE MOST REPEATED verse of the Quran, and CROWNS every chapter except Chapter 9? How can the number of its letters be considered anecdotal if there is possibility of a numerically structured system in the Quran? We should not forget that the importance and extraordinary nature of this system does not come merely from individual examples, but from the combination and interlocking nature of those examples. Remember that the Quran does not challenge disbelievers to bring one or several verses similar to its verses, rather it challenges them with more; in at least three verses. For the same reason, the word *aya,* in its singular form, is never used for the Quranic verses. On the other hand, its plural form, *ayaat,* is used for both, the Quranic verses and miracles, signs etc. Please check the 84 occurrences of the singular form *aya* and witness this unique Quranic semantic for yourself. Quran has a unique way of inserting details in conventional language. Besides, you will notice the grave mistranslation of the word *aya* (singular form) in verse 2:106, and the ignorance of those who claim the existence of abrogation in the Quran.

In summary, the number of letters in *Basmalah* is certainly true and significant.

LOMAX: From an overall examination of the techniques of counting used to generate the "miracle," we can see that what may be obvious in one example is undermined in another. Khalifa (and Yuksel, following him), would claim, again and again, that this count of 19 was completely obvious and that the assertions of scholars that there were additional letters were simply ridiculous. But when, in the next fact, the counts of the individual words in the Bismillah are given, what is counted for *BSM* is actually *ASM*. Where did this "A" come from? All I have stated is that there are other ways of counting letters than the one used by Khalifa, Yuksel, and others. Arabic orthography is, to a certain extent, arbitrary, and variations exist. We will come back to this issue: Is the Quran the written text, or is it the recitation? Since the word *Quran* means "The Recitation," I find the answer fairly obvious.

YUKSEL: Again, Lomax is confusing the matters. The first word of *Basmalah,* that is, BSM has three letters. However, without any prefix attached, the word is

ISM. There is no other way of counting the letters in *Basmalah*, since it has a unique spelling. None so far succeeded or even suggested to change this famous spelling. It is different to count the letters of a statement and counting the frequency of its words, since in the later case, you can isolate the word from attached propositions or contextual format.

The Quran means "The Recitation" or "The Book of Recitation." The Quran is both the written text and recitation. There are many other names (attributes) mentioned for the Quran: *Al-Kitab* (Book), *Suhuf* (Scripture), etc. Obviously, Lomax has not read the Quran carefully enough to see this simple fact. Here are few verses as reminder: 80:11-16; 56:78. Nevertheless, the Quran, ultimately, is neither a recitation nor a book. It is a revelation in the heart of those who possess knowledge (29:49).

LOMAX: There are, and have always been, variations in the written text, as well as in recitation. If one is going to count *alif* as a letter (some say that it is not properly a letter), then we must face the fact that it is written in various ways, and, for example, *Hafs* and *Warsh* differ in the use of *alif*.

YUKSEL: We are here discussing the number of letters in *Basmalah*, nothing else. So far, there are no variations in its spelling. Hafs, Warsh and even Marsh versions do not differ regarding the spelling of the *Basmalah*.[19] Muhammedan scholars who are masters of disagreement somehow failed to disagree on the spelling of *Basmalah* (or Bismillah). Again, Lomax is trying to confuse the issue by using a subtle implication.

LOMAX: Yuksel wrote: "If he is referring to the absurd arguments raised by modern Sunni scholars who hallucinate by counting non-existing letters in order to blind people to the mathematical miracle of the Quran..."

The comment that *alif* was not a letter is taken from a non-Muslim scholar of Arabic. Yuksel exceeds bounds in argument, and imputes motives with little evidence. If it is a hallucination, how then can it be claimed that *ASM* is a word in the Bismillah?

YUKSEL: The statement, "You're hallucinating my friend" has 26 letters and it contains the word "are." Nevertheless, whoever claims that the statement has 27 letters is hallucinating. This is especially true, if the statement has been deliberately and consistently repeated that way, as in the case of the Bismillah.

---

[19]   There is no such a version called Marsh. I meant that there was not a single version of the Quran manuscript that *Basmalah* was written with an extra Alif.

LOMAX: Yuksel has not bothered yet to respond to my much shorter document, the draft FAQ: 19, Study Problems, which explains why the existence of various ways of counting things is significant. I would have preferred to discuss that first, since it establishes principles of analysis that might avoid many useless words of argument.

YUKSEL: Well, I will deal with it later, God willing. I hope then you will not complain that it was a "draft," not a finished work.

LOMAX: Yuksel wrote: "I challenge him to show me "other ways of counting the letters" of *Basmalah*." This is truly bizarre. One would presume that Yuksel has read Philips on this subject, and further that he, himself, is capable of such an analysis. But, here goes.

This is a transliteration of the *bismillah* without *haraka* (vowel marks):

BSM ALLH ALRHMN ALRHYM As can be seen, this is **nineteen** letters.

Here is a translation with all the haraka (Hafs, Egyptian script): *bis.mi all:hi alr:H.mAni alr:Hiymi* (*Sukun* is written with a period, *shadda* with a colon.)

If I neglect the *sukuns*, and some would write additional *sukuns*, I come up with **29** letters.

The point is that to state how many letters there are requires a definition of "letter." It would seem an obvious precondition, but the Khalifites depend on such lack of definitions, because it allows them to shift definitions as needed to come up with "amazing" counts. One may argue that the short vowels are not letters. Okay, here it is without short vowels (and without *sukun*):

bsm all:h alr:HmAn alr:hym. **23** letters.

The *shaddas* (:) double letters. But perhaps we can eliminate them.

bsm allh alrHmAn alrHym. **20** letters.

Some of these letters are not pronounced. We can eliminate them.

bsmllh *lrHmAn lrHym* **17** letters.

The A in *al-Rahman* is pronounced but not written (by convention).

We can eliminate that too, for **16** letters.

157

One could also assert all these variations with an additional one, the elided *alif* in *bism*. This is a word which has been written in a special way; elsewhere in the Quran it does occur with the *alif: bi Asm*.

YUKSEL: Thank you Lomax for your exposition. You have proved that you are able to do anything and everything. An elementary level knowledge of Arabic alphabet is enough to notice the absurdity in your counting. It is one of the simplest and well known fact that Arabic has 28 (Twenty Eight) LETTERS in its alphabet. All Arabic dictionaries use these 28 letters. Kindergarten children in Arabic speaking countries memorize these 28 LETTERS. We are again discussing the number of LETTERS in Bismillah. I have challenged you regarding the LETTERS, not sounds, not *shaddas*, not *harakas*, not comas, not mumbles and bumbles.

This example is enough to demonstrate how Muhammedans are twisting the simplest facts in order to cast doubt on the mathematical miracle of the Quran. It is noteworthy that not a single Muslim scholar had a different count for the letters of *Basmalah* **before** the discovery of the code. Whoever mentioned the numbers of its letters acknowledged the simple fact: *Basmalah* consists of 19 letters. For instance, Molla Jami starts his Persian Divan by referring to the 19 letters of *Basmalah*. Fakhraddin Al-Razi, in his 30 volume commentary, et-*Tafsir al-Kabir*, links the 19 letters of *Basmalah* with 19 guardians of Hell. Furthermore, the Abjad (Gematrical) value of *Basmalah* is well known as 786 for centuries, which is the numerical value of its 19 letters. Many a Muslim still use this number on top of their letters, instead of *Bismillah*...

LOMAX: Nevertheless, it does stand that one of the most simple and obvious ways of counting, that of counting the unvowelled and unmarked traditional Arabic letters, as written in the earliest manuscripts, comes up with nineteen letters. But to assert that this is the only reasonable way to count is mere polemic. It would be more honest and courageous for Yuksel to put his energy into making explicit the definition of "letter" that produces the count.

YUKSEL: Lomax and those Muhammedans who cannot digest the message of the mathematical code of the Quran are wondering the "explicit definition of letter!" I cannot imagine a more ridiculous question than this one. Lomax has an obsession with "definition." Should I ask him to define himself, or to define "definition"? In order to show him the sun you need to define the sun. He will not see the sun if you take him out and show him. Thousands of elementary Arabic books, thousands of Arabic dictionaries and encyclopedias, and millions of Arabs know their alphabet. Even my Random House Webster's College Dictionary lists the 28 Arabic Letters in a table in the entry of "alphabet."

My 5 year old son, Yahya, discovered a word-game that amuses and sometimes frustrates me. For instance, if I tell him "You are cute," he asks, "What does 'cute' mean?" If I say, "It means 'charming,'" he asks "what does 'charming' mean?" In the end, I give up, "I don't know." He continues, "What does 'I don't know' mean?"

Instead of defining what is letter, I believe it is shorter, practical and more meaningful to list the Arabic letters. This way, I will try to deprive Lomax from nitpicking on the definition. Here they are: alif, ba, ta, tha, jim, Ha, kha, dal, dhal, ra, za, sin, shin, Sad, Dad, Ta, Za, 'Ayn, ghain, fa, qaf, kaf, lam, mim, nun, ha, waw, ya.

That's all I can do. If someone has planted his feet in the concrete, what can you do?

LOMAX: Yuksel quarrels with my use of the term "anecdotal" in reference to this fact. He has not quoted the parts of my paper which establish the meaning of this term in context. This means that it is a single measure. It refers to the problem of attempting to prove a hypothesis by referring to isolated events. For example, my wife went to the doctor, and she did not get better. Therefore medicine does not work. This is anecdotal evidence. In order to really confirm such a hypothesis, one needs to examine all occurrences of the same class of event.

A relevant way of doing that would be to examine the letter counts in all distinct verses of the Quran (that is, all verses which differ from each other in some way.) How are the letter counts distributed? Is there some preference for divisibility by 19 that is outside the normal distribution? Such a finding would be very significant. But that a single verse, even a verse which is repeated so many times, is divisible by nineteen is not evidence of a pattern. There are "facts" asserted which are not anecdotal. But they are not verifiable. I am referring to the counts involving *alif*. Others seem to represent multiple occurrences of a pattern, but, if one looks closely, no pattern produces more than a few hits without there being some change in the method of analysis. I have never seen any of these phenomena that are outside normal statistical variation.

YUKSEL: Lomax will never be able to see a miraculous pattern with this attitude. He has just demonstrated his method of evaluation regarding the numbers of letters in *Basmalah*.

159

6:104 Visible proofs have come to you from your Lord; so whoever can see do so for himself, and whoever is blinded, does the same. I am not a guardian over you.

10:39 No, they have lied about the things they did not have comprehensive knowledge of, and before its explanation came to them. Similarly, those before them denied, so see what the retribution of the wicked was!

LOMAX: "Anecdotal" is not a counter-argument to significance, by itself. It merely cautions against generalizing without actually analyzing the whole population of statistics of a particular kind. (And if there is only one member of the population, any fact is inherently anecdotal: If I know only one red-haired person, and he is very intelligent, I can still make no generalization about red-haired persons.)

YUKSEL: The number of letters in *Basmalah*, the frequency of the four words that it contains, and many other related features cannot be explained with mere accident. But, it is always possible to cover the truth with fabricated excuses, or extreme skepticism. What was the excuse of those who rejected Moses, or Jesus after they were provided with supernatural physical miracles? Contemporary ingrates are no different than medieval ones.

CLAIM 3

Every word in this verse is mentioned in the Quran in multiples of 19. a) The first word (Ism = Name/Attribute) is mentioned in the Quran 19 times, 19x1. b) The second word (*Allah* = God) is mentioned in the Quran 2698 times, or 19x142. c) The third word (*Al-Rahman* = Most Gracious) is mentioned in the Quran 57 times, or 19x3. d) The fourth word (*Al-Raheem* = Most Merciful) is mentioned in the Quran 114 times, or 19x6.

LOMAX: The counting of words in the Quran, to be simple, requires a clear definition of what words are to be included and what are to be excluded. To arrive at the counts reported by Dr. Khalifa requires arbitrary and inconsistent judgments, at best, and, at worst, errors or alteration of the text.

a) The count of Ism, as given, excludes the invocation itself. It also arbitrarily selects and rejects various forms of the word. There are many ways the words are written. For instance the word "ism" is written in many different forms adding to a total of 38 or 2x19. But, you eliminate half of it by arbitrary selection. The problem is obvious. "Word" is an undefined term. Later, we will see that, by the practices Dr. Khalifa follows with other word counts, the first

160

word of the invocation is "BISM," not "ISM."... He excludes the BSM in the invocation itself, whether found at the beginning of a chapter or elsewhere as well as the plural, which is formed by adding "A" (*Alif*) at the end.

b) The count of 2698 for Allah, as reported by Dr. Khalifa in VP does not include occurrences at 9:15, 25:68, 40:74, and 46:33. (The first three of these omissions involves reporting two occurrences as a single one.) He has counted an extra Allah at 22:60. The net result is a count of 2701. This count and the remaining two also involve the arbitrary exclusion of the other 112 initial invocations. Furthermore, he includes the word "Allah" with prepositions, such as, "*Lillah*" and "*Billah*." This is inconsistent with his counting method for the first word of *Basmalah*, that is, "*Bism*."

c) The count of 57 for *Al-Rahman* is accepted and verified.

d) The count of 114 for *Al-Rahim* involves the arbitrary inclusion or exclusion of various forms of the word, and completely omits the occurrence (which is a form otherwise included) at 9:128. The question of the exclusion of 9:128-9 from the Quran, as proposed in Dr. Khalifa's later publications, is definitely not simple. Dr. Khalifa did not even mention that this was necessary in VP. I certainly suspect that this was simply a reaction to his embarrassing and fairly easily discovered error. Further, we must note that we included in the count for *al-Rahim* the forms without the article, *Rahim*, and the forms with an additional alive at the end, *Rahiman*. The word al-*Rahim* occurs 116 times. However, it seems easy to exclude the one in verse 48:29, since the plural of "Ism" also were excluded. This leaves us with 115 occurrences. This is a different method of counting than was used for Ism where Ism was counted and Isman was not. Only one out of four counts ›s correct, and even that count requires an arbitrary decision (the exclusion of 112 invocations).

YUKSEL: Yes, there is a peculiarity regarding the counting of the word *Bism*. However, I assert that there is still a consistent method of counting words. The word *Allah*, *Rahman* and *Rahim* are counted according to the same method. I will discuss two reasons for the peculiarity in the counting of the word *Bism*: 1) Leaving a perceived loophole for arrogant disbelievers and hypocrites to find an excuse to reject the great mathematical system of the Quran; they don't deserve to see the miracle. 2) To teach us a linguistic lesson regarding the difference between the first letter of *Ism* (name) and *Allah* (God). Now, let's explain.

a) I believe this is a fair objection. However, this apparent problem can blind a person to impressive and extraordinary examples of the mathematical system by its priming effect. Our exposure to certain information in advance can change

161

our perception and cognition dramatically. If you start closing one of your eyes while watching a stereogram, you will not be able to see the three dimensional picture hidden among arbitrary-looking dots. Not only you need to keep both of your eyes open, but also you need to focus on the picture with a positive attitude. Otherwise, you will reject the existence of a three dimensional picture and make fun of those who claim to see it. You may even write a book trying to prove how those colorful dots do not have any pattern. Similarly, if you make up your mind based on a questionable criterion regarding the system of the Quran, you may disqualify yourself to see the extraordinary picture. If you don't see the picture you will not be able to understand the reason for that apparent problem. I will not speculate further on this point, but I leave it to your own choice. It is God's system to show His miracles to believers (2:118), not to fanatic disbelievers or hypocrites. I know the philosophical problem (circularity) with this argument. Too bad, since miracles are not presented as proof for ingrates. Therefore, I am not arguing this topic in order to convince someone who has already made up his mind; but to help those who have an open mind. I am not judging you. I think you are sincere and honest in your search. God knows, you may tomorrow witness this great miracle of the Quran yourself. I hope that your personal feeling towards Rashad is not creating a psychological mountain between you and the truth.

> 7:146   I will divert from My signs those who are arrogant on earth unjustly, and if they see every sign they do not acknowledge it, and if they see the path of guidance they do not take it as a path; and if they see the path of straying, they take it as a path. That is because they have denied Our signs and were heedless of them.

Therefore, I believe that if you want to reject this mathematical system, you will be provided with some deliberately arranged "loopholes" (3:7; 17:82). God distinguishes sincere believers and hypocrites in various ways (3:179; 74:31). We will find out the ultimate truth in the Day of Judgment.

However, I would like to remind us that we should not gullibly accept the claims regarding a "miracle," since attributing our speculation or wishful thinking to God Almighty is a very serious sin. This forces us to be extremely careful in accepting or rejecting ideas regarding the Quran:

> 29:68   Who is more evil than one who fabricates lies and attributes them to God, or denies the truth when it comes to him? Is there not a place in hell for the ingrates?

162

Curiously, Fuad Abdulbaqy too, in his famous Concordance of the Quran, *Al-Mu'jamul Mufahras*, follows the same method of counting. While categorizing the words, he consistently distinguishes the form of a word attached to a conjunction from the one without or separate from a conjunction. He also consistently distinguishes the regular form of a word from the contracted form, for instance, when a preposition is attached to a word starting with Alif, as it is the case with *Bism*.

However, he is not consistent with this method of classification in the counting of the word God. This inconsistency is curious, since this is the only method of classification that we can obtain 2698 (19x142) for the frequency of the word God, if we exclude 9:128-129 and un-numbered *Basmalah*s. (If you study this concordance you will find that the author separates the word Allah in three parts according to their last vowel points as an exceptional treatment. However, this is not because he considered them as different forms of the word. This exceptional treatment is to make it more convenient for us to find a verse containing the most repeated word in the Book.)

Here is how Abdulbaqy's concordance classifies the different forms of Ism. Ism or *Bism* (Name, in the Name. The three original letters of the word remains unaltered):

No	Sura:Verse
1	5:4
2	6:118
3	6:119
4	6:121
5	6:138
6	22:28
7	22:34
8	22:36
9	22:40
10	49:11
11	55:78
12	56:74
13	56:96
14	69:52
15	73:8
16	76:25
17	87:1
18	87:15

19   96:1

Bsm (In the name. The first letter is dropped):

1	1:1
2	11:41
3	27:30

Ismuhu (His name. A pronoun is attached to the original form):

1	2:114
2	3:45
3	19:7
4	24:36
5	61:6

Asma' (Names. The original form is changed):

1	2:31
2	2:31
3	7:71
4	7:180
5	12:40
6	17:110
7	20:8
8	53:23
9	59:24

Asmaihi (his names)

1	7:180

Asmaihim (their names)

1	2:33
2	2:33

Totaling to 39 frequencies.

No one can claim that Fuad Abdulbaqy cooperated with us by putting the number 19 under the word *Ism* as its frequency, since he completed his concordance in 1938 and died before the code was discovered in 1974.

The exceptional method applied to the counting of the word *Bism* might be due to its first letter, that is "A" (*Alif*). When this letter is omitted it changes its basic form, that is, ISM becomes SM. If we accept this fact as one of the reasons for a different method of counting, then, we have to accept that the first letter of "Allah," that is "A" (*Alif*) does not belong to its root. In other words, the first letter of "Allah" is "A" of the definite article "Al" (The). Many linguists have claimed that the word "Allah" is the Arabic word for "the god," which we write

164

as God, with a capital G. According to this understanding, omission in the first letter of "Allah" cannot be considered a change in the basic form of the word.

Again, we observe according to our current knowledge and sometimes learn new facts from our observation. Our observation in science occasionally forces us to modify our theories. Likewise, we study the mathematical structure of the Quran with our current knowledge; however, occasionally we may be forced to correct some of our preconceived ideas. An outsider may accuse the scientist of being inconsistent or cheating, but an insider will see it as a sincere search for a better understanding and explanation.

b) There were typing errors in the list published in VP (*Visual Presentation of the Miracle*). I really don't know how they occurred. However, the errors can be corrected by a careful comparison. As I have stated, I do not argue my position based on the authority of Rashad, but through verifiable and falsifiable physical facts. Therefore, speculating on his errors is not relevant here. So far, you have found few errors. I will give you the complete list of typos. My or your errors cannot change the number of the frequency of the word "Allah" (God) in the Quran. Independent researchers can find the real results. Here is the complete list of the errors in the count of the word *Allah* in VP with their corrections:

Index	Verse	Correction
565	4:69	(4:64)
784	5:47	(5:48)
828	5:57	(5:87)
1120 b	9:5	(add)
1176	9:46	(9:48)
1264	9:119	(9:118)
1271	9:127	(1272 index number)
1489	15:97	(96)
1567	17:39	(1576 index number)
1672	22:60	(subtract)
1792 b	25:68	(add)
2220 b	40:74	(add)
2306	46:23	(subtract)
2310 b	46:33	(add)
2457	58:4	(subtract)
2576	63:7	(6)

165

As you see above, we have 4 additions and 3 subtractions. That means, we have one extra word to add to the claimed result, that is, 2698 + 1 = 2699. This result includes the word "Allah" in 9:129.

Now, please tell me what is the result of your own counting? If you agree with 2699, then we can continue this argument. (You can use Rashad Khalifa's *Visual Presentation of the Miracle* or Fuad Abdulbaqi's *Al-Mu'jamul Mufahrasa Li-elfazil Quranil Karim* to check this number). By the way, it is becoming evident to me that your concordance is not accurate. You should not reject our counts based on your poorly edited concordance. I advise you to obtain Abdulbaqy's concordance, which is a well known work in Islamic circles. You will appreciate it after comparing it with yours.

Our counting excludes the 112 unnumbered *Basmalah*s. Had we include them, you would object again by labeling it as an "arbitrary inclusion of 112 unnumbered initial invocations which are merely repetition." Our exclusion is not arbitrary, since the other 112 *Basmalah*s are not numbered. You can see this fact in your own version of the Quran. Our method follows and justifies the well known difference between the other two *Basmalah*s and these 112 *Basmalah*s. It is a discovery which brings an explanation for this curious distinction. Though in the early manuscripts the verses were not numbered, they were ordered and separated from each other by dots which I believe justifies our numbering them. In fact, if there is a beginning and order of items of the same category, there is an implicit numbering in the structure. Therefore, early scholars were not wrong when they decided to number the verses as we know and use today.

c) The frequency of the word *Rahman* being 57, or 19x3, is obviously a significant fact regarding the numerical code.

d) As for the counting of "*Rahima*," the form without the article, I believe that your objection is very weak since we do not apply a different method of counting for the other three words in *Basmalah*. I agree that you can count a word in several different ways. The number of reasonable methods of counting this word is much less than nineteen ways. Thus, according to probability, it is still interesting to find one of them being a multiple of 19.

I believe that the method of counting that leads to a miraculous system is the most reasonable one, and we use it consistently. Furthermore, if we follow the same method of counting words, we find the word *shahr* (month) occuring exactly 12 times, and the word *yawm* (day) exactly 365 times. These two examples with their empirical implication, I believe, give us two important clues: 1) The author of the Quran has most likely attached meanings to the frequency of words. 2) While counting the words, the general method of counting should follow the same system that we discover in the frequency of these two words, which are related to their astromonical frequencies.

True, there are 115 occurrences of *Rahim* in current versions. But, we have always excluded the one mentioned in 9:128. Before we came to the conclusion that 9:128-129 is not from the original Quran, we had already and easily excluded it, since it is used for Prophet Muhammad. Therefore, the frequency of God's name *Rahim* is 114 (19x6) in the Quran.

LOMAX: Edip Yuksel wrote: "Yes, there is a peculiarity regarding the counting of the word Bism. However, I strongly believe that there is still a consistent method of counting words. The word Allah, *Rahman* and *Rahim* are counted according to the same method."

There is not just one peculiarity, there are at least two. However, Yuksel is reduced to saying "I strongly believe" because he is unable to state what the method is. I don't like to shout, but: WHAT IS THE METHOD OF COUNTING WORDS?

As will be noted from Yuksel's word counts from AbdulBaqi, 19 is the frequency of one particular form of *ASM*, and this is not the form found in the Bismillah. Other forms have other frequencies. Once again, I will note my amazement that Yuksel casually allows that there is an alif in BSM, so soon after he has called this a "hallucination" when it is asserted by "Sunni scholars."

YUKSEL: Again, Lomax is trying to confuse things. I have explained it before. The BSM of *Basmalah* does not have "A" (*alif*) in it. However, when we count individual words we can strip them off from attached propositions which are only relevant when they are in a sentence. I have briefly expressed the method of counting the words, and brought an explanation regarding the difference in the count of the first word, *Bism*.

LOMAX: I could infer from some of these facts a method of counting words. But it is a lot of work, which I am not going to exert right now. It is up to those who claim the miracle to state the experimental conditions. I have tried to do it in the past, and always found inconsistencies.

But, by varying the definition of "word" one can certainly increase the occurrence of any desired modulo 19 remainder. It is not difficult to do this with just four words, and this has been done with the Bismillah.

YUKSEL: Lomax was asking us the definition of "letter" too. He was confused on that simplest and most obvious linguistic unit. Knowing his great talent in confusing matters, I avoid wasting my time to bring a definition for "word." It seems that nothing can clear his fuzzy eyesight.

LOMAX: Yuksel wrote: "b) There were typing errors in the list published in VP (Visual Presentation of the Miracle). I really don't know how they occurred." Especially since these lists were supposedly produced by computer. Khalifa allowed people to assume that his counts were generated by analyzing a computer text of the Quran. It is likely that he did keypunch (yes, keypunch!) part, or perhaps all, of the Quran in the early 1970s. But his database was full of errors, and it is likely that his later work was done without the benefit of a computer-readable Arabic text. So his "computer" proof was actually just a manually-compiled spreadsheet. How did he come up with 2698?

YUKSEL: It really does not matter. I do not care how he came up with this number. I know for myself that there are exactly 2698 occurrences of the word "Allah" (God) in the numbered verses of the Quran, excluding the 9:127. Lomax might again marshal his Hafs and Warsh! By repeatedly referring to his versions of the Quran he reminds me his ancestors mentioned in verse 10:15.

LOMAX: He followed AbdulBaqi, who did not count 1:1 as part of the Quran, and who, I am told, reports 2698 occurrences of "Allah." Then he made his spreadsheet, and kept looking for errors until he had the same total as AbdulBaqi. It has been said that this was an error in AbdulBaqi, but perhaps he was following the Warsh reading, which does not number the initial Bismillah.

YUKSEL: AbdulBaqi DID count verse 1:1 as part of the Quran. His claim shows how shallow is his "research." If he has just checked his concordance for the other three words of *Basmalah* (Bism, *Rahman*, and *Rahim*) he would have realized that the omission of the word "Allah" in 1:1 is a typographical or a human error. The numbers indicating the frequency of the word "Allah" (God) is one less than the actual list. Therefore, he lists 2699 occurrences, including 9:127. Again Lomax' obsession with Warsh forces him to a wrong conclusion. AbdulBaqi did not care about Warsh version.

LOMAX: Yuksel could have saved me a great deal of trouble by sending me the list of errors in VP two years ago, when I first mailed him a copy of my paper. It met with complete silence then. I had to find those errors myself. At this point, I consider it unlikely that there are more errors, and tentatively accept that there are 2698 occurrences of "Allah" in the accepted Quran, plus 113 occurrences in the initial Bismillahs. I add the qualification "accepted," because

168

I have a copy of the Tashkent muSHaf, and I have found, on a fairly cursory examination, that there is a verse in it which does not have an "Allah" which is present in modern versions. Because it is "*innallaha*" and the subject of the following verb is obviously Allah, it does not change meaning. I would not be surprised to find other occurrences like this.

YUKSEL: Individual manuscripts might contain certain unintentional errors or omissions because of the human factor. If there was a "deliberate" omission of the word "Allah" in Tashkent copy it would not have escaped from public attention and hot debate among early scholars. There is no such a debate on that missing word, even in the books which argue the tiniest variations among different versions.

Lomax may hope to equate his "discovery" with ours. In the case of the spelling of *BasTata* we had clues and evidences to suspect the spelling of that word. Furthermore, we did not rely on Tashkent copy alone, we checked some other early manuscripts too.

LOMAX: Again, it would be useful if Yuksel would address the point mentioned in the draft FAQ: 19, Study Problems, "What is the Quran?"

YUKSEL: Here, I will briefly attempt to define the "Quran." Lomax may not like this unconventional definition. The Quran is a unique book and it has a unique (he may call it bizarre!) definition:

The Quran is a revelation of God to Prophet Muhammad in Arabic language containing 114 chapters, a number which equals the Gematrical value of God's attribute Jami' (The Editor). Its chapters, except the Chapter 9, start with a 19-lettered verse, Bismillahir*Rahman*ir*Rahim*. (Lomax may wonder: "what does 'chapter' mean? I will define it for him if he explains what does "mean" mean!)

It is a numerically coded book, detailed and explained by its author, easy to understand for believers, impossible to understand for hypocrites and disbelievers. With its 6346 verses (including unnumbered *Basmalah*s), it is complete and the only source of guidance. Disbelievers seek other sources besides it. It is a book of recitation which its preservation is guaranteed by God. It does not belong to Hafs, Warsh, neither the East, nor the West; it is God's light. Where can we find this Quran? It is in the heart of those who possess knowledge. Who are those who possess knowledge? They are those who know the Quran and accept its self-definition without doubt. How do they know the Quran? God teaches them. How can we know that God teaches them? If God

169

teaches you too! How do I know that God teaches me? You will not trade God's word with medieval fabrications and consensus of clergymen.

LOMAX: Yuksel writes: "Though in the early manuscripts the verses were not numbered, they were ordered and separated from each other by dots which I believe justifies our numbering them."

The Tashkent *muSHaf* is not numbered, nor is it dotted in any way. Dots were added later, and some of them were used to indicate pause. Yuksel ignores the fact that different readings of the Quran divide verses differently. If he really is interested in authenticity, he should probably be studying Warsh instead of Hafs, since it is likely that Warsh is closer to the Quran as recited by the Prophet (SAS).

YUKSEL: Lomax again is advertising his Warsh version. However, he has lost his credibility since he is the same guy who strongly advises people to accept the authenticity of hadith books which are collection of primitive stories. As for the Tashkent copy, I can see clear marks separating verses, even without my eyeglasses. Lomax probably has a newly discovered Warsh version of Tashkent manuscript ;-)

LOMAX: As to the remaining counts, since al-*Rahman* is used in only one form, I can state that it occurs, in Hafs according to modern numbering, 57 times in numbered verses. Since Yuksel has not stated the rules for counting words, I will refrain from confirming the Rahiym counts. It is also a bad sign that the necessary qualification "Hafs" and "numbered verses" needs to be stated by me. It should be part of the original claim, as should be the necessary definitions. Without those definitions, counting is impossible.

YUKSEL: Lomax appears to have a short memory. He had confirmed the frequency of Rahiym in his previous work. He could count 115 *Rahiym* (including 9:128) without trouble (see: Lomax' first round objection to Claim 3). With the exclusion of 9:128, the frequency of *Rahiym* comes to 114 which is a ZMN of 19. Lomax is learning so much that he has lost his ability for counting clear words ;-)

LOMAX: It is easy to define "word" in English, because we separate words with spaces. Arabic is not like that. Even so, in English we could run into problems if we try to count words. If I count the occurrences of the word "truck," should I also count "trucking" and "trucks?"

170

YUKSEL: Lomax should just count the word "truck," if he is counting only the "car" the "bicycle" and the "bus!" This is exactly what we did in the count of the words in *Basmalah*.

LOMAX: Yuksel is actually proposing that we should use the definitions which produce the multiples of 19. There is nothing wrong with doing this, as long as one recognizes the possibility that one is thereby creating the "miracle." Such a hypothesis (that the "miracle" is man-made) could later be disproven by showing that the results were of far higher significance than could have been created by such manipulation.

For example, I can decode the cryptogram in the newspaper by trying out certain letter conversions. If some of these conversions seem to produce a real word, I then look at all the occurrences of the translated letter, and see if other words appear as well. Because of the nature and complexity and redundancy of a genuine message, I can generally be very sure if I have found the correct code, because it produces a message with much greater coherency than can be explained by the relatively simply process of choosing letter conversions.

YUKSEL: Good job! Sometimes I wonder how Lomax can be among those who are blind to the mathematical miracle of the Quran.

LOMAX: But if I am allowed to do more than just substitute one letter for another, and if the spaces are considered part of the code as well, and if I can add positional coding (in other words, the translation varies with the position), then, if I am not limited in the complexity of the translation, I can convert any message into any other message of the same length. But the code will approximately as complex as the message decoded.

I find, on examining the claims of the Khalifites, that the decoding they do is as complex as the message it produces. They shift analytical methods as needed to produce multiples of nineteen. But to demonstrate this requires an overview.

YUKSEL: Lomax consistently tries to stigmatize us by labeling us as Khalifites. He is appealing to orthodox masses who excommunicate people by labeling them with names, ironically, in the similar way of how they call themselves: Hanefites, Shafiites, Malikites, Hanbelites, Wahhabites, etc.

After receiving my answer rejecting the claims that do not follow a consistent method of calculations, Lomax still insist to repeat his original criticism. He enjoys punching a straw-man. It is a tactic that works for politicians who appeal to the emotions of the masses.

171

LOMAX: Yuksel wrote: "True, there are 115 occurrences of *Rahim* in current versions. But, we have always excluded the one mentioned in 9:128. Before we came to the conclusion that 9:128-129 is not from the original Quran, we had already and easily excluded it, since it is used for Prophet Muhammad. Therefore, the frequency of God's name *Rahim* is 114 (19x6) in the Quran." Yuksel here shows how slippery all this is. Apparently, it is meaning that is being counted, not words. But this is not stated in the original "fact." True, Muhammad (SAS) is "*raHiyma*", merciful, but the word in the *Bismillah* is *al-raHiym* (The Merciful), so I would agree that it is a reasonable choice not to count this, but on the basis that it is an adjective, not a name or title. *Al-rahiym*, the title or name, occurs 34 times in the Quran. All Yuksel has demonstrated is that there is a way to count that produces a multiple of 19.

YUKSEL: In the Quran, all the attributes of God are adjectives. It is entirely arbitrary and baseless to divide them into groups of names and adjectives. Again, Lomax contradicts his previous count of *Rahiym*. All Lomax has demonstrated is that there is a way to distort and manipulate the facts that produces a result lacking a mathematical pattern.

LOMAX: Yuksel wrote: "I agree that you can count a word in several different ways. The number of reasonable methods of counting this word is much less than nineteen ways. Thus, according to probability, it is still interesting to find one of them being a multiple of 19."

Yuksel's knowledge of math is, well, deficient. Each possible method of counting reduces the significance of the result. Because it is possible to combine the results of various counts (particularly in this case, where words exist in many different forms), and also because of probability theory, one does not need nineteen different ways to make it likely that one of them is divisible by nineteen. In fact, with thirteen different ways, there is a fifty percent probability that one or more of them is divisible by nineteen, even without allowing combinations.

It is more complex to determine the odds with combinations, but a rough estimate would be that, if one may combine forms (as Khalifa did), four different forms is sufficient to make it likely (more than 50% probability) that one combination exists which is divisible by 19.

There are about seven forms of ASM, at least two forms of ALLH (Allah and Lillah), one form of ALRHMN, and four forms of ALRHYM. From this, one might expect more than one way to combine forms for ASM to produce a

172

multiple of 19, and, in fact, there is: The total of all forms of ASM is, in fact, 38, or 19x2, in the numbered verses in Hafs.

YUKSEL: Lomax generalizes my statement for the word *ISM* to the other three words in order to claim deficiency in my math. We have counted the word ALLAH, al-*RAHMAN*, and *al-RAHIYM* according to the clearest and most obvious method in a consistent fashion. The same method was employed by many independent researchers who did not have motivation to reject nor defend the mathematical code of the Quran. It is amusing to see many new methods of counting are being discovered by today's Muhammedans who are terrorized by the number 19. After their new discoveries they complain from confusion. In this endeavor, no wonder they end up losing the most common knowledge. They seriously ask what is a letter? What is the Quran?

Below is one out of many examples from previous works that do not demonstrate any problem in counting particular words. Though their count may not be exact, it will show that they have employed the same way of counting. Prof. Dr. Suad Yildirim, in his comprehensive book on the attributes of God in the Quran, "Deity in the Quran" (*Kuran'da Uluhiyet*, Kayihan Yayinevi, Istanbul, 1987) gives the frequency of the word Allah 2697 (p.101), the word *Rahman* 57 (p.115), the word Rahiym 114 (p. 124). Though Dr. Suad Yildirim is not interested in the mathematical code of the Quran (most likely he rejects it) he came up with the same counting method and with the same results, except he was one short for the count of "Allah."

LOMAX: Yuksel wrote: "I believe that the method of counting that leads to a miraculous system is the most reasonable one, and we use it consistently." This should be printed with every piece of Khalifite literature. It is an open acknowledgment that the methods of counting are selected to produce the "miraculous system." Now, if such a method is found, and it then produces consistent results applied to new data, it would truly be miraculous. But the Khalifites keep modifying the system as new data is presented, and, in fact, as will be seen, they do not use and system "consistently."

Now, their inconsistency is not a guarantee that there is no pattern. It merely means that they have not demonstrated one. The Khalifites are apparently not aware that the human mind is very, very capable of projecting pattern. It is a very useful capacity, but it has its limits. Perhaps our older readers will remember the "canals of Mars." Many astronomers drew them from observing Mars, but somehow they never showed up in photographs. No one draws them anymore, now those excellent photographs from close up fail to show them.

173

Gamblers believe that they can detect patterns in random variations, and they bet everything they have on this.

YUKSEL: Again Lomax is not dealing with my position. He can always find someone who will introduce some inconsistent or false counting. Instead of dealing with my argument, he is still looking for straw-men among what he calls "Khalifites". This attitude is irritating. As the readers have noticed I have occasionally lost my patience with Lomax.

Lomax obviously distorts my statement for his cause. Here I will re-word it to make it clearer: I believe that our method of counting is the most reasonable one, and we use it consistently. This method leads to a miraculous system.

LOMAX: I have challenged the Khalifites (and other students of "numerical miracle" in the Quran) to come up with a coherent statement of exactly what the miracle is, so that we could then determine if it exists in other books. So far, no takers. Until such a statement exists, it is really impossible to prove that the miracle does not exist, for exactly the same reason that an honest atheist would admit the general impossibility of proving the non-existence of "God." In order to disprove a proposition, the proposition must first be stated in a way that can be tested.

YUKSEL: An impressive and eloquent statement with no substance. I do not have any motivation to come up with a coherent statement of exactly what the miracle is for someone who is not able to see the simplest facts. You cannot help someone who stubbornly closes his eyes to the light and complain of not being able to see it. What if that person demands a coherent statement of exactly what the light is? Personally, I would not bother to bring a technical and comprehensive definition of the light for someone who hates the light and demonstrates all kind of blindness in the past.

CLAIM 4

The Quran consists of 114 chapters (Suras), which is 19x6.

LOMAX: True, but anecdotal. The number of chapters is not part of a pattern. I will make this "anecdotal" comment many times; I should explain it. Suppose I have a telephone book. I can generate a large number of counts from that book. In fact, the number of counts, the number of ways that the data can be analyzed, which can be generated from such a large collection far exceeds the number of atoms in the known universe. Approximately one out of 19 of those counts will be divisible by 19. By selecting and presenting only those counts, especially if I worked at it for fifteen years or so, I could show an amazing collection of facts, all of which would be undeniably true. I do not intend to do this; I consider it a recipe for insanity.

On the other hand, if I could show an extensive pattern, appearing with the application of a few simple rules, I would have found something truly interesting. Dr. Khalifa's early claims of the letter frequencies showed exactly such a pattern, which is why they interested me so much. Out of 14 initial letters, 8 showed total counts divisible by 19. There is substantially less than one chance in a million of that happening simply by coincidence. Unfortunately, even those counts already included what can only be considered deliberate modification of the data to create divisible counts. (To consider this kindly, Dr. Khalifa probably thought that he was discovering the "correct" way of counting or the "correct" text.)

YUKSEL: It is true, and it is very important. Why the number of chapters should not be a part of the pattern? The number of the Chapters being multiple of 19 is surely an important support for the importance of number 19 as the code of mathematical structure of the Quran. Yes, it is not difficult to compose a book with a certain number of chapters. However, if we have many examples of a mathematical pattern in a book and the number of chapters (the biggest element in a book) also supports that pattern, then, it becomes important. I am sure if the number of chapters were not multiple of 19, skeptics would repeatedly raise this question: "How can you claim that this book is constructed on a numerical code, and its chapters, the most obvious element, not a multiple of that code?"

There is no alternative counting for the number of chapters. For instance, if someone claim that the number of statements in a book which starts with letters B, C, E, M, R, T are each multiple of 7, then, I could object him by saying "It is just a coincidence. What about other 22 letters? You have one out of 7 chances to have such a case. Approximately 4 out 28 initial letters will be multiple of 7.

It is not significant to have 6 of them being multiple of the same number." But, I cannot object to the number of chapters by accusation or allegation of selective counting. The number of chapters has a unique importance in the numerical composition of a book.

Let me modify my example in order to get some insight regarding the mathematical structure of the Quran: If someone claims that the number of statements in a book which starts with letters Y, U, K, S, E, L are each multiple of 6, then, I would not rush in rejecting it as coincidence, if I knew that the author of that book is Edip Yuksel. I would raise some cautious questions: "How did you put these letters in this order? Why not K, L, Y, U, S, E or other pattern?" If I am provided with a reason for that order, then, I may consider the example to be intentional. For instance, "they are ordered according their first occurrence" or "they are ordered according to their frequencies, that is, Y occurs 6x50 times, U occurs 6x43 times, K occurs 6x12 times, so on so forth." Some other examples, for instance, the number of chapters and sentences may convince me regarding the intention of the author for using the number 6 as a code. In this case, in order to be convinced, I may need three times more examples than I need for a 19-based pattern, since 7 is smaller.

In order to claim a unique and extraordinary mathematical design in the Quran, we should first be convinced regarding the intention of the author. The number of chapters, I believe, is a crucial element regarding this conviction. Though the discovery of "intention" is important, it is not sufficient for to claim that the mathematical composition is authored by God. Extraordinary claims require extraordinary evidence.

Your example of a telephone book is deceptive for at least four reasons. First, a typical telephone book contains much more characters than the Quran, which contains approximately 300,000 letters. The white pages of Tucson's phone directory, excluding suburbs, contain approximately 8,500 characters in one page, totaling 5,200,000 characters. This includes approximately 200,000 different phone numbers with approximately 1,400,000 digits. (Quran, on the other hand contains 30 different numbers and 8 fractions, and they are repeated less than 300 times). This volume is obviously more than 17 times of the Quran. Therefore, in order to believe in a mathematical structure for a typical phone book, we need much more examples than we need for the Quran. Besides, the volume of a phone book is a discouraging factor for verification of any claim.

Second, you did not come up with any real examples from any phone book. You just made up a hypothetical claim. If any book shows similar pattern, I am ready to discard the mathematical miracle of the Quran. I am not talking about selecting fragmented patterns collected from a fraction of details from a vast number of possible elements, since I consider many so-called mathematical miracles as manipulation or selective calculations. I have written a lengthy article to distinguish between a genuine mathematical pattern and a fake one. I may include it to our argument later.

Third, it is absurd to look for a mathematical system in a phone book, since it has no chance to be beyond probability. Even if someone claims the existence of such a system, we lack motivation to verify or falsify those claims. However, it is relevant and reasonable to look for a mathematical pattern in a book which claims to be a "miraculous" message from the Creator of the Universe, the Greatest Mathematician. It is wrong to equate a probable and meaningful mathematical system in the Quran with improbable and meaningless "mathematical system" in a phone book.

Fourth, instead of a phone book, a novel or a non-fiction could be more relevant. Quranic verses and chapters have semantical relation and common theme. The numerical relation among elements can be supported by their meaning. Sometimes few examples are enough to give us a message. The frequencies of the words *month* and *day* in the Quran, for instance, are significant, since they correspond to their astronomical numbers, that is, 12 and 365, respectively. A phone book, on the other hand, contains a number of fragmented information without literal quality.

LOMAX: Edip Yuksel wrote: "There is no alternative counting for the number of chapters."

This is not true. I am quite sure that if there were 115 chapters, for example, Khalifa would have pointed out that there is no bismillah at the beginning of Sura 9, that it is really a continuation of Sura 8. It raises the question of how we know how the chapters are divided. In modern editions, there are distinctive markings at the beginning of chapters, but in the Tashkent Quran, there are no such markings. I have not looked yet at the division between 8 and 9 in the Tashkent *muSHaf*; it is a bit difficult to find. I will report in a future posting what is there, insha'allah. Further, apparently, the *muSHaf* of Ibn Mas'ud had 111 chapters, and that of Ubay bin Ka'b had 116 chapters, as did that of Ibn Abbas. (Ibn Mas'ud excluded the Fatiha and the last two chapters, apparently considering them merely supplications, and the two additional chapters of the others are also known as supplications.) However, my statement that this number is an anecdotal fact still stands. It is not based on the application of a pattern.

YUKSEL: Second-guessing hypothetical claims can be deceptive. I can claim similar thing for Lomax: I am quite sure if he was living during Moses and had witnessed one of his miracle he would have claimed that he was doing magic. Lomax again has found something useful in his holy hadith books, books that he can find whatever he wishes (68:35-38). Knowing the credibility problem of his sources, he craftily avoids giving reference for his claim regarding the manuscript of Ibn Mas'ud and Ubay bin Ka'b. The original reporters of this claim are *Bukhari* and *Ibn Hanbel*, which are full of fabricated narrations. According to the same books there was a verse legislating stoning to death and it was abrogated by a hungry goat after Muhammad's death. Half of the moon fell down on Ali's backyard after it was remotely split by Muhammad's index finger. Monkeys practiced Sunna by stoning a couple of adulterer monkeys to death. Muhammad possessed sexual power of 30 men. Muhammad negotiated with God to reduce the number of prayers from 50 times a day (every 28 minutes) to 5 times a day, after getting advice from Moses who was residing in the sixth heaven, and many more nonsense. According to Lomax, the number of chapters in the Quran is doubtful based on the books that narrate these kinds of stories! What an argument!

It is noteworthy that Lomax, who does not hesitate to ignore Quranic verses for the sake of the consensus of scholars, ignores the consensus about the number of chapters. Lomax again proves that he is ready to fight against the mathematical code by any means possible. This time he resorts to the collections of medieval lies.

LOMAX: Yuksel wrote: "For instance, if someone claim that the number of statements in a book which starts with letters B, C, E, M, R, T are each multiple of 7, then, I could object him by saying 'It is just a coincidence. What about other 22 letters? You have one out of 7 chances to have such a case. Approximately 4 out 28 initial letters will be multiple of 7. It is not significant to have 6 of them being multiple of the same number.' But, I cannot object to the number of chapters by accusation or allegation of selective counting. The number of chapters has a unique importance in the numerical composition of a book."

Every fact is unique. However, some facts are related to others as applications of a pattern. If, for example, I were to assert that every initial letter occurs in its chapter an exact multiple of 19 times, this is a single description which would cover many facts, and it would truly be miraculous, if it were true. Other facts are isolated and do not prove anything. If my house number is divisible by 19, or is not divisible by 19, what would that prove?

YUKSEL: Every element in the Quran is obviously related to each other since they are parts of the same book. Sure, some elements are more important and obvious than the others. If your house number was divisible by 19, and it had 19 windows and 19 pillars, and it had a lot number divisible by 19, and had 19 peach trees in its backyard, and had a phone number divisible by 19, then we could seriously think about a deliberate arrangements or extraordinary coincidences. If additionally, it had a long welcome note engraved on its door starting with a 19 lettered headline and containing a "vague" statement: "on it is nineteen" then we could talk about a deliberate arrangement.

If a skeptic had come up with an objection such as "There is no relation between windows and trees or between house numbers and telephone numbers" we would tell him: "Yes, they are generally not related but here (in this house) they are related, since they can be listed as elements of a single category, that is, the house." If the critic challenged us with a smirk on his face, "Define the window. What is the difference between windows and doors? In fact, I can claim that there are 21 windows (including the two doors). Why do you distinguish between doors and windows? They are all holes in the walls. You separate them to manipulate numbers and create an impressive pattern. You are a charlatan!" You would not probably waste your time to bring a "coherent" and "satisfactory" definition for windows. If you came up with any definition, you are sure that the critic is smart enough to find a vague point that can confuse doors with windows. He could reject your definition by saying that people can enter from windows if they want, or, some windows can extend down to the floor, etc. He would dance on that fine line repeating his claim: "You see, you cannot define what is window without excluding doors." Furthermore, if he could not find a fine line to dance on, he would resort to a fictional book written by a former resident claiming that his house number was 20 instead of 19!

You cannot count anything with such a person; even the number of eyes in his head. He would invite you to define the "eye." In case you gullibly accepted the invitation he would dwell on your definition and find a way to confuse the eye with ear or nose or with some allegorical concept! This is not a hypothetical imagination, but how the mind of a knit-picking extreme skeptic works. Personally, I can do what Lomax is doing. Here is my challenge for Lomax: "if you believe that you have two eyes in your head you are wrong! There are different ways of counting eyes. First of all, you should provide us with the definition of 'eye.'"

LOMAX: What the Khalifite claim boils down to is that significant counts of various phenomena in the Quran are divisible by 19. What Yuksel does not face is that these phenomena are part of a larger class of phenomena, and only those members of the larger class have been presented for examination, which are divisible by 19. Yuksel claims that the chapter count is particularly significant. Perhaps it is; perhaps, indeed, it is the most obvious thing one could count: it is certainly the easiest, and, indeed, it is, in all modern texts, divisible by 19.

YUKSEL: Here again Lomax is appealing to the tribal emotions of orthodox Muslims by labeling and insulting me as Khalifite. This is a psychological tactic called "projection." Lomax again twists his argument. After so many words he acknowledges the fact that the number of chapters is divisible by 19. This is a pattern in Lomax' argument: claims full of "maybes" and "perhaps" occasionally ending up with "certainly" and "indeed," or vice versa.

LOMAX: But I have no problem with the concept that nineteen has something to do with the structure of the Quran. In fact, it appears to me that a certain level of nineteen-related significance has indeed been woven into the Quran, enough to make a fitnah, a trial, for us. This is how I understand 74:30-31. Satan's temptation to us, however, is to say what we do not know, which, in this case, is to exaggerate the significance of nineteen and to create, from the threads which exist, a much more elaborate structure than is justified by the evidence.

YUKSEL: What a lousy logic: "God arranged some 19, but not enough to make a miracle. This 'certain level' of 19 is a test for those who can see a miracle!" The Quran repeatedly mentions those who do not see the obvious miracles. God portrays them as *musrifun murtab* that is, extreme skeptics.

Besides, Lomax is repeating his false claim regarding the meaning of 74:30-31 without answering my criticism. I have challenged his (mis)understanding in the first round argument. Unfortunately, he repeats himself like a broken record!

LOMAX: It is a very big step from the discovery of certain nineteen-related facts to asserting that enough of the pattern has been demonstrated that one is justified in excluding two verses from the Quran. Dr. Khalifa's early work proves beyond doubt one thing: it is possible to find a perfect pattern in imperfect data. He claimed that his early data proved the exact preservation of the Quran. When errors were found, he did not abandon his hypothesis; he merely re-analyzed the data with new criteria, and again claimed perfect preservation. When more errors were found, he preferred his hypothesis over perfect preservation. These "facts" that Yuksel is presenting are the result of extensive analysis and re-analysis of the data using the criterion that he admitted earlier: if it is divisible by 19, keep it; if it is not, keep looking for the "correct" way to count.

YUKSEL: This is a general accusation based on biased evaluation of a meticulous and arduous research that needed time to be completed. You can undermine and ridicule many scientific discoveries with the same criticism. They can be seen the same way through the eyes of dogmatic or extreme skeptics.

Lomax falsely attributes a "criterion" to me by chopping and distorting my words. Let me reciprocate with the same style: The criticism that Lomax is presenting is the result of extensive analysis and re-analysis of the data using the criterion that he admitted earlier: "if it is not divisible by 19, keep it, if it is, keep looking for 'another' way to count. If still there are some patterns left after trying all imaginable ways then you should claim that the irreducible minimum is a divine test for those who try to understand and appreciate the pattern."

LOMAX: NEVER have we seen from a Khalifite an honest statement about this, such as "We examined 430 facts and found that 24 of them were divisible by 19." They just present the 24 facts and it looks very impressive.

YUKSEL: Well Lomax, we use our time intelligently. We know that out there, there are many Muhammedans biting their fingers out of rage because of this number. We leave the examination of those 406 "facts" for them. As in the case of Lomax' house number metaphor, we leave it to them the counting of bricks, studs and nails in the walls (*harakas, shaddas, sukuns,* etc.), or the number of weeds in the backyard (differences in Warsh version), or investigating the history of the house from fictional books (Ibn Masud's and Ubay b. Ka'b's personal manuscripts reported by Bukhari and Ibn Hanbal), or the invisible ghosts that reside in the house after dark (invisible letters in *Basmalah*), etc.

Lomax, who is shouting with majuscule NEVER, should listen to the proverb "never say never." If Lomax considers me a KHALIFITE (a cheap label frequently used by him despite my frequent reminder) then I have given "an honest statement" whenever it was relevant. We cannot find "430 facts" regarding the four words of *Basmalah*, or 430 methods of counting for the simple count of its letters, or 430 alternatives to examine the statistical probability for the frequency of "day" as 365, or 430 manuscripts regarding the number of its chapters, etc.

181

However, whenever we felt a relevance or necessity for such a revelation, we tried to convey that. For instance, in the count of the attributes of God and their Gematrical values we have stated that out of more than 120 names we examined we found only four of them have a ZMN frequency and again only four of them have a ZMN Gematrical value.

I want to mention another criticism for this fact. In his DRAFT FAQ: 19 Lomax claims: "Some years ago, confining myself to names of God found in the Quran, I found: At 17:42, there is &y |L@r$ (possessor of the throne), which is clearly a name of God. The value is 700 + 10 + 1 + 30 + 70 + 200 + 300 = 1311, or 19x69. I did not find any others, but I was using a computer only to aid in the calculations, not to find the names; I cannot testify that there are no others." With his bizarre transliteration he refers to "Zil'arsh" (Possessor of the throne). This claim indicates his level of Arabic. You cannot mention "Zil'arsh" as the name of God by itself. It should be "Zul'arsh," since "Zil'arsh" is a grammatical form that can be correct only in the context of a sentence as an object. I will try to use an analogy in English: "Yahya's brother Matine drools on everything he likes." Now, what is the name of Matine's brother? Lomax's answer would be "*Yahya's*" when in fact, the boy's name is "Yahya."

Therefore, the Gematrical value of "Zul'arsh" is not 1311, but is 1307, and is not divisible by 19.

LOMAX: Yuksel wrote: "Extraordinary claims require extraordinary evidence." Exactly. To claim a miraculous pattern requires more than a few important (in the sense of obvious, like chapter count) facts divisible by 19. Such a pattern could easily occur by chance. The only statistical calculations I have seen from Khalifites have been seriously defective, overstating the probability of a single count by a factor of 10,000 (What was, in fact, 1 in 19 was reported as 1 in 190,000, by Arik.)

YUKSEL: Well, the number of chapters being 114 and the name of God *Jami'* (the Editor) having exactly a Gematrical value of 114 (3+1+40+70) cannot be discarded easily as a coincidence. In order to see the mathematical system of the Quran, Lomax should stop treating primitive story books as the second source of God's religion. Then, he should go and study elementary level Arabic Alphabet in order to see that there are 28 letters in Arabic language and there is no question regarding the definition of letters. Then, he can start learning how to count words from an expert, such as Fuad Abdulbaqi.....

LOMAX: Yuksel was not happy with my mention of the possibility of finding patterns in a telephone book. He complained that I did not actually count words or whatever in a phone book. I will respond that if he will provide a description of exactly what the miracle of the nineteen is, in terms that are testable against any book, I could then attempt to find a similar pattern in another book. I must note, however, that it took almost twenty years of work, counting and recounting, analyzing and reanalysing, for Khalifa to come up with his final statement of the "miracle;" it would be arrogant for me to think that I could invent a similar "miracle" in a few days. It takes work to find these things. I could certainly do it faster than Khalifa, because I have better access to computing power, and, further, an explicit knowledge of how to generate these statistics; I could automate the process. But it would still be, probably, many man-months.

YUKSEL: There is again two contradictory statements concocted in a single paragraph. If Lomax is sincere in his need for "a description of exactly what the miracle nineteen is" then how can he invent a SIMILAR miracle in many man-months? How can someone imitate something if he does not know what that thing really is? I am sure that Lomax can invent impressive statistics in less than many man-months. I am sure they will be similar to the lousy, foggy and incoherent pattern he sees in the Quran. It is always easy for a legally blind to imitate the portrait of Mona Lisa.

LOMAX: Milan Sulc is working on a description of the prime rank phenomena he claims to have found in the Quran. It has been suggested that he offers a very substantial reward for anyone who can show a similar phenomenon in another book. If he can find the description, it will be possible to test it. Until then, all these claims are unprovable.

Khalifa claimed that this "miracle" was "scientific." The potential of disprovability is the essence of science. Khalifa's hypothesis has never been stated with sufficient clarity to be disprovable if it were not true. (Actually, this is not quite true. Short summaries of the "miracle" are quite easily disprovable, as we have seen with the "claims." In order to test the claims, one needs additional definitions and qualifications....)

YUKSEL: Again the same pattern. First, say "never" then correct it! I call this idiosyncratic style "rickety-wobbly arguments" As if it were not Lomax who was rejecting many claims as "false." How can someone reject something as "false" if that thing is nonfalsifiable? Here is the reality: Lomax always found an easy way to reject our claims. If there is no way to refute it then state an alternative way of counting, no matter how improbable it is. If there is no alternative way of counting, then exclaim that the claim is not clear, it needs definition and clarification. You can cut every rope with this scissors in your hand.

CLAIM 5

The first verse, which can be considered as the foundation of the miracle, occurs 114 times despite its conspicuous absence from Sura 9 (it occurs twice in Sura 27) and 114 is 19x6.

LOMAX: True, but anecdotal. The count of *Basmalah* is not part of a pattern. Based on common recitation practice, it could also be said that the Quran contains only two *Basmalah*s (at 1:1 and 27:30), rather than 114 times.

YUKSEL: Not only is the claim true, but it is important. How can you say that the count of *Basmalah* (or Bismillah) is not part of a pattern? Are we not discussing a pattern which involves the number of its letters, the frequency of its words, etc.? Why should the frequency of the whole formula should not be considered as part of this pattern?

You claim that there are two *Basmalah*s in the Quran. If you follow the Hanafite sect (I know, you call it Hanafite School of Thought!), you are left with only one *Basmalah*s, that is 27:30. Hanafites do not recite *Basmalah* of al-Fatiha, since they do not accept it as the first verse of the Quran. However, none of the Orthodox Sunni scholars deny the existence of other *Basmalah*s. The argument is about whether they are independent verses or just repetition based on revelation. The prominent view is that *Basmalah* is an independent verse in 2 cases and a repetition in 112 cases. No one claims that these repetitions are man-made. Thus, the total of all *Basmalah*s come to 114, which is multiple of 19. Chapter 9 is well known for the absence of *Basmalah* in its beginning. This fact makes it clear that other 113 Chapters do HAVE *Basmalah*s.

In fact, the role of *Basmalah* in the mathematical structure of the Quran reflects this peculiarity. Now we know why there were arguments regarding the status of *Basmalah*s. We have discovered that the unnumbered 112 *Basmalah*s participate in the count of letters and the total number of verses of the Quran. However, only the two numbered *Basmalah*s (1:1 and 27:30) participate in the frequency of words in the Quran. This is an observation guided by the theory of 19. The theory is supported by this possible method of counting. As in natural sciences, ambiguity is solved based on clear and certain facts. Similarly, if clear and unquestionable parameters convince a person regarding an intentional design, then he/she can easily discover and see the intended design in ambiguous cases.

184

One point to reflect on: If every Chapter had *Basmalah* as their opening statement, we could not cite the number of *Basmalah*s being multiple of 19, since it would be redundant. It would follow from the number of Chapters. However, the author of the Quran, by deliberately violating this pattern in Chapter 9, pulls our attention to one missing *Basmalah*. While looking for the 114th *Basmalah* He shows us His intention by restoring it 19 Chapters later with an extra *Basmalah*. The verse number of this extra *Basmalah* is also significant, since it indicates the semantical relation between 74:30 and *Basmalah*.

LOMAX: Discussing "Claim 5," Yuksel defends the assertion that there are 114 Bismillahs in the Quran. However, in counting words earlier, he excluded all the Bismillahs except for two. Now, he objects to my comment that it is possible to say that there are two. Which is it? This is a variation on the question "What is the Quran?" which, so far, has not been answered by those who claim to see a numerical miracle in it. If the envelope of the Quran can be contracted and expanded as needed to produce facts, they are not as significant as they would be if a constant definition were used, and, perhaps, they are not significant at all.

YUKSEL: Lomax is a perfect example how people pretend to argue while instead they are performing a monologue tirade. I have explained the reason behind this different treatment of *Basmalah*. If 112 of *Basmalah*s are part of the Quran but not numbered, then it is very reasonable not to treat them like other *Basmalah*s which are numbered. This Quranic distinction is reflected in the mathematical system of the Quran. A simple question for Lomax would be enough to blow out his balloon: Are all those 114 *Basmalah*s have the same position in the Quran? Is there any difference between the *Basmalah*, say, in the beginning of Chapter 2 and the *Basmalah* in the middle of Chapter 27? I hope Lomax's answer will not repeat his "rickety-wobbly" style of argument.

LOMAX: Are the vowel marks part of the Quran? Are the verse numbers part of the Quran? How about the meanings we associate with the words? Where is the limit? What, exactly, are we studying? If vowel marks and verse numbers are included, which version do we follow, since there is variation in these things. In fact, with any characteristic, it may be relevant which text we study, since it appears that the 'Uthmanic copies may not all have been identical in every respect.

YUKSEL: Vowel marks are not part of the Quran. An elementary knowledge of Quranic archeology is enough to learn this fact. The verse numbers are the part of the Quran, since the verses of the Quran are not jumbled but positioned in ordered chapters. Numbering verses is acknowledgment or expression of this existing order. Besides, the Quran describes itself as "Kitabun Marqum" (Numerically Structured Book) in verses 83:9, 29. (I know that Lomax will have problem with the meaning of "marqum").

Here Lomax is becoming a philosopher by wondering the "meanings we associate with words." He is not aware that he is digging his own grave by obliviously quoting this philosophical question. He does not wonder the "meaning we associate with HIS own words, and the meaning he associates with OUR words" He wants to toss our argument based on the Quran to oblivion by this apparently innocent question. He wants to cast doubt on the QURAN ALONE. Yes, the Quranic words have meaning and believers can associate correct meanings with Quranic words by grace and guidance of God, who is the teacher of the Quran (55:1-2).

LOMAX: (Some of my scholar friends confirmed this, others are offended by the statement. There is a dogma under scrutiny here. I suggest that the Quran which is preserved is the recited Quran, and, from the beginning, variations were allowed. 'Uthman (RA) tried to settle on a single text, by consensus, and nearly succeeded.)

YUKSEL: Lomax thinks he is defining his Quran by saying "the recited Quran." Who is reciting it? The one that I do or Lomax? This claim is a sure way to reject any mathematical miracle, since it is not possible to witness a mathematical system coming from the mouth of Lomax or Abdulbasit. If the Quran is only a recitation, then Muhammad and his companions made a big mistake by putting it in a book. How can God who orders believers to write down their financial transaction before live witnesses not choose the Quran to be written down? Does He consider His words less important than financial transactions?

CLAIM 6

From the missing *Basmalah* of Sura 9 to the extra *Basmalah* of Sura 27, there are precisely 19 chapters.

LOMAX: True, but anecdotal. Further, it must be recognized that there are other positions which would produce a ZMN fact. For example, if it were Sura 28, instead of 27, which had the extra invocation, we could make exactly the same statement, and it would still be "true."

YUKSEL: The claim is true and important. Again, so many parameters on *Basmalah* encounter us with the number Nineteen. We observe that there are 19

186

chapters between the missing *Basmalah* and the extra *Basmalah*. This reduces the probability of coincidence dramatically if evaluated TOGETHER with your other so-called "anecdotal" examples. You are trying to diminish the power of accumulated evidences by isolating them from each other. Each evidence can be minor, but a number of related minor evidences can create a very strong thread of evidence. This example is related to *Basmalah* and decreases the probability not by addition of 1/19, but by multiplication of 1/19.

As for different "other positions," you give only one example. I really wonder how you will come up with other reasonable positions. Your suggestion is not better than the actual position we observe in the Quran, since it would have justified a more serious objection.

LOMAX: Yuksel wrote: "You are trying to diminish the power of accumulated evidences by isolating them from each other. Each evidence can be minor, but a number of related minor evidences can create a very strong thread of evidence. This example is related to *Basmalah* and decreases the probability not by addition of 1/19, but by multiplication of 1/19."

Yes, "a number of related minor evidences *can* create a very strong thread." I emphasize the word CAN. If the evidences are related, and all examples which satisfy the relation are given, and all of them are divisible by 19, then a strong thread would be created. But if each fact is independent of the others, sharing only the fact that they are counts of objects in the Quran, then the whole population of similar counts must be studied (or a representative sample) to show significance. I'll give an example from a phone book. Suppose I look in a phone book, and I find a list of fifty people, all named Smith, whose phone numbers are divisible by 19. Is this significant? To answer, we would need to know how many Smiths were in the phone book, and whether or not the sample was biased.

YUKSEL: It is true that 50 Smiths in a phone book may not show any significance. The whole population of similar counts should be considered. However, it is not true to that the facts that I have listed in the end of the first round argument can be criticized by this phone book metaphor. We have discarded the claims that suffer from selective method of counting. Please note that this criticism is an answer to my defense for Claim numbber 6. Please read the claim and my defense again. There is no relevance of FEW selected Smiths from a large phone book, since the entire population of *Basmalah* is considered. Lomax is expressing a meaningful criticism for an unrelated case.

LOMAX: We already know that the sample shown to us by Yuksel is biased. I have done a lot of counting in the Quran. 19 does not show up particularly often; I would suggest that it shows up about 5% of the time.

YUKSEL: Remember our house metaphor. You can equate basic and clear elements with trivial and vague elements and come up with 5% probability. You can reject the periodic table with the same mentality. You can count many other aspects of atoms and suggest that periodical table of elements is an arbitrary and biased arrangement. Well, my friend, even if it is biased, I see an obvious pattern and system in the periodic table.

LOMAX: "This example is related to Bismillah." But we can generate many, many facts from the Bismillah, and all of them are related, but not all of them are divisible by 19. For example, take each word in the Bismillah, exactly as it is spelled. In the whole text considered by Khalifa to be the Quran (i.e., excluding 9:128-129), BSM occurs 115 times, ALLH +LLLH, 2810 times (I don't know the breakdown), ALRHMN 169 times, and ALRHYM 146 times. None of these numbers is divisible by 19. It is not that all of Khalifa's counts are wrong (in his later work, errors are more rare), rather it is that equally significant counts which are not divisible are not mentioned.

YUKSEL: You are contradicting your own previous counts. Again you start with a strong objection ending up "not all of Khalifa's counts are wrong..." etc. Another example of "wobbly argument"...

LOMAX: Of all the Khalifite apologists, Yuksel is the most coherent. But, apparently, he still has not understood what it would take to really show a scientific correlation. I highly recommend an introductory course in statistics.

YUKSEL: How did he come up with this conclusion? What is his population of study? How many "Khalifite apologists" has he seen in order to make such a superlative claim about me? I think I know what he is trying to do. He wanted to bring me to the top floor and then push me out the window. He deserves credit, since he is pretty consistent in following his "wobbly" pattern.

LOMAX: I note that Yuksel acknowledges the lack of an explicit definition of "word" in Arabic. This is refreshing, given the heaps of abuse piled on me by Azhar Khan for pointing this out in earlier discussions. Fuzziness in definitions allows the creation of 19-divisible statistics. As far as I have seen, Khalifa was NOT fully consistent in his choice of word divisions. There are examples in Philips.

188

YUKSEL: I have refrained using few examples that involve the count of total words in a text (not a particular word). The mathematical miracle of the Quran is so vast that it does not need those few examples which can be rejected by skeptics.

CLAIM 7

It follows that the total of the Sura numbers from 19 to 27 is 342, or 19x18. This total (342) equals to the numbers of words between the two *Basmalah*s in Chapter 27.

LOMAX: Partly true, partly not verifiable, and anecdotal. That the sum is ZMN simply follows from fact 6, and, without a definition of "word," the word count is not verifiable.

YUKSEL: The claim is possibly wrong, difficult to verify, but interesting. I agree with you that "the sum is ZMN simply follows from fact 6 (the previous claim)" since it is a mathematical property. The total of every 19 consequent numbers will give a sum which is divisible by 19. However, if you have noticed, we don't mention this as a part of the miracle. That is why Rashad began this claim with the qualifier "it follows." The number of words between the two *Basmalah*s in Chapter 27 being multiple of 19 and matching exactly the total of the Sura numbers from 19 to 27 is what we count.

I agree with Lomax regarding the problem in counting all the words contained in a text. I had verified the number of those words as 342 for myself many years ago. To be sure, I counted them again according to the consistent definition of words implicit in Rashad's counting. After several attempts I came up with 344, excluding the intial combination of letters. This is 2 words more than the claimed 342 (19x18). It would be an interesting feature if the number of these words were multiple of Nineteen.

However, I have an unanswered question regarding the definition of "word." Inshallah, I will explain this later. Still, it is significant, if we come up with this number according to a consistent counting system.

[I did not receive the second round response from Lomax regarding the claims 7 and 8].

CLAIM 8

The total number of verses in the Quran is 6346, or 19x334. There are 6234 numbered verses and 112 unnumbered verses of *Basmalah*s, and 6234 + 112 = 6436. Note that 6 + 3 + 4 + 6 = 19

LOMAX: Anecdotal, arbitrary, false. There is substantial disagreement on how to divide verses. Without specifying an edition of the Quran, it is impossible to verify the "6234 numbered verses." However, totaling from the table of contents to Madinah, I get 6236 numbered verses. Provided that I have calculated correctly, this "simple fact" requires the exclusion of two verses, presumably 9:128-129. Note also that the ZMN total is arbitrarily created by the inclusion of the 112 unnumbered invocations, which were excluded from counting in fact 3.

The digit count sum is arbitrary as it depends on numerical representation and number base. This works with base 10 "Arabic" numerals (which actually came later from India and were not used by the Arabs of the time of the revelation.) I am quite sure that if such a total did not come out to 19, it would not be mentioned.

YUKSEL: The claim is true (if 9:128-129 is excluded) and becomes significant with other related parameters. The ZMN total for the absolute value of digits is arbitrary. It is true that the numbering system was adopted from India two centuries after the revelation of the Quran. However, the Quran uses the base 10 system. Numbers are represented by names and also by alphabet letters called Gematrical system. In fact, you can notice the base 10 system in the Arabic name of numbers. Since modern numerical representation is based on the base 10 system, it is neither difficult nor unreasonable for an Omniscient author to employ such a representation for future generations, that is, us. If I see enough examples of this kind of calculation then I will not have any problem for accepting it. On the contrary, such a pattern will be strong evidence regarding the divine nature of the book. Personally, I am convinced regarding the relevancy of this method of calculation. The example that convinced me is the absolute value (digit count sum) of the frequency of "HaMim" letters. They exhibit a marvelous mathematical pattern. Inshallah, I will present that evidence later.

CLAIM 9

The famous first revelation (96:1-5) consists of 19 words.

LOMAX: False unless qualified. Until a better definition of an Arabic "word" comes along, I'll follow this: a "word" is a unit of meaning, as found in Arabic dictionaries. I.e. "rabbuka" is two words, "rabbu" (Lord) and "ka" (your). Evidence for this is that it is impossible to translate such a phrase with less than two words into English without major loss of meaning. There are further refinements which would be necessary to be thoroughly consistent: I'm not going to work them out now.

By this definition, 96:1-5 consists of 28 words. I have counted the definite article as a separate word. I am quite aware that this definition is arguable, but it is incumbent on one who wishes to argue to state an alternative. We could then look at these "facts" and verify them. Dr. Khalifa, as far as I know, never specified his method of counting words, and his counts do not seem to follow a consistent pattern.

YUKSEL: The first revelation consists of either 19 or 20 words. By just looking at Rashad's counting I was able to understand what his definition of "word" was, since he demonstrated several lengthy examples of his counting one by one. He did not need to further verbalize this well illustrated task. For instance, if I claim that there are 99 words in this paragraph, it will give a good idea regarding my definition of words. If I give several more examples of my word count, then, by comparison you can have a very clear understanding of my method of counting.

Since I am not a linguist, and English is my fifth language, my lexicon is not sufficient to define our method of counting words. I believe you can do this, since you did a very good job in defining the simple fact regarding the count of letters in *Basmalah*. Obviously, Rashad is accepting a word as a monogram. If prepositions, prefixes, and suffixes are attached or dependent to the main word, he accepts all of them as one word. Furthermore, by his non-verbal definition, a word should have at least two letters. He does not consider a single letter as a word as it is the case with conjunction "W" (and) in Arabic.

I believe that our method of counting words is virtually physical (does not require extensive grammatical knowledge and linguistic speculations), clear (for those who are not blinded by antagonistic passion), and the most reasonable one. Many researchers, independent of Rashad, employed the same method in the counting of words. For instance, Dr. Mahdi Bazargan who later became a prime minister in Khomeini's first cabinet, in his great statistical work on the size and topic of verses according to the chronology of revelation, gives the same count, 19, for the first revelation. (Sayr-i Tahawul-i Quran, -Process of Quranic Evolution-, published by Book Distribution Center, P.O. Box 22933, Houston, TX 77027, in 1974).

I found your claim bizarre: "evidence for this is that it is impossible to translate such a phrase with less than two words into English without major loss of meaning." This statement is both arbitrary and baseless. I can translate "Rabbuka" (your Lord) by one word into Turkish as "Rabbin", by one word to Persian as "Khudayet" or by two words as "Khuday-i Tu" How can you claim that the English translation of Arabic words is the criterion for defining Arabic words. With your reasoning, an Arab should reject the word count of my Microsoft Word Processor Program by translating one English word with two or three Arabic words!

Finally, I find Rashad's method of counting for the sum of words consistent, relevant, and reasonable with the exception of two problems: First, he does not count "Ma" (both negating adverb and pronoun) as a word. "Ma" is the only exception with its two letters. I do not see any grammatical or physical justification for this exceptional treatment. However, I find the results of counting based on this exception impressive. Curiously, this exceptional treatment is independently committed by other researchers. Maybe you and I should study this subject more thoroughly.

Second, probably the most important problem: he relies on narration to establish the chronology and the size of early revelations. Narration regarding the early revelations can be considered more reliable than the other narration (hadiths) which were shaped and filtered by dubious and diverse motivation and agenda. But, personally, I do not feel comfortable with any mathematical pattern based on those conjecture, since those information is not the part of the Quran, at least.

LOMAX: Yuksel wrote: "I believe that our method of counting words is virtually physical (does not require extensive grammatical knowledge and linguistic speculations), clear (for those who are not blinded by antagonistic passion), and the most reasonable one."

If it is "virtually" physical, then it should be possible to describe the algorithm so that we could use a computer to count words. Yuksel pleads ignorance of English for his inability to state explicitly what standards are used to divide words. Sorry, his English, while it is obviously not his native language, is quite good enough.

YUKSEL: My vocabulary of the linguistic terminology is limited. Therefore, I wanted to avoid granting you errors to dwell on. I believe that there is no consensus about the counting of all words in a text. But, I found Rashad's method of counting the most reasonable one. The same method is also used by many independent scholars BEFORE him. I found only one problem which I mentioned earlier.

LOMAX: But we already know how to divide words: divide them so that the counts of certain important chapters or pieces of text come out divisible by 19. The more that one does this, however, the pattern becomes fixed, and additional counting will only work out one out of nineteen times. But one can still get a few good "facts" out of manipulating the word divisions.

Yuksel acknowledges that "maa" is two letters but is not treated as a separate word. He can find no justification except the fact that it produces ZMN (zero modulo 19, i.e., divisible by 19) counts. How many does it produce? I noticed early on that while other statistics continued to pile up, where methods of analysis could be varied indefinitely, these word counts slowed to a trickle.

YUKSEL: Enough said on this.

LOMAX: Yuksel also acknowledged the problem that Khalifa relied on hadith to make claims about sequence of revelation, whereas he elsewhere denied hadith. It would be very useful if, before we proceed, Yuksel would direct his attention to the Draft FAQ: 19, Study Problems, which has been posted and also e-mailed to him. (It is available on request). Unlike the material to which Yuksel is replying, which is over two years old, this represents a more mature analysis of the problems in these studies. Most or all of what Yuksel continues to claim is without the kind of foundation necessary to establish statistical significance. Otherwise we stay bogged down in hordes of details with no standards for determining what is coincidence, what is the result of manipulation, and what is, possibly, a genuine miracle.

YUKSEL: I agree with Lomax regarding the necessity of foundation to establish statistical significance. However, ultimately, we have to deal with "hordes of details." He was the one who first dealt with those details and published it with his copyright stamp on it. I believe that we will agree on almost all of the statistical foundation, but we will still have differences in evaluating particular examples. I found Lomax occasionally abusing those statistical foundations when they were not relevant.

CLAIM 11

The word God (Allah) is mentioned 2698 times as shown above, 19x142, and when we add the numbers of the verses wherever the word "God" occurs, the total comes to 118123, which equals to 19x6127.

LOMAX: False unless qualified, not simple, anecdotal. I have not verified the verse number totals. It is an arduous task to find such a sum without having a comprehensive verified database, which is not accessible to me. I suspect that such a database was not available to Dr. Khalifa, either; otherwise he would not be repeating such errors as the count of the word Allah. Therefore I have little confidence in the total 118123. Remember, too, the numbering of verses is not an obvious "fact," unless the edition is specified. But at best, this is anecdotal.

YUKSEL: While I was translating the Quran to Turkish, I recorded the cumulative frequency of the word God on the left bottom of the page, and the cumulative sum of verse numbers where God occurs on the right bottom of the page. I came up with the same result, 118123, using my calculator. I agree that verification is an arduous task, and I am still not certain about my result regarding this big number. It is still possible, though with low probability, that I have made a mistake which brought me to the same result. Therefore, I say, "it is true unless it is proven wrong." You can verify (or falsify) the accuracy of this number (118123) by checking the third revised edition of Rashad's translation. You can find the cumulative sum of verse numbers where the word God occurs at the right bottom of each page. The burden of disproof is on you, since we have demonstrated our count page by page. If the result is true, I think it is more than anecdotal, since it is a reasonable part of an interlocking mathematical pattern regarding the most important word in the Quran, that is, God.

[Lomax did not continue the second round debate on Claims 11, 12, and 13.]

194

CLAIM 12

The word, "Quran" is mentioned in the Quran 57 times, or 19x3.

LOMAX: Vague, arbitrary, anecdotal. I haven't looked at all of them, but my concordance shows 70 occurrences. Some of them have the definite article, some not, some have affixed pronouns, etc. Perhaps there is some narrowing of the specification which will produce the magic 57; is it worth checking?

YUKSEL: Clear, arbitrary, and anecdotal. There are 68 occurrences of the word Quran. Quran (58), Quranan (10). The other two occurrences are in form of "quranahu" and is not a name, but a verb. To be consistent with our method of counting, I find 68 occurrence of the Quran. One of them is not used for the Quran, that is as a proper name for God's word, but referring to its literal meaning, i.e., a quran (book of recitation) other than the Quran (10:15). Therefore, the total comes to 67, which is not multiple of Nineteen. It is clear that Rashad did not include Quranan (with an Alif in the end, indicating its grammatical position in the sentence), which occurs 10 times. I find this method of counting inconsistent with others and reducing the significance of other counts.

CLAIM 13

The word "Quran" is mentioned in 38 suras, 19x2.

LOMAX: True but anecdotal. Since we do not agree on the number of mentions of "Quran," it is astounding that my concordance agrees on the number of Suras. Either the extra "Quran"s are only found in suras with the "real" occurrences, or Dr. Khalifa's friends have changed their method of counting. Nevertheless, this is very much anecdotal: this is the first time that we encounter this method of counting.

YUKSEL: True but anecdotal. I agree with what are said by Lomax. Unfortunately, in this case the method of counting is changed.

CLAIM 14

Sura 96, first in the chronological order, consists of 19 verses.

LOMAX: True, but anecdotal. Well, it is true if we exclude the invocation, which is traditional. Nevertheless, we have established no pattern that the verse counts of suras is a multiple of 19.

YUKSEL: Since Sura 96 is not just any chapter, but it is chronologically the first chapter, it does not need a pattern among other chapters to be significant. This can justify a unique treatment. Furthermore, if every chapter had 19 or multiple of 19 verses, the mathematical code of the Quran would not have remained hidden in a Chapter called "The Hidden One" for centuries. If the purpose of this mathematical miracle of the Quran was to provide the computer generation with a mathematical miracle, then it is very appropriate for it to be hidden from previous generations (10:20).

The position of Chapter 96 is also interesting: it is the 19th Chapter from the end of the Quran.

LOMAX: Nevertheless, within the parameters of our use of the term "anecdotal," the term still applies. Certainly it is possible to justify the unique treatment of anything, but if all we have is a collection of unique treatments, which may be chosen according to whatever shows the pattern of 19, there is no way to distinguish between a "miracle" and the result of a series of choices.

YUKSEL: What if we are convinced that those choices are intended in the Quran? What if we can predict or explain many things based on those unique treatments? In the end of the first argument I provided some examples.

LOMAX: It is to be noted that it is not impossible that Allah intended there to be an increase over the normal occurrence of 19-divisible statistics, without there being an absolute code or pattern. This would be consistent with the description of this number as a "trial" in 74:31. However, even this increase has not been demonstrated by Yuksel, Khalifa, or any other worker in this field. In general, the statistical problems have been ignored.

YUKSEL: Previously, Lomax claimed that there was a little pattern that functions as a trial for desbelievers. Now, he makes that position looser by making it a mere possibility. Another twist, in other words, wobbly pattern.

LOMAX: Yuksel wrote, "The position of Chapter 96 is also interesting: it is the 19th Chapter from the end of the Quran."

There is no end to specially-constructed facts. Note that a chapter has four ways to have a 19-divisible count of this kind, because one may choose to count from the beginning or end, and one may count cardinally or ordinally.

YUKSEL: There are only two ways of counting the position of chapters. And we found it significant that the 19th chapter from the end of the Quran has 19 verses and is unanimously believed as the FIRST revealed chapter. One does not need to take an introductory nor a doctorate level of statistics to see the significance of this simple fact.

CLAIM 15

Sura 96 consists of 304 Arabic letters and 304 equals to 19x16.

LOMAX: Arbitrary, unverifiable, anecdotal. Some hamzas are counted, some not. An example of an uncounted hamza is sitting on the alif in the first word, Iqra', and one counted is in Raahu, the 25th "word" shown in VP. The text used is not specified (although Dr. Khalifa gives the words for this: some uncounted hamzas are shown). Hamzas are ordinarily not considered letters, but rather marks. While is not unreasonable to count them, the choice is arbitrary. If it depends on pronunciation, we have abandoned our rule that it is the written language that is being studied. In any case, again, we have no pattern that the letter count for suras is ZMN.

YUKSEL: Not arbitrary, verifiable, interesting, but most likely false. It appears that your lack of knowledge in Arabic is the reason for your arrival at arbitrariness regarding this count. Both "Iqra" the first word, and "Raahu", the 25th word, have 4 Arabic letters in current manuscripts. If you had looked carefully to their spellings you would find a difference justifying this count. In fact, you can tell this difference also from their pronunciation. Dr. Khalifa uses a known Uthmani text, which is the official text of the manuscripts published in Saudi Arabia and all of the Arabic speaking world. Turkish, Pakistani and other non-Arab Muslims use a manuscript that contains extra "Alifs" in many words in order to help people to read easily and accurately. I came up with the same count which includes *Basmalah*, 304 (19x16). The count can easily be verified by a person who is little bit familiar with Arabic calligraphy. Therefore, I have no doubt that Rashad's counting is consistent, verifiable, and leads to an interesting pattern in the "chronologically first" Chapter.

However, I believe the count of letters should be less than 304, because early manuscripts did not contain "hamza." Hamza is a scriptural innovation intended to help unsophisticated readers when many non-Arab nations converted to Islam. Though dots were also invented later during the reign of Uthman, the Arabic language had 28 letters even before the revelation of the Quran, and native speakers could differentiate among different letters written in almost identical appearance, such as, B, T, S, Y, N. When I study the Tashkent Copy, I see the difficulty of reading the early manuscripts written without dots and other paraphernalia. It requires a good command of Arabic and a high IQ to be able to read them. Reading those manuscripts is like solving quadratic problems. While trying to read a word you are required to distinguish and identify several identical characters. You eliminate some candidates based on your familiarity with the shape of words. Then, you narrow down your choice to one specific combination of letters based on the context and position of the word. It is sometimes possible that you may have two legitimate readings and meanings for a word. If their meanings do not contradict the overall context, you may end up with two equally accurate but different meanings for a single statement.

Anyway, the count in early manuscripts is less than 304. I do not have time to search other early copies, nor do I have the time to deduce the counting from the Tashkent Copy that I have. The copy is not complete.

LOMAX: I have learned a great deal of Arabic since writing the document Yuksel is responding to; it is over two years old. But my comments still stand. In the case of the hamza on the initial alif of Iqra, some styles of writing do not use it, some do. Arabic scholars have also asserted that Khalifa's counting of hamza is arbitrary, and examples can be given of contradictions in his counting.

YUKSEL: I hope you will learn more Arabic. However, and more important than Arabic, you need to learn the fact that those who trade God's word with medieval fabrications and follow other sources besides God's word (Hadith, Sunna, Consensus of Ulama, etc.) are handicapped from witnessing the miracle of the Quran. Miracles are divine blessings for believers and for those who sincerely search for truth.

LOMAX: Yuksel wrote, "Both 'Iqra' the first word, and "Raahu", the 25th word, have 4 Arabic letters in current manuscripts. If you had looked carefully to their spellings you would find a difference justifying this count. In fact, you can tell this difference also from their pronounciation. Dr. Khalifa uses a known Uthmani text, which is the official text of the manuscripts published in Saudi Arabia and all of the Arabic speaking world."

First of all, the text used by Khalifa in Final Testament is, as Yuksel claims, a so-called Uthmani text, following the Hafs reading. But so are the Warsh texts, and they differ substantially in the use of hamza. Since Warsh is preferred in quite a few Arabic-speaking countries, it is not true to call this text "official" for all the Arabic-speaking world. It appears, moreover, that Warsh is probably closer to the dialect of the Prophet, SAS.

YUKSEL: Another Hafs-Warsh mumbo jumbo. I am very doubtful about the truth-value of your endorsement. By using SAS after "the Prophet," that is Muhammad, you prove, besides your acknowledgement, that you are a dedicated Muhammedan. You have mentioned God's name frequently without any additional praise words. SAS is used for the purpose of praise, since anyone who does not use that phrase after Muhammad's name is considered disrespectful to Muhammad. However, like other Muhammadans you feel obligated to add the fabricated abbreviation SAS (Sallallahu Alayhi Wasallam) each time you mention his name or refer to him. Anyone who is not blinded by Hadith and Sunna can see that the meaning of "Salli Ala" (support) in 33:56 is distorted by Muhammad worshipers in order to exalt him more than they exalt God. In order to see the distortion, please compare verse 33:56 with 33:43 and 9:103. You will see the same expression is used in all of them. I have a lengthy argument on this traditional deliberate distortion, but here is not the place.

LOMAX: Yuksel wrote, "Turkish, Pakistani and other non-Arab Muslims use a manuscript that contains extra "Alifs" in many words in order to help people to read easily and accurately."

This is true. But there are many other variations besides this.

YUKSEL: I wish Daniel Lomax could give some examples. He made same claim about the number of letters in *Basmalah*, and later his examples demonstrated the absurdity of his claim.

LOMAX: Yuksel wrote, "I came up with the same count which includes *Basmalah*, 304 (19x16). The count can easily be verified by a person who is little bit familiar with Arabic calligraphy."

Note that the Bismillah is included in the count. Elsewhere, when counting the verses and words, the Bismillah is excluded. But, in this case, it makes no difference, since, as Khalifa counts, the Bismillah contains 19 letters; if it is 19-divisible without the invocation, it will also be so divisible with it. But this does, once again, demonstrates how the Khalifites do not use consistent methods of counting.

199

YUKSEL: I have given my answer for this tirade before.

LOMAX: Yuksel wrote, "Therefore, I have no doubt that Rashad's counting is consistent, verifiable, and leads to an interesting pattern in the "chronologically first" Chapter.

Khalifa was apparently not consistent; and, as Yuksel notes below, hamza was not written in the early ms. of the Quran. To be verifiable, a text must be given and rules for counting specified: we have only anecdotal evidence as to Khalifa's rules for counting.

YUKSEL: As I said before, I do not really consider this example as centerpiece of the system. Indeed, I accepted it as false considering the "hamzas" in old manuscripts.

LOMAX: Yuksel wrote: "Lomax quotes the third paragraph of my answer in the first round mentioning the calligraphy of the early manuscripts." Actually, this is true for unvowelled Arabic as well, though the problem is not so severe. But to one who already is familiar with the Quran and Quranic language, it is not so difficult to read the Tashkent text.

YUKSEL: I agree.

LOMAX: Yuksel wrote, "Anyway, the count in early manuscripts is less than 304. I do not have time to search other early copies, nor do I have the time to deduce the counting from the Tashkent Copy that I have. The copy is not complete." I suspect that we have the same copy, or at least parts of the same copy. the original is obviously in poor condition. The edges of some pages are missing, and one finds holes in the text. But enough is there to raise serious questions about any attempt to count letters and words in the Quran. There are numerous variations from the modern conventional text!

YUKSEL: I don't know which copy you have either.

CLAIM 16

When we add the numbers assigned to the verses wherever the word "the Quran" is found, the total comes to 2660 or 19x40.

LOMAX: Arbitrary, anecdotal. I have not verified this; notice that, presumably, we are only looking at the word with the definite article. We have no pattern that words occur in verses which add up to a multiple of nineteen. Of course, about one out of nineteen common words will have such a total.

YUKSEL: Verifiable, yet anecdotal. I have not verified this either. If the count of "the Quran" is multiple of 19 and the verse numbers wherever they occur add up to a multiple of 19, it might be considered interesting. Sure, one out of nineteen common words will have such a total. However, if we find that most of the key words exhibit this pattern, then we may exclude other common words from our statistics. Obviously, it will be a matter of dispute which ones are the key words, and how many of those words we have. I think none will dispute the fact that the word "Allah" (God), and "The Quran" (The Book of Recitation) can be distinguished as "key" words. But, in order to asses their significance in the mathematical system we need to have an approximate idea regarding the members of this category. Therefore, it is very difficult, if not impossible, to objectively verify the significance of this count in the mathematical system of the Quran.

LOMAX: Yuksel is beginning to acknowledge the difficulties. I would point out that if one decides what are the significant words in the Quran *after* finding out which ones have allegedly special numbers associated with them, then the process is open to the charge that the results have influenced the selection. It is a major step forward that Yuksel recognizes that the "miraculous" phenomena are selected out of a larger group of phenomena. So far, his claim that there is a miracle here seems to be based on nothing more than an impression produced by looking at a lot of 19-divisible statistics, which were largely preselected. To convert such an impression into objective knowledge, at least in a scientific sense, requires that a *comparative* study be done.

Milan Sulc saw this right away, when I pointed it out. So he has set for himself the task of describing the numerical phenomena he has found in a way that they can be studied comparatively. I do not think it will be an easy task. All this talk about the 19 began in an atmosphere of puffery and hype, where only evidence on one side of the question was considered. We find a few numbers and they are divisible by 19 (or have prime rank linkages, in Sulc's case) and we get very excited. This is addicting. I know; I believed it for about twelve years.

YUKSEL: An important acknowledgment that explains why Lomax is so paranoid about this mathematical system! Lomax, "believed" it for about twelve years. That is the difference between himand me. I never believed in the miracle. I saw it and examined it from the very beginning. I did not swallow everything presented by the discoverer. Lomax, on the other hand, got addicted and excited without sufficient knowledge and examination. He just believed it as many people do. Later, certain things forced him to examine the system. I can only guess what has triggered his suspicion after twelve years of gullible faith! He found himself in a dilemma, like many other Muhammedans did: either accept the Quran alone, or follow orthodox teachings promoted by Muslim clergy. He chose the orthodox teachings.

CLAIM 17

If we take the above number (2660) and add the number of suras where "the Quran" occurs to it, the total becomes 2698, or 19x142, which is exactly the frequency of occurrence of the word "God."

LOMAX: False, anecdotal. When two large numbers, produced by separate calculations, are identical, we are impressed. However, note that, since these two numbers are already presumably established as ZMN, we are really saying that 140 plus 2 equals 142. This is not nearly as impressive, and, in fact, suggests a method for generating such impressive statistics. Unfortunately, the count for the word "God" has not been shown to be 2698; it is likely 2701.

YUKSEL: Redundant, anecdotal. It is an obvious mistake to count this as an example of the mathematical system of the Quran, since it is simply a mathematical property, as Lomax indicated. However, the count for the word "God" is 2698 without 9:127-128, and Lomax's number 2701 is surely false. I have discussed this before, and if needed I can present the list of the 2698 (19x142) occurrences.

LOMAX: As I have acknowledged before, my early examination of the count of "Allah" was complicated by numerous undiscovered errors in Khalifa's published work. Philips had found some errors in it; I (and he, apparently) assumed that there were no more, which was incorrect. As I have pointed out, there are still problems with this count, but it is apparently true if properly qualified (i.e., if the ms. is specified and exceptions noted).

YUKSEL: We have noted the exception. We did not need to specify the manuscript, since we thought it was obvious for people. Now we are learning that we need such a specification because of Lomax who apparently became an expert of finding odd and non-existing manuscripts, such as the ones that contain less or more chapters.

Probably, it is better to specify the manuscripts by negation: our manuscript is not the non-existing manuscripts which are reported by hadith books. Also, our manuscript is not the Warsh version which is esteemed by Lomax.

CLAIM 18

When we add the Sura number, plus the number of verses, plus the sum of verse numbers (1+2+3+...+n, where n = number of verses), the grand total of all suras is 346,199 or 19x1822.

LOMAX: False if unqualified, arbitrary method. Following the Madinah edition (which is not the only way of numbering the verses), this statistic is false. However, if verses 9:128-129 are excluded, it is true. Why that particular combination of figures? The answer is simple. Because the answer comes out as a multiple of nineteen. The kicker is the sum of verse numbers, a completely arbitrary, out-of-the blue statistic, which is not an independent variable (it equals $(n*n + n) / 2$). There are many such dependent variables which could be chosen; all we need to do is to find one that comes up with a ZMN figure.

Note that if n is ZMN, then 1+2+...+n is necessarily ZMN, so the inclusion or exclusion of the sum of sura numbers from the calculation makes no difference (since there are 6x19 suras). It merely makes the calculation seem more complex. This calculation is also redundant to Facts 34 and 35; that is, it is necessarily true if they are true.

YUKSEL: The claim does not have any mathematical value for the reasons well explained by Lomax.

CLAIM 19

The Quran is characterized by a unique phenomenon -- twenty nine suras are prefixed with "Quranic initials." There are 14 letters which make up 14 sets of initials and these occur in 29 suras. The total of these numbers is 14 + 14 + 29 = 57, that is 19x3.

LOMAX: True, anecdotal.

YUKSEL: True, anecdotal.

CLAIM 20

Sura 50 is entitled Q (Qaf) and prefixed with the initial "Q". The letter Q occurs in this sura 57 times, or 19x3.

LOMAX: True, anecdotal. The occurrences of the letter "Q" in suras 42 and 50, and the identity of their sum with the number of suras in The Quran is one of the interesting true facts Dr. Khalifa discovered. I keep returning to God's explanation of the Nineteen. For some people, these true facts (and even the false assertions) may lead to an increase in faith; others are led down endless pathways of delusion, obsession, and worse.

YUKSEL: The claim is true and significant. Examples of the frequency of other letters that prefixes Chapters provide a very strong evidence for their role in the mathematical system of the Quran based on the code Nineteen. The true facts can only increase the faith of those who understands and accepts them.

CLAIM 21

The only other sura prefixed with the Quranic initial "Q" is Sura 42. This sura also contains 57 Q's, or 19x3.

LOMAX: True, anecdotal. This is anecdotal because the method of analysis does not apply to any other Quranic initial. Originally, Facts 20 and 21 were indeed part of a pattern; as I mentioned above, in Dr. Khalifa's early publications, 8 out of 14 totals for the abbreviated letters were ZMN. However, after suffering corrections, Dr. Khalifa's figures now show only 3 ZMN totals, and two of those require changes to the text (See below, Facts 23 and 24).

YUKSEL: The claim is true and very significant. Since the total of the letter "Q" is 114 (19x6) in the chapters where it prefixes, the method of analysis do apply to other Quranic initials. The general method of counting is this: if we see an initial letter or combination of letters in the beginning of a Chapter we count all the occurrences of those letters in ALL chapters where they prefix. If some letters have extra significance, such as the letter Q, there is no reason not noticing it, especially if we are provided with a good reason for that peculiarity.

LOMAX: Yuksel is still avoiding the problem with "good reasons." It is possible to generate a "good reason" for many different ways of counting. I believe I have done a study on applying the method of analysis which Yuksel proposes; it does not at all give consistent results. I will follow up with a copy of that, inshallah.

CLAIM 22

"Q" stands for Quran, as the total of all Q's in Sura 42 and Sura 50 is 114, the same as the number of suras in the Quran.

LOMAX: Redundant. Again, because we have preselected for ZMN, this is equivalent to saying 3 plus 3 equals 6. The assertion that "Q" stands for "Quran" is speculative, though not unreasonable.

YUKSEL: The claim is not redundant but an interesting and meaningful speculation. If the total was 152 (19x8) or another ZMN, then it would be redundant to mention this again. The claim is about the relation between the total frequency of "Q" in the Chapters where the letter "Q" prefixes and the number of Chapters in the Quran. If it is a reasonable speculation to find a relation between "Q" and the "Quran," it is also reasonable to see the relation between the frequency of "Q" and 114 chapters of the Quran.

Another point: You are evaluating these claims with a double standard. For instance, you have labeled the claims 20 and 21 as anecdotal by saying that their "method of analysis does not apply to any other Quranic initial," that is, they are counted separately. Now, you reject the total count which shares the same method of analysis applied to other Quranic initials. This time your excuse is "redundancy." Objecting something on the basis of redundancy implies that you have accepted the previous one. Otherwise, you should not object to this claim on the basis of redundancy. In brief, your argument is fallacious and it is like saying "X is insignificant because of Y, and Y is insignificant because of X."

In order to eliminate your objections I will re-arrange the order of the claims. Please consider the Claim 22 before the Claim 20 and 21. Let's see:

20. There are 114 (19x6) letter "Q" in the two chapters that start with the letter "Q".

22. The frequency, 114, is equal to the number of chapters in the Quran. This is numerically meaningful when we consider that the letter "Q" stands for the Quran.

21. It is a special feature of the letter "Q" to occur 57 (19x3) times in both Chapters. With this arrangement, the claim 22nd becomes significant, since it is analyzed according to the same method used for counting other initials. You cannot ignore others as redundant either.

LOMAX: I am glad to see Yuksel using the word "speculation" regarding the significance of these claims. Remember that Khalifa et al have called these claims "Absolute Proof."

YUKSEL: I do not call the mathematical system as the "proof." They are divine evidences *ayaat* for those who can see them.

LOMAX: Yuksel wrote, "Another point: You are evaluating these claims with a double standard. For instance, you have labeled the claims 20 and 21 as anecdotal by saying that their "method of analysis does not apply to any other Quranic initial," that is, they are counted separately. Now, you reject the total count which shares the same method of analysis applied to other Quranic initials. This time your excuse is "redundancy." Objecting something on the basis of redundancy implies that you have accepted the previous one." No, it merely means that the claim is redundant, at least to some degree, to another claim. The two claims are really not two separate phenomena. If one is allowed to multiply claims by simple restatement using different words, one may create an appearance that 19-divisible statistics are more common than they actually are.

YUKSEL: Great. Does this mean he has accepted the claim 22 as "another claim" and the two previous claims "the redundancy?

LOMAX: Yuksel wrote, "Otherwise, you should not object to this claim on the basis of redundancy. In brief, your argument is fallacious and it is like saying 'X is insignificant because of Y, and Y is insignificant because of X.' " Yuksel is not without a point here. Really, the situation is that Claim 20 and 21, considered together, because they use the same method of analysis, are a little more than anecdotal, and the two claims, added together, produce another interesting result (cited in Claim 22) which equals the number of Suras of the Quran. But we have already used the number of Suras of the Quran as Claim 4, and, since we are now dealing with a selected population of 19-divisible statistics, this is equivalent to saying that there are 3 blocks of Q in sura Q, and 3 more blocks in another Sura, and 3 + 3 = 6, which is the number of blocks of Suras in the Quran. It does sound a little less impressive stated that way, doesn't it? So, let us say that there is a little less here than enumerated. How much less depends on definitions; in the absence of a clear definition of a "numerical miracle," we could argue for thousands of years....

YUKSEL: Lomax is playing with numbers to make it less impressive. He seems to be not aware of the statistical trick he is employing. Obviously, matching with a smaller number (6) is less impressive than a bigger number (114). In other words, surely, the total of Qs being equal to 114 (the number of real chapters) is much more impressive than being equal to 6 (the number of supposed chapters). Lomax cannot devalue this fact by suggesting us to pretend fake Qurans.

LOMAX: Yuksel wrote, "In order to eliminate your objections I will re-arrange the order of the claims. Please consider the Claim 22 before the Claim 20 and 21. Let's see: (20) There are 114 (19x6) letter "Q" in the two chapters that start with the letter "Q". (22) The frequency, 114, is equal to the number of chapters in the Quran. This is numerically meaningful when we consider that the letter "Q" stands for the Quran. (21) It is a special feature of the letter "Q" to occur 57 (19x3) times in both Chapters. With this arrangement, the claim 22nd becomes significant, since it is analyzed according to the same method used for counting other initials. You cannot ignore others as redundant either."

A consistent method has NOT been used for analyzing initials, as we will see.

YUKSEL: We will see, inshallah.

CLAIM 23

Sura 68 is initialed with "N" (Noon), and contains 133 N's or 19x7.

LOMAX: Deliberate deception. It is incumbent on the claimant to specify the text upon which this assertion is based, and further to state whether or not this is the same text upon which the other counts are based. It is obvious upon reflection that the freedom to pick and choose between texts would allow one to fabricate ZMN totals.

All Qurans available to me show N as an abbreviated letter, with 132 of them in the sura. As I mentioned above, Dr. Khalifa, in VP, crudely altered the text in his photocopy to expand N to NwN, so, I assume, he did not possess any such original.

YUKSEL: You do not need to assume, it is true that Rashad did not posses any such original. "NwN" is one of the 4 corrections made based on the mathematical system of the Quran. Have you thought why it is just 1 (one) less in your Quran? A proofreader can correct my spelling errors based on conventional language and the context of my statements. Similarly, we can find errors committed by scribes. If I am convinced about the mathematical structure of the Quran, I can make some corrections based on compelling evidences. Obviously, scribes were humans and they did commit some errors while writing. Besides, the accuracy of the mathematical code is proven with a prophetic correction. I will explain this later when we discuss the number of "Saad"s.

By the way, I do not think our "alteration" of the text is crude! Though this is an irrelevant and a silly argument, I will tell you a secret regarding this correction. While I was assisting Rashad at the Masjid Tucson, it was me who corrected it. In order to do this correction I followed a sign from my Lord. I cut the last part of "Zan-NuN" in verse 21:87 and pasted it there. Inshallah, in the end of this defense I will explain my reason for this apparently bizarre transfer.

LOMAX: It is true that one could make corrections in a text which is covered by a code. But first it must be established that the code exists and is sufficiently precise to justify the corrections. So far, Khalifa has managed to demonstrate that a code will appear in false data. In other words, that a code can be found does not guarantee that the data is correct. The converse is also true. If there is a "defect" in the code, it does not mean that the data is incorrect. It is possible, for example, that the method of analysis is incorrect. Since Khalifa varied the method of analysis as necessary to produce 19-divisible results, and since, in many cases, no clear standards exist for counting, changing the text *without notice* is very, very shaky, especially as part of a demonstration that the code exists.

This is like a scientist who is convinced that his theory is true, so he alters his data to make his results more convincing. After all, if the theory is true, that particular experimental result must have been wrong anyway.

YUKSEL: Another monologue eloquently stated. It blends a general common sense with his biased evaluation. I quote this to be fair with him. I believe that Lomax is doing a great job by providing excuses for those who do not want to see the miracle. No wonder he became a hero on Muhammedan news groups. It is a blessing to witness miracles. May the Possessor of Infinite Bounties guide us.

LOMAX: Yuksel wrote, "By the way, I do not think our 'alteration' of the text is crude! Though this is an irrelevant and a silly argument, I will tell you a secret regarding this correction. While I was assisting Rashad at the Masjid Tucson, it was me who corrected it. In order to do this correction I followed a sign from my Lord. I cut the last part of "Zan-NuN" in verse 21:87 and pasted it there. Inshallah, in the end of this defense I will explain my reason for this apparently bizarre transfer."

I was a typesetter and layout artist. I described the change as "crude" because it was clearly visible. One can see, in Visual Presentation of the Miracle, two defects: first, there is some extraneous ink on and to the right of what would have been the original Nuwn; and, second, the added letters are hanging into the right margin. It can be clearly seen, unlike the change done elsewhere with the letter Sad. It is true that this is not an argument against the "miracle," since it is not required that Messengers be layout artists. But, somehow, I think they would not resort to this kind of deception, and it *is* deception to change the text of the Quran and use it to prove a numerical phenomenon which is then the excuse for changing the text. Only if the phenomenon is first shown, on strong evidence, to exist in a received text, would it become legitimate to, with notice, restore the original text.

YUKSEL: There is a misunderstanding. I was referring to the Rashad's transalation of the Quran that contains the original text. I do think that it was a mistake not to inform the reader the change in Visual Presentation of the Miracle. The size of the mistake depends on the circumstances and intention of the person who committed it. I will not condemn Rashad for this. If I did not know him personally, it could be easier for me to accuse him of deliberate deception as well.

LOMAX: This does not prove that the change is not correct, just as the alteration of data by our dishonest scientist does not prove that the original data was correct or that his or her theory is wrong.

However, Yuksel has now acknowledged that the errors of Khalifa are also his own errors; in fact, this particular one was done with his own hand. I fear for him, in this life and the next.

YUKSEL: Ditto. Thank you for your concern about my salvation.

LOMAX: One more point. If N is spelled out, why is it considered an initial letter?

YUKSEL: Would it be a word? If a letter is spelled out, it is "a spelled out letter."

CLAIM: 24

Three suras are prefixed with the initial "S" (Saad), and the total occurrence of the letter "S" in these suras is 152, or 19x8.

LOMAX: Deliberate deception. In his early publications, Dr. Khalifa shows chapter 7 as having 98 S's, 19 as having 26, and 38 as having 28. This adds up to the ZMN total of 152. He emphasized the miraculous nature of this total by pointing out the unusual spelling of "bsTatan" as "bSTatan" in 7:69 as necessary to preserve the miracle. (He made a similar point about the spelling of Makkah as Bakkah in 3:96.)

However, it was discovered that his total of 28 for chapter 38 was incorrect, the true total being 29. In order to preserve the miraculous total, it was necessary to eliminate a letter.

In his presentation of the count of S in VP, at 7:69, "bSTatan" has been altered to "bsTatan" by drawing the one letter in place of the other, reducing the count in chapter 7 to 97 and restoring the ZMN total. In this case, the change was done with sufficient skill as to be undetectable without comparing with another text.

Note also that there is a chapter initialed with S alone (chapter 38). If the chapter initialed with Q alone were ZMN in Q (as it is), and the chapter initialed with N were ZMN in N (as it is, only one short), then, if there is a pattern, we might expect the chapter initialed in S also to be ZMN in S. It couldn't be further away from that.

Neither of these textual changes alter the pronunciation, as the name of the letter N, which is how it is read, is NwN, and bSTatan was always pronounced bsTatan.

YUKSEL: The claim is true, the objection is based on suspicion, lack of knowledge, and fallacious argument. Since Lomax is coming with a very big accusation, I will give a relatively longer reply to this objection.

Dr. Rashad Khalifa put the Quran into the computer at Monsanto Company, while working as a biochemist, in 1969. He did not have a particular expectation regarding a mathematical code. At the time, he was interested in the meaning of "Huruf al-Muqatta" (Initial Letters) which were considered as mysterious by virtually all Muslim scholars. In 1973, he published the results of his research in a book titled "Miracle of the Quran: Significance of the Mysterious Alphabets" (Islamic Productions International, Inc. St. Louis, 1973. The book was revised and endorsed by Virgil I. Moss, Ahmad H. Sakr, Sulayman Shahid Mufassir, and Mujahid Al-Sawwaf). In this book, he pointed to some significant statistics, since at that time he was not aware of the hidden code 19. He had discovered that the frequency of initial letters were very high in chapters that they prefixed. For instance, the frequency of letter "N" in Chapter entitled "The Pen" (Al-Qalam) was 133 and was among the top six chapters with the highest percentage of letter "N". Another example: he found that the two chapters initialed by letter "Q" contain total 114 "Q"s matching the total number of chapters in the Quran.

In early counting there were some errors. Some were typing errors, and others were caused by vague or primitive method of counting. It is very interesting that in the earliest computer data (also in the book published in 1973) the count of the letter "N" in the chapter that starts with "NwN" was 133 (19x7). This was one extra from the manuscripts used and accepted by Dr. Khalifa. In other words, he got 133 by "mistake," years before he decoded that 19 was the common denominator and years before he noticed the traditional error in the spelling of the initial letter "N". Similarly, the frequency of the letter "S" (Saad) in three chapters containing this initial was 152. In fact, it was 153 in the manuscript he used.

The third "error" which led Rashad to the discovery of the pattern of 19 was in the count of "Allah" when he was not aware of the big problem generated by 9:128-129. These early "errors" actually helped us to see the miracle. With my own personal experience, I have no doubt that this was a divine wisdom and mercy; through our "mistakes," God, the possessor of infinite bounties, led to us to discover the truth and see the real "errors." Had we not committed those errors, neither Rashad nor the rest of us would have seen or accepted the mathematical miracle of the Quran.

As I mentioned above, our early count of "S" (Saad) despite our error was 152, and this number was a multiple of 19. None can claim that Rashad deliberately played with numbers in order to deceive people, since it is an unquestionable fact that he was not aware of the significance of 19 at the time of his early publication. If he had such a secret agenda, we should have witnessed such manipulations in the frequency of ALM (Alif, Lam, Mim) in those publication before 1974. Two publications, The Perpetual Miracle of Muhammad and Miracle of the Quran: Significance of the Mysterious Alphabet, raise a serious doubt for such an accusation.

We also repeated the error

In the discovery of the code 19, the importance of 152 was paramount. It was one of the few numbers that inspired Rashad to see the common denominator of the frequency of initials, that is, code 19. If it was 153, instead of 152 (19x7), probably Rashad would not have noticed the mathematical system. Even if he could, he would have had hard time to publish it.

This is not a mere conjecture, since I had almost the same experience. I repeated two crucial errors exactly as Rashad committed in his computerized work. I will not go to the details of how God's great signs prove His control and blessings in our life.

I was representing Muslim Turkish Students in an International Conference held in August 1980, when I first heard about the mathematical code of the Quran. Ahmad Deedat, president of the Islamic Propagation Center in South Africa, summarized Rashad's work with passion. It was a prophetic surprise for me since I had a political-personal experience with that number without knowing its importance in the Quran in 1979-1980 (FT/19). I decided to translate Deedat's book, *Al-Quran: the Ultimate Miracle*, after checking its argument and data. However, I was arrested few days after this conference because of my political activities and the publication of my articles promoting Iranian-style Islamic revolution in Turkey. The military regime sentenced me to six years in prison and two years of banishment.

In prison, I spend several months to check the data presented in Deedat's book. Since my childhood I was always questioning the things I hear or read. I was in a Turkish military prison when I verified Rashad's claims summarized by Ahmad Deedat. I used Fuad Abdulbaqy's concordance to check the frequency of words. I did not have a computer to verify the count of the letters. (Indeed, it was a luxury to have books in the prison at that time. Several months later, personal books and magazines were banned by prison administration.) I mobilized my inmate friends who happened to be the members of my political organization. I assigned the same page to five different groups who were able to read the Quran. We counted all of the initials except the letter ALM (Alif, Lam, Mim), since I knew the problematic nature of letter "A" (Alif). I would compare their count of letters verse by verse. If there was a difference we would re-count

212

and check the result. It is still a mystery for me how we also came up with 152 "S" (Saad) in the three chapters starting with that letter. Years later I was surprised when I learned that both Rashad's count and our's were one short. In retrospect, I see this "error" as a blessing from God, the Most Wise.

Otherwise, I would not have translated that book and would not have believed in the mathematical structure of the Quran. I kept the list of our letter count as a memento. Later, when Rashad discovered that his early counts were one short, I checked my list again. I found that we had forgotten to count the letter "S" (Saad) in verse 38:41. The compensating letter came from "*BasTata*" of 7:69.

If you check a Quran manuscript you will see that the word "*BasTata*" in verse 7:69 has a unique spelling: it is written with letter "S" (Saad) having a letter "s" (Sin) written on top of it. Furthermore, you may find in many manuscripts an interpretation written in tiny Arabic letters "yuqrau bil sini" (it is pronounced with "sin"). There was another, less popular interpretation of this peculiar spelling: the word could be written in two different spellings without change in its meaning. However, there were several hadith supporting the first interpretation. The mathematical code of the Quran finally solved this problem: The correct spelling of the problem word in 7:69 was with "s" (Sin), not with "S" (Saad).

Bankruptcy of traditional interpretation

Before noticing an extra "S" (Saad) in our count which was caused by an error in spelling of "*BasTata*" of 7:69 in current manuscripts, we found the unique spelling of "*BasTata*" interesting and shared the traditional interpretation based on several hadiths. According to a hadith narrated by Ibn Ebi Davud in Kitabul Masahif p. 108, and several other sources, Prophet Muhammad reportedly said: "Gabriel revealed this word specially spelled with letter "S" (Saad)." We found that this word with its unique spelling of "S" (Saad) had a significant meaning in a chapter that starts with combination of letters A.L.M.S. which includes "S" (Saad).

However, after the discovery of an extra "S" (Saad), Rashad suspected the very word that he initially found so interesting. I was shocked when I first noticed the correction made by Rashad in his book Visual Presentation of the Miracle. I think it was irresponsible of him not to indicate this important change with a footnote, probably with some explanation. This change shocked me for two reasons: I felt that Rashad was manipulating the text. Second, I had a lengthy evaluation regarding the important role of the word "*BasTata*" in the count of

213

letter "S" (Saad) with its unique spelling, in my book "Kuran En Buyuk Mucize" which became a best seller in Turkey for several years.

I went to the Sulaimaniya library and Topkapi Palace Museum to see the oldest available manuscripts. I was relieved when I saw in several manuscripts the word "*BasTata*" written with "s" (Sin), not "S" (Saad). Those manuscripts, including the Tashkent copy did not contain a "specially" spelled "*BasTata*." I took the photograph of the verse 7:69 of the Tashkent copy, and later presented it to Rashad. We put it in the first appendix of his 1989 translation as a historical support of the miracle. The mathematical system of the Quran had predicted a spelling error, and it was confirmed by early manuscripts.

Now, I can comfortably say this: The word "*BasTata*" in verse 7:69 of today's Quran manuscripts in circulation contains a spelling error. The hadiths defending this error is obviously a fabrication and they illuminate the early arguments on the spelling of this word.

The Deliberate Deception of the Supporters of "Saad"

According to both mathematical system of the Quran and the earliest manuscripts, the word "*BasTata*" should be written with letter "s" (Sin). Now let's speculate on what have happened in the past. The duplicated manuscripts of the Quran were not error free despite the extraordinary care of the scribes. Since there were no printing or photocopy machines it was eventual that human errors would creep in the text of copies of copies. One of the early copy of the Quran misspelled the word "*BasTata*" of 7:69 by writing it with "S" (Saad) instead of "s" (Sin). A careful proofreader or just an ordinary reader found the scribal error and corrected it by writing a tiny letter "s" (Sin) on top of the letter "S" (Saad). Unfortunately, later scribes who duplicated that copy did not realize the correction; they copied the word with both its spelling error and correction. Since then the word became famous with its peculiar spelling, that is, a "S" (Saad) carrying a "s" (Sin) on its top.

Of course, there were copies with correct spelling. Most likely an argument erupted between pro-Saad and pro-Sin groups. When those arguments caught public attention, the pro-Saad group resorted to fabricating hadiths to support their position. Finally, the pro-Saad group won the day and the Quran manuscripts containing spelling error in 7:69 became popular in Muslim world. Your Quran version most likely contains the extra "S" (Saad) which was supported by fabricated hadiths. Now, it is up to you to correct the spelling error, which is a product of deliberate deception, with an ink pen. This will show that you have witnessed the mathematical miracle of the Quran, the preservation

214

of the Quran by a divine system and you are not following your ancestors blindly.Furthermore, this will teach you how hadith books serve the worst deceivers in the history of the world.

LOMAX: Note that Yuksel is presenting information about the state of mind of Khalifa in 1969, long before he knew him, as if that were a fact known to him. This is a man who hates "hadith"!

YUKSEL: No Lomax, I will not let you get away with this accusation. As I have expressed in the beginning of this argument, the mathematical code of the Quran can be examined and witnessed independent from its discoverer or his early work. However, your personal accusation based on Rashad's early work forced me to mention the other side of the story. I reject "hadith" as a source of my religion. I do not invite anyone to accept the mathematical miracle of the Quran based on stories from Rashad. Nevertheless, if a person like you comes up with a personal accusation of "deception" regarding my close friend and teacher, then I am compelled to tell the truth of the matter. This is especially critical since, the accusation is coming from a person who repeatedly claims personal acquaintance with Rashad in order to gain credibility. No Lomax, you cannot falsely accuse my close friend and get away with it.

Besides, it can be inferred by ANY non-paranoid person that Rashad Khalifa did not have any particular expectation regarding a mathematical code. His early books and articles in magazines clearly indicate this fact. Especially, the errors in his early works are solid proof that he did not know about the code when he first printed his computer results. Thus, my defense of Rashad's state of mind is not a personal or subjective testimony, but it is a strong argument based on his early works published through various media, including periodicals.

LOMAX: Yuksel wrote, "The third "error" which led Rashad to the discovery of the pattern of 19 was in the count of "Allah" when he was not aware of the big problem generated by 9:128-129. These early "errors" actually helpedus to see the miracle. . ."

In other words, it was false data which led to the theory. Now, Khalifa said, in the last issue of Submitter's Perspective before his assassination, that "A False Messenger is a Messenger of Satan." Is this not a warning? Does this not indicate that one should check such information and the reasoning process used to analyze it very, very carefully?

YUKSEL: You intentionally confuse the expression "false messenger" with "unintentional personal errors." What a hideous tactic to falsify the truth!

LOMAX: Yuksel wrote, "As I mentioned above, our early count of "S" (Saad) despite our error was 152, and this number was a multiple of 19. None can claim that Rashad deliberately played with numbers in order to deceive people, since it is an unquestionable fact that he was not aware of the significance of 19 at the time of his early publication."

Yuksel has mistaken my comment of "deliberate deception" to mean that all of Khalifa's work was deliberate deception. No, I think that Khalifa himself was deceived. However, my reference was specifically to the alteration of the text in Visual Presentation of the Miracle without mentioning that it had been altered. He knew he was altering it; he knew that his readership would have been unhappy with alteration; so he did not mention the alteration. Yuksel was there. In this case, these "claims" came from Sam Khalifa, who is himself responsible for what he has claimed, since I would expect that he too knows these "details."

YUKSEL: Again Lomax is wobbling. Now he is trying to switch the meaning of "deliberate deception" with "self-deception." However, again, he goes back to his original accusation.

LOMAX: Yuksel wrote, "If he had such a secret agenda, we should have witnessed such manipulations in the frequency of ALM (Alif, Lam, Mim) in those publication before 1974. Two publications, The Perpetual Miracle of Muhammad and Miracle of the Quran: Significance of the Mysterious Alphabet, raise a serious doubt for such an accusation."

Yuksel is responding to a charge that I did not make. The comment "deliberate deception," I will repeat, refers to the publications of the Nuwn and Sad counts.

YUKSEL: My response is valid for your specific charges too. If a person is choosing "deception" as a method, then we should be able to see some traces of deception in his work where he could employ his deception easily. My knowledge of his work stops me from such an accusation. People sometimes make mistakes during their research and in its presentation. I can only blame his passion for "one of the greatest" miracles. Who is error free? If you were a contemporary of Muhammad, most likely you would have rejected him as a false messenger, since he committed serious errors. (22:52-55; 33:37; 80:1-12)

LOMAX: Yuksel wrote, "We also repeated the error In the discovery of the code 19, the importance of 152 was paramount. It was one of the few numbers that inspired Rashad to see the common denominator of frequency of initials, that is, code 19. If it was 153, instead of 152 (19x7), probably Rashad could not have noticed the mathematical system. Even if he could, he would have had hard time to publish it."

Yuksel has made this argument before, on the telephone. Essentially, it is an argument that a series of mistakes was the will of God. This is not to be denied. But Khalifa did not make mistakes just in his data. He also followed an implicit theory of probability, which was itself fallacious. He never faced the problem that it was possible to create a "miracle" like this through selection of data and analytical method. The history described above proves the point. When the data was found to be false, the theory which rested on the data was not changed. Rather, the data was re-analyzed to fit the data [theory? sic]. Any scientist who deals with statistical data is cautioned against this kind of analysis. They teach about this in first-year statistics.

YUKSEL: A person with one eye and not being able to see the 3-D holographic picture can accuse us of "self-deception" or remind us of first-year statistics course. The existence of the code is so obvious, even an ardent enemy such as Lomax is compelled to acknowledge that it serves as a "trial." What he does not realize is: this: it is he who is on "trial." This is "thought" in the first-verses of "The Spider" (Chapter 29 of the Quran).

LOMAX: This is not a mere conjecture, since I had almost the same experience. I repeated two crucial errors exactly like Rashad committed in his computerized work. I will not go to the details of how God's great signs proves His control and blessings in our life.

I also report similar experiences. In fact, I started studying the Quran again because it had been "proven" by Khalifa that it was literally the word of Allah. And I found miraculous patterns in the Quran that disappeared when the data was examined more closely.

YUKSEL: And Lomax later traded the Quran again with collection of hadith books. What an unfortunate circle.

LOMAX: Yuksel wrote, "I assigned the same page to five different groups who were able to read the Quran. We counted all of the initials except the letter ALM (Alif, Lam, Mim), since I knew the problematic nature of letter "A" (Alif). I would compare their count of letters verse by verse. If there was a difference we would re-count and check the result."

This is an excellent method; however, there are potential pitfalls, especially if those counting can read the words. It is possible that there is a particular word in which the spelling deceives people according to what would be a common misreading. But what Yuksel did was better than what Khalifa himself did. I would expect Yuksel to come up with far fewer errors than Khalifa, using the method he described. If this method were applied to create a computer-readable copy of the Quran, one could be pretty sure that it was without errors.

Yuksel wrote, "It is still a mystery for me how we also came up with 152 "S" (Saad) in the three chapters starting with that letter. Years later I was surprised when I learned that both Rashad's count and ours were one short. In retrospect, I see this 'error' as a blessing from God, the Most Wise." Or a trial. What Khalifa did, I am fairly sure, once he had developed the theory of nineteen-divisibility, was to keep looking for errors until counts came out to nineteen. Once he had a 19-divisible count, he stopped looking. This is why it took so long to find all the errors, if, indeed, they have all been found.

YUKSEL: Yes, as a "trial" too. Curiously, Lomax tends to ignore the other objectives of the mathematical code. What he sees is "Hell" and "Trial." Nothing else! But, the verse 74:31 mentions two positive objectives of the number nineteen. Why does Lomax keep forgetting them? The answer is simple: his meaningless 19 cannot function that way, except in case of few anomalies.

LOMAX: Yuksel wrote, "Otherwise, I would not have translated that book and would not have believed in the mathematical structure of the Quran [. . .]The mathematical code of the Quran finally solved this problem: The correct spelling of the problem word in 7:69 was with 's' (Sin), not with 'S' (Saad)." Yuksel does not know or does not mention that what he has said is according to the Hafs reading. Warsh has a Sad and reads it as Sad; in fact, it is difficult to pronounce a sin just before an emphatic Ta. It is actually more likely that the pronunciation as a sin was inauthentic, though it is one of the accepted readings.

YUKSEL: Lomax again is thumping on his Warsh version. Whenever he exhausts all the excuses to reject the miracle of the Quran he resorts to his Warsh version!

LOMAX: Yuksel wrote, "Bankruptcy of traditional interpretation Before noticing an extra 'S' (Saad) in our count [. . .] We found that this word with its unique spelling of 'S' (Saad) had a significant meaning in a chapter that starts with combination of letters A.L.M.S. which includes 'S' (Saad)."

Khalifa, before his error in counting Sad was known to him, used to cite the use of a Sad as one of the miracles: if the word had beenspelled as pronounced, the

218

19-divisibility would not have been preserved. When the error was found, he no longer made that claim.

YUKSEL: See the rest of my response.

LOMAX: Yuksel wrote, "However, after the discovery of an extra 'S' (Saad), Rashad suspected the very word that he initially found so interesting [. . .]The mathematical system of the Quran had predicted a spelling error, and it was confirmed by early manuscripts."

Confirmed by one early manuscript, yes. But, as we will see, there are other variations in those manuscripts which demolish other major pillars of Khalifa's theory, such as the count of "Allah." One again, we come back to one of the central problems: what is the Quran? If we can pick and choose pieces from manuscripts, we can choose them so as to create or amplify a numerical pattern. Together with variations in analytical method, this is a powerful tool for forging a "miracle," so powerful, in fact, that it can work to a certain extent without conscious intent.

YUKSEL: As I noted before, I did not just look at one manuscript, but several. Manuscripts are not error-free. A comparative study is required for questionable cases.

LOMAX: Yuksel wrote, "Now, I can comfortably say this: The word '*BasTata*' in verse 7:69 of today's Quran manuscripts in circulation contains a spelling error. The hadiths defending this error is obviously a fabrication and they illuminate the early arguments on the spelling of this word."

It would be necessary, though not quite sufficient, before claiming this, to establish that the pattern of 19-divisibility is necessary. But the Khalifites have never done the work to demonstrate that. Instead, they rely on bluster, as Yuksel is not doing.

YUKSEL: First, Lomax knows that I reject to be called a Khalifite. Perhaps, he is calling me this way to justify his being a Muhammedan. A psychological projection! Lomax ignores that this particular case is one of the examples that establishes the pattern of 19-divisibility. Lomax is repeating this for each individual pattern that he is forced to accept as correct. Here is the caricatured demonstration of how his argument works.

Premise:	There are 4 main features of "X." They're A, B, C, D.
Objection:	Though they are basic features, there are some others. According to "W" version there are 5 or 6 features.
Claim:	A is divisible by 19.
Objection:	Anecdotal. You did not establish a pattern yet.
Claim:	B is divisible by 19.
Objection:	Anecdotal. You did not establish a pattern yet.
Claim:	C is divisible by 19.
Objection:	Anecdotal. You did not establish a pattern yet.
Claim:	D is divisible by 19.
Objection:	Anecdotal. You did not establish a pattern yet.

He fragmentizes or isolates each parameter of the pattern and rejects it in the name of that pattern. What is evident is that Lomax has established a pattern of denial with the mantra, "you did not establish a pattern yet."

LOMAX: Yuksel wrote, "The Deliberate Deception of The Supporters of 'Saad' [. . .] Since then the word became famous with its peculiar spelling, that is, a 'S' (Saad) carrying a 's' (Sin) on its top."

All of this neglects the fact that the word is *pronounced* in Hafs with a sin. The Quranic manuscripts were only aids to recitation. However, it is certainly possible; in fact it is certain, that there have been spelling errors or variations in texts of the Quran. In this case, it is more likely that the sin of the Tashkent was idiosyncratic. Yuksel examined a copy of the Tashkent text. There is another reputed Uthamic [Uthmanic, sic] recension at Topkapi. What does it show for this verse? What do other very early copies show? Yuksel's analysis is reported from a point of view of one who already knows the "correct" answer; he also apparently does not know that Warsh has no "sin" written above the letter.

YUKSEL: I did not check the copy in Topkapi Museum which has fake blood on it. But, I checked several other ancient manuscripts and found that they confirm our prediction based on the mathematical code of the Quran. If your Warsh version does not have "sin" on the suspected word, it only shows that it is not corrected by a careful editor.

LOMAX: Yuksel wrote, "Of course, there were copies with correct spelling. Most likely an argument erupted between pro-Saad and pro-Sin groups. When those arguments caught public attention, the pro-Saad group resorted to fabricating hadiths to support their position. Finally, the pro-Saad group won the day and the Quran manuscripts containing spelling error in 7:69 became popular in Muslim world. Your Quran version most likely contains the extra 'S' (Saad) which was supported by fabricated hadiths."

Now Yuksel, without any appropriate evidence, is stating that hadith are fabricated. Perhaps he is right. But how is he so certain? What, precisely, is the method by which this "numerical code" can be used to "correct" the text? Of course, it would first be necessary to prove that the code is real. How, precisely, can we know that?

YUKSEL: Again Lomax repeats his pattern with different wordings: you did not establish a pattern yet. I will not repeat my answer anymore.

LOMAX: Yuksel wrote, "Now, it is up to you to correct the spelling error, which is a product of deliberate deception, with an ink pen. This will show that you have witnessed the mathematical miracle of the Quran, the preservation of the Quran by a divine system and you are not following your ancestors blindly. Furthermore, this will teach you how hadith books serve the worst deceivers in the history of the world."

We already knew that hadith can be incorrect; this is not even controversial. Surely Yuksel is drawing a much bigger conclusion than is justified from all the evidence seen so far. He has not, in 24 claims, shown anything more than a few interesting number correlation. None of these correlation warrant the conclusion that there is a numerical code sufficiently pervasive to allow correction of the text. His comment here is merely a lapse into his habitual polemic and diatribe against hadith. One must look for emotional causes to understand this. It is not in the data, nor in any logical or rational analysis.

YUKSEL: Thanks to Lomax for his psychological analysis. Have you seen a bald doctor prescribing medicine for baldness? Or have you seen someone who lives in a glass house stoning others?

CLAIM 25

Sura 36 is prefixed with the initials "Y.S." (Ya Seen) and the total occurrence of these two letters in this sura is 285, or 19x15.

LOMAX: True, interesting. If I may accept totals not rejected by Philips [op.cit.], this total is accurate, and one of three such suras where the initial letter totals are definitely ZMN. If we did not know the totals, we could predict that, about one time out of thirty, such a coincidence would occur (there is a probability of .035 that three or more events would occur, each of which has 1 chance in 19 of occurring, out of a set of 28 trials.) Coincidences of this level are everyday experience.

YUKSEL: True and more than interesting. As we had demonstrated and we will continue to do so, the mathematical system of the Quran is well beyond probability.

LOMAX: There has not been one single demonstration in this discussion of a result which is wildly improbable! Yuksel is, again, lapsing into habitual polemic. This phrase, "Beyond Probability," is from a book by Abdullah Arik, who has successfully demonstrated in it his utter ignorance of statistical theory. Yuksel is not showing any better understanding.

YUKSEL: Lomax demonstrates the wisdom of knowing that the probability of using the words "beyond" and "probability" together without quoting from a book is zero! ☺ Who is demonstrating utter ignorance of statistical theory? (By the way, I do not necessarily endorse all of Abdullah Arik's calculations).

CLAIM 26

The letters K (Kaf), H (Ha), Y (Ya), 'A ('Ayn), and S (Saad) occur in the "K.H.Y.'A.S."- initialed sura (Chapter Mary) 798 times, or 19x42.

LOMAX: True, interesting. See Fact (claim) 25.

YUKSEL: True, and more than interesting. See claim 24, 25 and the following claims.

LOMAX: Note that this claim is a "within the sura" count. Some counts are like that, some require combination across suras. It is really necessary to look at tables showing all the combinations, and then to apply the different methods of analysis, consistently, to these tables. Whenever I have done this, I have found the number of 19-divisible statistics to be within normal variation. But by varying the analytical method with each "fact," it is possible to find some "interesting" result for each initial letter or initial letter combination. This is the functional anatomy of a miracle.

YUKSEL: There is no varying analytical method, except in the count of T.S.M. which I will mention later. Lomax fails to see the simple method we use. All chapters that share the same initials are counted together. For instance, there is a SINGLE letter "S" that starts the Chapter 38, thus, we count all the cases where this single letter is an initial letter. The same with H.M. and other initials... Since

K.H.Y.'A.S. of Chapter 19 is not repeated anywhere it is consistent to count it in that chapter alone.

CLAIM 27

The total occurrence of "H.M." (Ha Mim) in their 7 suras (40-46) is 2147, or 19x113.

LOMAX: Probably true, interesting. I have counted a few of these suras. The count given requires the inclusion of the invocation each time (i.e., each letter in the invocation is counted seven times), contrary to the "Important Rule" announced earlier by Dr. Khalifa, and still followed with his word counts. Further, his totals have changed from earlier reports. Always including the invocation for each chapter (the invocation has 2 H and 3 M):

	c. 1976 (old xerox)			c. 1980 (SNC)			1982 (VP)
count of H:							
sura 40	64	+0	→	64	+0	→	64
sura 41	58	-10	→	48	+0	→	48
sura 42	53	+0	→	53	+0	→	53
sura 43	45	-1	→	44	+0	→	44
sura 44	16	+0	→	16	+0	→	16
sura 45	31	+0	→	31	+0	→	31
sura 46	37	-1	→	36	+0	→	36
count of M:							
sura 40	389	-9	→	380	+0	→	380
sura 41	276	-2	→	274	+2	→	276
sura 42	308	-9	→	299	+1	→	300
sura 43	317	+6	→	323	+1	→	324
sura 44	145	+5	→	150	+0	→	150
sura 45	200	-1	→	199	+1	→	200
sura 46	227	-2	→	225	+0	→	225
totals	2166	-24	→	2142	+5	→	2147
totals modulo 19:							
	0			14			0

It is apparent from these figures that Dr. Khalifa's early work was extremely sloppy. My suspicion is that, initially, he merely counted the letters once and accepted his numbers if they produced the results he wanted. If they did not, he recounted until he had found sufficient errors to produce the desired result. The

223

large drop in the counts from the first to the second publication is odd. He may have changed some critical measure. At the second publication, he still hadn't gotten it right; some letters were still overlooked. I am aware of no errors in the third publication.

In doing the counts myself (I have counted two of these chapters), depending on my state of mind, I have made as many as 10 errors in counting a chapter, always errors of omission. My procedure was as follows:

I xeroxed the pages of the Quran in question. I marked with distinctive colors the letters under consideration. I then checked my counts by comparing them with Dr. Khalifa's, page by page totals (in VP). If a count disagreed, I reexamined the page, looked for missing letters.

I estimate as high as one chance in 30 of missing a letter by this method, and a much lower chance of counting non-existent letters. The chance that two persons counting will miss the same letter is perhaps 1 in 900, although it could be argued that some letters are particularly easy to miss (such as the m in bma), especially for a non-Arabic speaker like myself.

Ultimately the way to produce solid, reliable counts is to have multiple persons count the chapters, reporting totals by page, carefully examining any discrepancies, proving that each discrepancy from the final count was produced by an error. Even better (and easier in the long run) would be to have multiple persons key the Quran into a computer, then to compare the results letter by letter, which can easily be automated, producing a reliable database.

Given that many persons have critically examined Dr. Khalifa's work, it may be reasonable to accept the counts in VP. On the other hand, he produced so many other relatively easily discoverable errors that it may be that the other researchers simply gave up at this point.

Most significant is the fact that with all three sets of data he claimed ZMN perfection in the total. How did he do this in SNC? Simple (actually, not so simple; in fact, quite clever). He included the invocation once only, with the count of H and excluded it entirely (violating his rule) from the count of M. This results in the removal of 6x2, or 12, H's and 7x3, or 21, M's. The original remainder of 14, less 12, less 21 results in a ZMN total of 2109.

In other words, he created the HM total in SNC by a combination of errors.

YUKSEL: The claim is true and more than interesting. Though Lomax feels obligated to accept this fact in his first statement with a dubious "probably," he tries his best to undermine this fact by speculating on early errors of Dr. Khalifa's work. After a lengthy demonstration of how he is careful and meticulous and how Dr. Khalifa was sloppy, he ends up with an entirely negative remark. An "interesting" fact in his first statement transforms to "a combination of errors" in his last statement. Instead of accepting the truth he tries his best to loose it among early errors committed by a human like him. (By the way, Lomax made an error while correcting Dr. Khalifa's earliest report. The count of M for chapter 45 was correct and the correction -1 was not needed.)

The question is simple: How many H and M letters are there in chapters starting with the initials of H and M? Through our research and observation we have learned that unnumbered *Basmalah*s participate in the counting of initial letters. We consistently include them in the counting of letters. However, in the count of words we observe a clear pattern that requires the exclusion of unnumbered *Basmalah*s. This method is consistent and also meaningful in distinguishing unnumbered verses from numbered verses.

Furthermore, I would like to introduce a recent discovery made by Milan Sulc regarding the frequency of HM letters in seven chapters initialized by them:

The frequency of H (Ha) and M (Mim) letters in those 7 chapters are 2147, that is 19 x 113. The probability of this fact is 1/19. There are only four subgroups whose combinations result in a multiple of 19. (Two of them depend on the other two). This can be seen as an arbitrary arrangement or coincidence, since there are 112 possible combinations out of the seven chapters. However, if you add the digits of each frequency you will end up with 113, which is exactly equal to the multiplication factor. This is valid also for all the subgroups. The digits of the individual occurences of these letters in the individual suras always add up exactly to the coefficient of the number 19 in the total group as well as four determining sub-groups, as shown below:

Chapter	$\hat{H}$	$\hat{M}$	Absolute Values											SUM
40	64	380	6	+	4	+	3	+	8	+	0	=	21
41	48	276	4	+	8	+	2	+	7	+	6	=	27
42	53	300	5	+	3	+	3	+	0	+	0	=	11
43	44	324	4	+	4	+	3	+	2	+	4	=	17
44	16	150	1	+	6	+	1	+	5	+	0	=	13
45	31	200	3	+	1	+	2	+	0	+	0	=	6
46	36	225	3	+	6	+	2	+	2	+	5	=	18
TOTAL	2147(19x113)		26	+	32	+	16	+	24	+	15	=	113
		▲	D_1		D_2		D_3		D_4		D_5		▲

Now let's study each subgroup of seven chapters where the combined frequencies of ḤM are divisible by 19. In the previous pages, we learned that in each of the four subgroups the frequencies of ḤM were divisible by 19. Please note the multiplication factors of 19. For instance, note that the factor in the first group containing 1121 ḤM letters is 59, since 1121 = 19x59. The multiplication factor in the second group is 54, in the third 55, and in the fourth 58.

Chapter	Ḥ	M	Absolute Values									SUM
40	64	380	6	+	4	+	3	+	8	+	0	21
41	48	276	4	+	8	+	2	+	7	+	6	27
42	53	300	5	+	3	+	3	+	0	+	0	11
TOTAL	1121 (19x59)		15	+	15	+	8	+	15	+	6	**59**
		▲	D_1		D_2		D_3		D_4		D_5	▲
43	44	324	4	+	4	+	3	+	2	+	4	17
44	16	150	1	+	6	+	1	+	5	+	0	13
45	31	200	3	+	1	+	2	+	0	+	0	6
46	36	225	3	+	6	+	2	+	2	+	5	18
TOTAL	1026 (19x54)		11	+	17	+	8	+	9	+	9	**54**
		▲	D_1		D_2		D_3		D_4		D_5	▲
41	48	276	4	+	8	+	2	+	7	+	6	27
42	53	300	5	+	3	+	3	+	0	+	0	11
43	44	324	4	+	4	+	3	+	2	+	4	17
TOTAL	1045 (19x55)		13	+	15	+	8	+	9	+	10	**55**
		▲	D_1		D_2		D_3		D_4		D_5	▲
40	64	380	6	+	4	+	3	+	8	+	0	21
44	16	150	1	+	6	+	1	+	5	+	0	13
45	31	200	3	+	1	+	2	+	0	+	0	6
46	36	225	3	+	6	+	2	+	2	+	5	18
TOTAL	1102 (19x58)		13	+	17	+	8	+	15	+	5	**58**
		▲	D_1		D_2		D_3		D_4		D_5	▲

227

The probability of this phenomenon for each subgroup is 1/9. Therefore, the combined probability of these phenomena occurring altogether is 1/(19 x 9 x 9 x 9 x 9 x 9), or 1/1,121,931.

I hope Lomax will not rush into falsely labeling this extraordinary mathematico-literary pattern as a mathematical property.

LOMAX: Yuksel wrote, "The claim is true and more than interesting. Though Lomax feels obligated to accept this fact in his first statement with a dubious "probably," he tries his best to undermine it by speculating on early errors in Dr. Khalifa's work."

I said "probably," because I have not verified the counts personally, and I know of no non-Khalifite who has verified the numbers (and, really, only one Khalifite: Yuksel himself.) But I do consider it likely that the counts are accurate, as far as they go. The speculation about errors becomes relevant when one realizes that, through a host of errors, one thing remained constant: Khalifa always reported a multiple of 19 in his counts.

YUKSEL: Lomax should learn another lesson: never use ALWAYS. He is not aware that his passion for denial of the miracle of the Quran leads him to exagerate and falsely accuse the proponets of the miracle.

Dr. Khalifa did not always report a multiple of 19 in his counts. For instance, he did not claim that the number of all verses in the Quran is multiple of 19 until he discovered that the last two verses of Chapter 9, Ultimatum was not part of the original Quran. Again, he did not claim that the sum of the verse numbers that contain the word God is multiple of 19 until that discovery. He discarded his claim regarding the frequency of 9 words in the last two verses of Chapter 9. (See Muslim Perspective, March 1985.) As with many human, he had sometimes rushed and commited errors and inconsistencey in counting.

LOMAX: Yuksel wrote, "After a lengthy demonstration of how he is careful and meticulous and how Dr. Khalifa was sloppy, he ends up with an entirely negative remark. An 'interesting fact' in his first statement transforms to 'a combination of errors' in his last statement."

Yuksel has completely misunderstood what was being discussed. The "combination of errors" reference was not to the final figures presented later, and which Claim 27 is about, but to the numbers shown in a paper from Khalifa called Secret Numerical Code, or SNC. This was fully referenced in the original paper.

Yuksel wrote, "Instead of accepting the truth he tries his best to loose and hide it among early errors committed by a human like him. (By the way, Lomax made an error while correcting Khalifa's earliest report. The count of M for chapter 45 is correct and the correction -1 is not needed.)"

The correction of -1 refers to what is necessary to change the original number to the number in SNC. If one looks at the table, one will see that this is balanced by a later change of +1 to get the number in *Visual Presentation*. I made errors in the writing of the original paper, but this was not one of them.

YUKSEL: Thanks for this clarification; but, we were dealing with Claim 27. Why should we realy care? Are we here discussing how Rashad did this or that or are we verifying the claims ourselves? I thought we were doing the latter. Either you are incapable of doing your own research, or you are so obsessed with Rashad that you keep looking for a particular spelling or a data-entry error in his work. Were we conducting a historical or archeological study here? By sifting through the early draft of Rashad's work, you managed to escape from the present issue, which is the claim number 27! You just demonstrated a perfect example of redherring!

LOMAX: Yuksel wrote, "The question is simple: How many H and M letters are there in chapters starting with the initials of H and M? Through our research and observation we have learned that unnumbered *Basmalah*s participate in the counting of initial letters. We consistently include them in the counting of letters. However, in the count of words we observe a clear pattern that requires the exclusion of unnumbered *Basmalah*s. This method is consistent and also meaningful in distinguishing unnumbered verses from numbered verses."

However, Khalifa did not always follow this rule. Further, there are other rules which must be stated, otherwise one would not know why this particular series of chapters was chosen. Yuksel has still not responded to the concerns in the Draft FAQ: 19, Study Problems, apparently preferring to focus on the old writing in my paper, written over two years ago, and before a lot of new data was available to me, including the Tashkent copy of the Quran. Yuksel claims that the method is consistent. If this is true, why does he not simply state the method used to determine that there is a miracle here? In other words, answer the questions in the Draft FAQ. But he knows, I think, that such a document would expose how complex the selection criteria are: I assert that they are complex enough to account for the patterns which they reveal.

YUKSEL: Lomax is not telling the truth regarding Rashad's treatment of Bismillahs in the count of words and letters. The only exception is the word

229

"BSM" and we have explained the possible reason behind this exception. Here we are talking about a simple counting of two simple letters: H.M. Lomax's criticism is not relevant here.

I have stated the counting method. Instead of criticizing my method, Lomax repeats his chronic complaint: "If this is true, why does he not simply state the method used to determine that there is a miracle here?"

Lomax pontificates: "There are other rules which must be stated, otherwise one would not know why this particular series of chapters was chosen." What rule? Is it a difficult thing to infer that we count initials together in the chapters where they are found? This miracle is neither for morons, nor for extreme skeptics.

LOMAX: Yuksel wrote, "Furthermore, I would like to introduce a recent discovery made by Milan Sulc regarding the frequency of HM letters in seven chapters initialized by them: The frequency of H (Ha) and M (Mim) letters in those 7 chapters are 2147, that is 19 x 113. The probability of this fact is 1/19." I have suggested to Mr Yuksel that he study probability theory. His statement of probability does not state the experimental conditions sufficiently to judge that probability.

I could say that the probability that a random number is divisible by 19 is 1/19. But if the number has been preselected by someone, who is only presenting multiples of 19, I could say that the probability that such a number would be divisible by 19 is very close to 1. So the history of how the number came to be the number under consideration is very, very relevant.

YUKSEL: At first I suspected that the discovery was a mere mathematical property. I was almost sure about it. However, when I checked it extensively, I realized that it is not a pattern that I know. Therefore, I took it for evaluation to a professor in number theory at the University of Arizona. He worked on it for a while, and finally found a formula for it I lost the formula, but it was based on number nine.

[After this debate, I found the formula: $9 (D3) = (D1 + D4) + 2(D2 + D5)$, where D stands for the sum of each digits in the frequency of H and M in the set. For instance, let's apply the formula to the frequency of main set of H.M. initialed 7 chapters:

Chapter	H	M	Sum of the digits
40	64	380	6 + 4 + 3+ 8 + 0
41	48	276	6 + 8 + 2 +7 + 6
42	53	300	5 + 3 + 3+ 0 + 0
43	44	324	4 + 4 + 3 + 2 + 4
44	16	150	1 + 6 + 1 + 5 + 0
45	31	200	3 + 1 + 2 + 0 + 0
46	36	225	3 + 6 + 2 + 2 + 5

TOTAL 2147 (19x113)   26+32+16+24+15 = 113

$$9 (D_3) = (D_1 + D_4) + 2 (D_2 + D_5)$$

D1 = 26
D2 = 32
D3 = 16
D4 = 24
D5 = 15

$$9 (16) = (26 + 24) + 2 (32 + 15)$$

This formula similarly works for all other four subsets.]

LOMAX: Yuksel wrote, "There are only four subgroups whose combinations result in a multiple of 19. (Two of them depend on the other two). This can be seen as an arbitrary arrangement or coincidence, since there are 112 possible combinations out of the seven chapters."

Now, I am glad that Yuksel mentions the number of possible combinations, because the very fact that there are such subgroups has been cited by other Khalifites as a miracle in itself, when, given a total divisible by 19, it is actually completely normal that there would be such subgroups.

Yuksel wrote, "However, if you add the digits of each frequency you will end up with 113, which is exactly equal to the multiplication factor [. . .] Therefore, the combined probability of these phenomena occurring altogether is 1/(19 x 9 x 9 x 9 x 9 x 9), or 1/1,121,931."

This probability is overstated. The variables are not independent variables, for there are relationships between numbers and the sum of digits, and the relationships are stronger when the numbers involved are divisible by 19. For example, if the sum of a series of seven numbers is divisible by 19, it is not unlikely that the multiplier of 19 will be equal to the sum of digits of the individual numbers, when the numbers are within certain ranges. It would take more time than I now have to give to this problem to show why, and to quantify this. But note that the probability is stated without any substantive comment as to how and why it is appropriate to multiply the probabilities. Multiplication of probabilities is only valid when each event is independent, and sums of digits are not independent from the digits. I'll give a more concrete example. Take any series of numbers, the sum of which is divisible by 19. Take the sum of digits of the series of numbers. If the result has two or more digits, add them together, and repeat this process until there is only one number remaining as the result of the process. Do this same process with the sum of the series. The single digit resulting will be identical with that from the sum of the individual numbers. And it will also be identical with the repeated sum from the multiplier of 19. These are not independent variables.

YUKSEL: You are explaining a well-known property of the number 19. Knowing this property, initially, I discarded the pattern as a mathematical property. However, my later examination showed that they are independent variables. Lomax could examine it for himself by changing the numbers while keeping their sum as multiple of nineteen. After a score of trials, he will relaize that he has rushed to the conclusion despite my warning.

LOMAX: Now, because there is more than one way to come up with this sum, it is not rigidly determined that the non-repeated sum is identical. But if the orginal numbers are of a certain average size, and the list is of modest length, it is not particularly unlikely. I do not have time for a rigorous analysis here.

YUKSEL: If this is not "gobbledygook" nothing is.

LOMAX: Yuksel wrote, "I hope Lomax will not rush into falsely labeling this extraordinary mathematico-literary pattern as a mathematical property." But there are certainly mathematical properties involved. Yes, the pattern is interesting, like most of Milan Sulc's work. But it is not conclusive. For one thing, once it is really true that computers are being used to find these patterns, it becomes less and less remarkable that one can find a way to analyse a body of data that produces results like this. But a really good code will not have a probability of one in a million of being produced by chance: more likely, the probability will be smaller than one in the number of atoms in the universe. What is the probability that the letters in this paragraph came together merely by chance? Honestly, we might as well say that it is zero; the difference between it and zero is insignifigant.

YUKSEL: I expected from Lomax to study this particular pattern. However, he is not facing it. Rather, he hides behind his rhetoric, speculations and generalizations.

CLAIM 28

The total of "'A.S.Q." in their sura (42) is 209, or 19x11.

LOMAX: Probably true, interesting. Dr. Khalifa has been faulted for separating 'A.S.Q. from the preceding HM, but it is not an unreasonable analysis, as they are separately numbered verses. The effect of this choice is merely to reduce the significance.

YUKSEL: True, more than interesting if evaluated with other patterns of the mathematical system. Again, Lomax inserts a sly "probably" in order to reduce its significance.

LOMAX: As before, "probably" is inserted because the count has not been independently verified. That does NOT reduce the signifigance; signifigance is reduced by idiosyncratic selection criteria. In this case, there is not other split set of initial letters such that one could know whether or not a pattern exists. Thus the phenomenon is anecdotal. What is missing from Yuksel's analysis of the initial letters is a comprehensive statement which covers all initial letter analyses. In fact, the analyses vary with the initial letter under analysis. It is precisely this variation which makes it possible to assert the possibility that these counts have been deliberately chosen for 19-divisibility, out of a population of possible ways of counting which are equally as reasonable.

YUKSEL: I repeated several times a comprehensive statement that covers all the cases except one (T.S.M.). And I strongly believe that the exception is due to our incomplete and erroneous count. A more accurate count will show that the method of counting of initials is comprehensive.

Lomax is a very good example of how a good mind can be blinded with sectarian agenda and personal vendetta. I really feel sorry for him, since he confuses the simplest facts and cannot see the clearest pattern.

Have you noted the one whose god is his ego? Consequently, God sends him astray, despite his knowledge, seals his hearing and his mind, and places a veil on his eyes. Who can guide him, after such a decision by God? Would you not take heed? (45:23)

CLAIM 29

The initials "A.L.M." (Alef Lam Mim) occur in 6 suras, and the total occurrences of the three letters in 6 suras are as follows: [6 ZMN totals for chapters 2, 3, 29, 30, 31, and 32 omitted)

LOMAX: Unverifiable, at least partly false. Dr. Khalifa's method of counting A (alif) is not based on any known or stated system. Sometimes upright fetha is counted, sometimes not. Sometimes hamza is counted (as A -- which is arbitrary, @ would make more sense); , sometimes not. Because upright fetha and hamza are common, it is quite difficult to untangle his counting rules. By varying slightly the rules, it is possible to manipulate the counts.

Philips [op.cit., p. 21] points out that Dr. Khalifa has missed an L (Lam) in 30:21. He further shows how the hamza in "Lain" has been counted in 3:158 and not counted in exactly the same word (and written the same) in 30:51.

Dr. Khalifa changed his counts of A, L, and M greatly over the years:

Count of A
Sura	2	4592	32 →	4624	-122 →	4502
Sura	3	2578	21 →	2599	-78 →	2521
Sura	7	2572	18 →	2590	-61 →	2529
Sura	10	1353	5 →	1358	-39 →	1319
Sura	11	1402	10 →	1412	-42 →	1370
Sura	12	1335	12 →	1347	-41 →	1306
Sura	13	625	-1 →	624	-19 →	605
Sura	14	594	10 →	604	-19 →	585
Sura	15	503	9 →	512	-19 →	493
Sura	29	784	9 →	793	-19 →	774
Sura	30	545	22 →	567	-23 →	544
Sura	31	348	12 →	360	-13 →	347
Sura	32	268	0 →	268	-11 →	257

Count of L
| Sura | 2 | 3204 | -2 → | 3202 | 0 → | 3202 |
| Sura | 3 | 1885 | 6 → | 1891 | 1 → | 1892 |

Sura	7	1523	5 →	1528	2 →	1530
Sura	10	912	1 →	913	0 →	913
Sura	11	788	3 →	791	3 →	794
Sura	12	812	-1 →	811	1 →	812
Sura	13	479	1 →	480	0 →	480
Sura	14	452	0 →	452	0 →	452
Sura	15	323	0 →	323	0 →	323
Sura	29	554	0 →	554	0 →	554
Sura	30	396	-3 →	393	0 →	393
Sura	31	298	-1 →	297	0 →	297
Sura	32	154	1 →	155	0 →	155

Count of M

Sura	2	2195	-1 →	2194	1 →	2195
Sura	3	1251	5 →	1256	-7 →	1249
Sura	7	1165	-3 →	1162	2 →	1164
Sura	13	260	0 →	260	0 →	260
Sura	29	347	-3 →	344	0 →	344
Sura	30	318	-1 →	317	0 →	317
Sura	31	177	-4 →	173	0 →	173
Sura	32	158	0 →	158	0 →	158

Somehow it still continues to astound me that Dr. Khalifa never wavered from his claim that his numbers proved the perfect preservation of the Quran, even when he began changing the Quran to make his results neater. Each change in the numbers brought a new method of analysis which recovered the "miracle."

In the light of the above figures, I look at a statement in the April 1990 Submitters Perspective: "The Quran's mathematical composition is so vast and intricate that we do not expect these people who cannot even count simple letters to go beyond the opening verse in the Quran."

Who can't count letters? Who publishes extravagant claims before doing the work to back them up? This fact, if verifiable, by itself, would bring the Nineteen to the edge of certainty. It is the simplest rule that produces the most hits. With facts 30-32, we have all thirteen A initialed suras producing (it is claimed) ZMN totals within each sura. The chance of this occurring without a code or other underlying cause is about one in 42 followed by 15 zeros.

Unfortunately, the most likely explanation is manipulation of the count of A (Alif) to produce the ZMN totals. In fact, the very simplicity and depth of this claim, compared with many of the others, is evidence that the "Miracle" is indeed an artifact of the investigation process. Where is not necessary to resort to arbitrary combinations (because the letter counts themselves can be arbitrarily altered due to the ambiguities around A), no such combinations are needed to find ZMN.

YUKSEL: I have always found the count of A.L.M. letters problematic. However, I do not think that it is unverifiable. By a comparative and analytical study of earlier manuscripts, I believe, it is possible to obtain an accurate count of Alifs. I strongly believe that the count of ALM of six chapters will produce a ZMN total. I do not expect them to have ZMN totals within each sura, since the pattern seen in other initials does not require it. Dr. Khalifa became aware of the errors after they were indicated by critics. We were planing an extensive research on the count of A (Alif). However, he could not fulfill the plan since Sunni fanatics brutally assassinated him in January 1990. Inshallah, one day we will fulfill this research.

LOMAX: Yuksel wrote, "I have always found the count of A.L.M. letters problematic."

Yuksel is the only dedicated Khalifite who has ever admitted this to me. (Milan Sulc believes in numerical coherency in the Quran, but is not, most definitely, a Khalifite.) But this is obvious to anyone who actually tries to count alif and verify Khalifa's counts.

YUKSEL: Lomax, somehow, is sparing Milan Sulc from his derogatory label. This way, he put himself in a position of a fair judge.

LOMAX: Yuksel wrote, "However, I do not think that it is unverifiable. By a comparative and analytical study of earlier manuscripts, I believe, it is possible to obtain an accurate count of Alifs."

Having examined the Tashkent copy, which is one of the two earliest known manuscripts of the Quran, I am very skeptical. For one thing, it appears that the early manuscripts were not identical. For sure, the Tashkent copy does not confirm Khalifa's counts. As Yuksel has noted, hamza was not used in the Tashkent Quran, and it was variation in the count of hamza which allowed Khalifa to assert 19-divisibility. Is it not odd that one finds the most perfect pattern of 19-divisiblity, a pattern which repeats itself across many Suras with no change in the apparent rule for counting, where it was possible for Khalifa to modify his counts pretty much at will?

236

YUKSEL: After seing the problem in the count of A.L.M., Lomax does not hesitate to call it as "the most perfect pattern of 19-divisibility" I wonder whether he would say similar things about the counts of H.M letters or K.H.Y.'A.S. or Q, or Y.S., if some errors were found in them.

LOMAX: Yuksel wrote, "I strongly believe that the count of ALM of six chapters will produce a ZMN total. I do not expect them to have ZMN totals within each sura, since the pattern seen in other initials does not require it. Dr. Khalifa became aware of the errors after they were indicated by critics."

Khalifa's method was generally to count and keep recounting until the count came out to a multiple of 19. Then he would publish. But Yuksel has essentially acknowledged that this pattern has not yet been verified. And this makes the claims that it is proof of a miracle false and misleading.

Yuksel wrote, " We were planing an extensive research on the count of A (Alif). However, he could not fulfill the plan since Sunni fanatics brutally assassinated him in January 1990. Inshallah, one day we will fulfill this research."

Someday, someone will find a way to combine fragments of the early manuscripts with more modern texts in such a way that a pattern of 19 will be reported with more substance. But since there is no way to authoritatively determine what the original manuscript was for the Quran (there were, it is very likely, variations between the 'Uthmanic copies in the writing of alif, which often does not affect pronunciation -- which makes it very difficult to check), this will mean nothing.

This is why an overall pattern, combining all occurrences of an initial letter set, is such a poor coding. It is true that it is very sensitive to errors: even one error destroys the pattern. But a good coding will remain even if single bits are destroyed; enough of the pattern will be left that the original message can be reconstructed. Now, this possibility is truly exciting, and it is what attracted me initially to the "miracle of the nineteen." But it is not that kind of pattern, and if it is, it must be said that it has been applied by people who do not understand the laws of probability and the danger of creating the pattern out of selective analysis.

YUKSEL: A good lecture. And it is the right place to repeat the rhetoric once more. After witnessing Lomax' attitude towards the clearest and basic facts I really do not much hope that he can ever see the miracle of the Quran.

CLAIM 30-32

The initials of seven suras initialed with ALR, ALMS, or ALMR, occur in their respective suras ZMN times.

LOMAX: Unverifiable (30-32), false (31). See Fact 29. Note that Dr. Khalifa's totals for R and S have changed over the years, and, as mentioned above (Fact 24), he has altered the count of S in chapter 7 by changing the text.

YUKSEL: Verifiable (30-32). Again, a comprehensive comparative research on older versions can verify the count of these initials. The "alteration" in the count of S in chapter 7 is in fact a correction directed by the mathematical system and supported by the oldest versions of the Quran. The peculiar spelling of the word "*BasTata*" of 7:96 in modern versions is another evidence for the justification of this correction. It is Lomax and the majority of Muslims who have a version of the Quran with an altered text! We have discussed this issue in our defense on claim 24 in detail.

LOMAX: Yuksel's statement that these claims are verifiable is a mere insertion and shows no appreciation for the difficulty. Further, it is dishonest to claim the counts of Sad in ALMS without noting the textual variation which is followed. IF one has shown that there is a pattern in initial letter counts which is outside of normal statistical variation, then it might be possible to assert that the use of S in the modern texts is an error. But Yuksel has exaggerated the case, I suspect. I know of one ancient text which has a Sin in the place of Sad in the passage in question. It is possible that others exist; but it is my understanding that there are other ancient texts with a Sad there. This is a textual variation; it is not the only such variation.

YUKSEL: This is the only textual variation that is acknowledged in a unique way: an idiosyncratic spelling of the word in the versions that contain the error. This peculiarity explains everything regarding the variation. Somehow Lomax is unable see this clear sign.

LOMAX: Yuksel wrote, "The peculiar spelling of the word "*BasTata*" of chapter 7 in modern versions is another evidence for the justification of this correction. It is Lomax and majority of Muslims who have a version of the Quran with an altered text! We have discussed this issue in our defense on claim 24 in detail.

I have the Tashkent Quran showing this variation. But it also has other variations which demolish the Khalifite claim of 2698 occurrences of Allah in the Quran. Once again, if one can pick and choose pieces out of texts, it becomes possible to create, deliberately or otherwise, this fairly fragile "miracle." Further, the texts of the Quran, quite obviously, are not what was protected by Allah, for they all contain variations. Rather, it is something else. Perhaps it is the recited Quran, though variations exist with that, as well. My own view is that it is the message which is eternally preserved; individual letters are not that important. The reports that the Prophet (SAS) allowed recitation in dialect confirm this view....

YUKSEL: I do not want to repeat my response. There are only few instances that we referred to ancient copies of the Quran, and with a good reason. Lomax, as usual, exaggerates it.

Lomax, with his theory of the Quran, cannot defend any claim regarding the preservation of the Quran. It is a historical fact that the original Quran, the Quran written by Muhammad disappeared while Muhammad's grave, cloth, teeth and hair were well preserved! The original Quran reportedly was burned by Marwan, the Umayyad ruler. If someone asks: "how do you know that this verse or that verse is from the Quran?" what will be his answer? "My friend, or father-in-law recites it this way?"

CLAIM 33

The initials "H" (Ha), "T.H." (Ta Ha), "T.S." (Ta Seen) and "T.S.M." (Ta Seen Mim) occur in the suras (19, 20, 26, 27 & 28) a total of 1767 times, or 19x93.

LOMAX: Arbitrary, anecdotal. First, it should be noted that the "Ha" here is different from the "Ha" in "Ha Mim." Dr. Khalifa and his followers use a non-standard system of transliteration which obscures some of the differences. My system is also non-standard, but is designed to allow the distinction of letters. Here we are discussing "h", "Ha Mim" uses "H".

Second, and perhaps most important, there is no visible logic to the lumping together of these particular initials, except that the result is ZMN. Suras 20, 26, 27, and 28 at least share one common initial as the first initial. But "Ha" is not an isolated initial. Sura 19 is prefixed with KHY'AS. Why count h and not KY'A, and S? Why include Sura 19 at all?

Third, I note again that the counts of these letters changed repeatedly over the years, yet each change simply brought a new method of analysis to find a ZMN total. Either the invocation was included with each sura, or not included at all, or just included once for the total and not for each sura. The suras were combined in different ways as well.

YUKSEL: The claim appears to be arbitrary and anecdotal as Lomax said. I wish I had time and resources to count those letters using the earliest manuscripts. Lomax is unfair by criticizing Dr. Khalifa for non-standard system of transliteration. He could see that Dr. Khalifa distinguishes Arabic letters in VP (Visual Presentation).

LOMAX: Dr Khalifa was not "criticized" for using a non-standard system. It was merely mentioned to point out that the letter h and H were not the same. As I have pointed out, my own system is also non-standard, being designed to allow ASCII characters to designate unique Arabic letters, without using digraphs.

YUKSEL: I found your system much more obscure and complicated.

CLAIM 34

When we add the sura number, plus the number of its verses, plus the sum of verse numbers in each sura, we find the grand total for the 29 initialed suras is 190122, or 19x10007.

LOMAX: True, arbitrary. This Fact should be considered together with Facts 18 and 35.

YUKSEL: The claim is true, but arbitrary. Adding sura numbers, plus the number of its verses, plus the sum of verse numbers in each sura is an arbitrary arrangement, since it is one out of many possible arrangements.

LOMAX: Once again, Yuksel is unique among Khalifites for acknowledging much of what is clearly true. But he does not see the degree to which arbitrary selection of counting methods penetrates Khalifa's claims.

YUKSEL: We will see together inshallah!

CLAIM 35

The same calculations for the 85 un-initialed suras produces 156066, or 19x8214.

LOMAX: False if unqualified. If Claim 34 and 35 are true then Claim 18 is true. (ZMN plus ZMN always equals ZMN.) Therefore Fact 18 is redundant.

Using the Medinah edition, this Claim is only true if one or more verse counts are changed. One possibility is to eliminate any two verses from chapter 9. Presumably this is what has been done. However, there are many other ways to make this total come out ZMN. The elimination of one verse from any chapter having a number of verses which is 11 modulo 19 (has a remainder of 11 when divided by 19) will produce this result, in particular, 37:182, 52:49, 62:11, 67:30, 89:30, 93:11, 100:11, and 101:11. The same result, of course, would appear if a verse were merely combined with the preceding or following verse, as many are in recitation. Thus there are 335 places where a small change which might have been invisible to the main guardians of the Quran (the reciters from memory) would have produced a ZMN total.

Similarly we could add a verse (or split a verse, which is also common in recitation) where there are 5 modulo 19 verses. 4:176, 53:62, 59:24, 97:5, 105:5, 111:5, and 113:5 fit this criterion, or another 282 verses.

The addition of a verse from 21:112, 54:55, 83:36, 86:17, or the elimination of a verse from 8:75, 39:75, 49:18, 64:18, 74:56 will have no effect on the remainder modulo 19. Basically, the statistic in this Fact is a poor guardian against change. Most of the other statistics are sensitive to changes of one word or one letter or one verse, this one is not (and the same is true for Fact 18 and Fact 34).

What is hidden by the complexity of this statistic is that it is simply dependent on the pattern of verses. Why is the statistic not simply the sum of the verse totals? That answer is obvious: because it does not come out zero modulo 19. So our diligent researchers simply look for another measure that does. There is no end to this.

YUKSEL: I agree with Lomax that the statistic in this claim is a poor guardian against change. However, we should always consider all examples as a whole. Though one of the three claims is redundant, the chance for the two cases to produce ZMN total is 1/361. The three claims can be summarized in one claim: "When we add the sura number, plus the number of verses, plus the sum of verse numbers in each sura, we find the grand total for the 29 initialed suras 190122, or 19x10007. The same calculations for the 85 un-initialed suras produces 156066, or 19x8214." Again, I see an arbitrary selection of combinations in this calculation.

LOMAX: Once again, I will object to the statement of a probability without stating the experimental conditions. It is true that there is one chance in 361 for two independent random numbers to both be divisible by 19. But both these numbers have been selected out of a much larger universe of possible ways of counting suras and verses, and, further, the sum of verse numbers is not independent of the number of verses.

YUKSEL: Lomax has a point here. As I expressed earlier, this claim does not have significance.

LOMAX: It follows that if the original number is divisible by 19, and it is also divisible by 2 (i.e., it is divisible by 38), then the sum will also be divisible by 19. Another way of putting this is that half of all numbers which are divisible by 19 will show an integer sum which is also divisible by 19.

YUKSEL: This is not true. For instance, 2698 is divisible by 38, that is, 19 and 2, but the sum of its integers are not divisible by 19.

CLAIM 36

The numbers mentioned in the Quran are 1, 2, 3, 4, 5, 6, 7, 8, 9, 10, 11, 12, 19, 20, 30, 40, 50, 60, 70, 80, 99, 100, 200, 300, 1000, 2000, 3000, 5000, 50000 and 100000. The sum of these numbers is 162146, or 19x8534.

LOMAX: Anecdotal. I have not verified this statistic. It is not clearly specified. Is the sum the sum of the numbers mentioned, each one being counted once? (No, that does not equal 162146.) Does it include all uses of the dual, which is the normal way in Arabic to indicate that there are two of something? In fact, most nouns clearly indicate whether one is speaking of one, two, or more than two, so the singular is definitely present in the language as well. This is not an easy statistic to check without the kind of publication that Dr. Khalifa produced in VP.

YUKSEL: The claim is interesting, and significant if other related cases are considered. I have verified this statistic. Each number is counted once and they add up to 162146. Lomax most likely omitted the number 50000 while trying to verify (indeed, falsify!) this claim, since I corrected his quotation. The claim is not vague and cannot be clearer. The numbers have nothing to do with dual or plural forms of words. This is one of the easiest statistics to check, since it is based on the sum of all the different numbers mentioned in the Quran. To falsify this claim you need only to show a Quranic number which is not mentioned in the claim, or vice versa.

I would like to mention here another claim made by a Syrian author which I have not verified yet. All the numbers mentioned in the Quran, with repetitions, occur 285 times, that is, 19x15. The total of all these 285 numbers is claimed to be also multiple of 19.

LOMAX: I checked my copy of Submitter's Perspective from which this claim was taken. 50,000 was also missing from the list there. With the restoration of this number to the list, the sum is indeed 162146. The claim that it is significant if "other related cases are considered" is empty unless the principle for relating cases is stated.

YUKSEL: Well, I checked mine and saw 50,000 was not missing (Muslim Perspective, September 1988). Probably, Lomax has another issue, or just has obtained a Warsh version of Muslim Perspective ;-) Here, I would like to quote a comment made by one of the followers of the debate:

> Sadruddin wrote: "Br. Yuksel has put it very aptly. These numbers do add up to 162146. I do not understand how br. Lomax can say otherwise. If it were untrue, one only needs to point out which number does not appear in the Quran, or state any other number that does."

LOMAX: Yuksel misquoted my original list of numbers, which was taken from the list of claims printed in Submitter's Perspective. That original list was missing one of the numbers. Yuksel did note that he had "corrected" my list, but then he speculated that the error was mine and made out of eagerness to refute. But I specifically stated that I had not verified the list.

Many, perhaps most, of the numerical claims turn out to be true if the method of counting is properly qualified, in other words, if the choices which might be made in counting are precisely specified.

YUKSEL: Again another "wobbly" argument. Lomax appears to be accepting the numerical structure. He just wants us to specify things (some of the elementary facts too) more clearly.

LOMAX: What must be understood, however, is that this particular claim is anecdotal; i.e., it is not part of a clear pattern.

YUKSEL: Again, Lomax parades with his pattern: "it is not part of a clear pattern." Aren't we examining the importance of the number 19 in the mathematical system of the Quran? What can be clearer than to check the numbers mentioned in the Quran? What is more reasonable than to examine their total?

LOMAX: There are many different ways to count the occurrences of numbers in the Quran. This particular way, the brother has verified as totaling a multiple of 19. Probably most of the other ways do not. This actually proves nothing.

YUKSEL: Again, Lomax is talking about "different ways" to count numbers in the Quran. How many different ways? We witnessed how he counted the letters of Bismillah in different ways, and we saw how he counted the number of chapters in different ways! Obviously, he can come up with many different, but false ways of counting things.

Yes, the number of all numbers is important and it is not multiple of 19, since there are 30 non-repeated numbers in the Quran. However, there are only 8 different fractions mentioned in the Quran. The total of all numbers, without repetition again is divisible by 19.

LOMAX: One can find a practically infinite number of 19-divisible statistics in any large collection of random numbers. But to qualify as a clear code, there would need to be something very different from these anecdotal statistics. For example, if every sura initialed by the initial letters, contained those letters an exact multiple of 19 times, then there would be a very strong evidence. Now, there are several ways to understand what I have written, and by applying different understandings, one might be able to say, yes, there are these multiples: but I am referring to a single method of counting.

Khalifa did report such a pattern with the suras involving alif and a few others; if these reports were verifiable, the kind of evidence needed to assert a code with confidence would have already been discovered. But the alif counts are not verifiable, for reasons which have already been presented in detail. As to the other suras, it is true for Sura Qaf; for Nun, it is necessary to use a non-standard spelling. Khalifa also reports 19-divisibility for the intial letters of Sura 19. If this latter report is true, there are two Suras that have the pattern reported in the received Hafs Madina text. This is not statistically significant. Nor would be particularly significant.

YUKSEL: It is incredible to see how Lomax craftily ignores the pattern in initials of Chapter 36 (Y.S.), of seven chapters starting with H.M., the three chapters starting with letter S., etc. As I have explained earlier, all the counts of these chapters follow the same method of counting and are divisible by 19. We should listen to the warning of our Creator, the Lord of the Day of Judgment:

> 2:42   Do not confound the truth with falsehood, nor shall you conceal the truth, knowingly.

CLAIM 37

Why 19? Nineteen is the Gematrical value of the Arabic word "WAHD" which means ONE. "WaHd" has a Gematrical value of W(6) + A(1) + H(8) + D(4) = 19.

LOMAX: False. Causation has not been established. The question "Why 19?" is essentially the question asked by the unbelievers and those in whose hearts is a disease (74:31), and the reasons are given there. This is not one of them. Further, the value of this word depends on how it is spelled. If it is waHd, then, indeed, the value is 19. But if it is spelled wHd (with upright fetha on the w), then the value is 18. Now, the edition of the Quran that Dr. Khalifa has reproduced in VP (which generally matches Madinah) shows the word as wHd at 41:6, 42:8, and 38:65. I haven't seen it spelled as waHd in his text. But it appears from his counts of "A" that he may be counting the upright fetha as "A". Once again, what text is he using?

When Dr. Khalifa's critics count upright fetha in claiming that the invocation doe not have only 19 letters, his follower says "these people ... cannot even count simple letters." From 2:13, "When it is said to them: "Believe as the people believe, they say, shall we believe as the fools believe? Are they not indeed the fools? But they do not know."

YUKSEL: The Abjad (Gematrical) value of "waHd" is 19, if it is spelled with "a" (Alif). Lomax is right in his criticism regarding the different spelling of this word. In order to answer this criticism conclusively I need to have a research on the topic. The Gematrical value of "*waHd*" (One) being 19 is a common knowledge for both Arabs and Jews. Here I would like to quote from Dr. Cesar Adib Majul, former dean of Institute of Islamic Studies, University of Philippines:

> "Incidentally, that 19 is the numerical value of Wahid in terms of the Abjad is well known to persons knowing both the Abjad and the "Names" of Allah, especially those enumerated in the so-called "Ninety-nine Names of Allah." Such knowledge has also been in the possession of many persons in parts of the Muslim world who applied it for astrological, magical and other superstitious purposes. Relations between the Abjad and Names of Allah can also be found in Western literature of the last century. (See for example, the article on "*Da'wah*" in Thomas Patrick Hughes, Dictionary of Islam, London, 1885, pp. 73-76; and *Ja'far Sharif, Islam in India: Qanun-i-Islam*, translated by G. A. Heklots, pp. 255-258.) To some students such correlation was

245

simply utilized to distinguish such names in terms of their numerical values. For example, J. Redhouse used the Abjad to distinguish the Name *Wajid* (numerical value of 14) from *WaHid* (numerical value of 19). (See J. W. Redhouse, "On 'The Most Commonly Names,' i.e., The Laudatory Epithets, or the Titles of Praise bestowed on God in the Quran or by Muslim writers." *Journal of the Royal Asiatic Society.* Volume XII, 1880, p. 65). But all this is a matter of numerals; it was Dr. Khalifa who first pointed out that the number 19 is not just a number serving merely as a base for the mathematical structure of the Quran, but that it has a reference to the Quranic argument that Allah is One." (*The Names of Allah in Relation To the Mathematical Structure of Quran*, Cesar Adib Majul Ph.D., Islamic Productions, Tucson, 1982, p. 13).

The "causal relation" between 19 and *WaHid* (One), is very clear for me. Dr. Majul's work establishes the relationship between 19 and God's attributes with a very strong argument. I will give an example of that argument at the end of this defense, God willing.

Lomax does not see any relation between the number 19 and God's attribute "*WaHd*." Obviously, Lomax is not taking the entire Quran in regard of the meaning of 74:31. A careful study of the Quran puts God's oneness in a very close relation with the objectives mentioned in 74:31. Those unbelievers and hypocrites who block themselves from understanding the meaning and implication of 19 after it is revealed by God; they are the ones who are not happy with ONE GOD as the ONLY source of power and religion. Those who are not satisfied with God and His word ALONE, those who associate fabricated medieval narration (Hadith and Sunna) and sectarian jurisprudence issued by Muslim clerics (ulama) are handicapped from seeing this great mathematical structure. To witness the relation between 19 and the authenticity of the Quran and its prime message, that is, oneness of God, requires some qualifications. The Quran repeatedly mentions those qualifications.

LOMAX: Yuksel wrote, "The Abjad (Gematrical) value of "waHd" is 19, if it is spelled with "a" (Alif). Lomax is right in his criticism regarding the different spelling of this word. In order to answer this criticism conclusively I need to have a research on the topic. The Gematrical value of "waHd" (One) being 19 is a common knowledge for both Arabs and Jews."

Yes, but the point is that this is not the Quranic spelling of the word. Rather, it is based on the non-Quranic language. I might add, however, that, long before I heard this argument from Khalifa as to why 19, I was asked the same question by Abdul Aziz Said at American University in Washington, D.C. My answer was, immediately, "wahid." I then checked the abjad values, and, indeed, they came out to 19. I did not even know all the abjad values at that time, and I still do not. I had to look them up.

YUKSEL: I am cautious about making such a universal claim regarding the spelling of "waHd" before an extensive and comperative study.

LOMAX: I will not repeat the quotation from Dr. Majul, whom I knew from Tucson before I knew Dr. Khalifa, and whom I also recently had the pleasure of seeing again. I will say that he was absolutely opposed to the kind of extreme conclusions to which Khalifa took his work on the 19. Like many, he initially supported Dr. Khalifa's work, then warned Khalifa of extremes of interpretation, extremes which were not supported by the data. The 19 in the Quran is quite interesting; regardless of what subsequent research may show. But we did not suspect that Khalifa had falsified data, and it appears that he did, whether deliberately or not is open to interpretation. The strongest case may be made with the alif data, but it can also be seen that he was less than forthcoming in many other situations where his data was based on special interpretation or textual variation. We have already discussed this in the case of the extra Nuwn at the beginning of Sura Qalam. It was altered in Visual Presentation (by Yuksel, he now tells us), without any notice that there was anything unusual. In the context of a stand-alone "proof" of the miracle, this is nothing less than deceptive.

YUKSEL: I know Dr. Majul too and occasionally exchange ideas with him. He is cautious by nature and feels free to criticize Rashad whom he still calls as "ustad" (teacher). He communicated with Rashad until his assassination. Dr. Majul is well convinced about the miraculous mathematical code.

The real "deception" is trying every means possible to fight against "one of the greatest" signs of God.

CLAIM 38

The Word "WAHD" (ONE), referring to God occurs 19 times.

LOMAX: False. I find the following 21 occurrences: 2:133, 2:163, 4:171, 5:73, 6:19, 9:31, 12:39, 13:16, 14:48, 14:52, 16:22, 16:51, 18:110, 21:108, 22:34, 29:46, 37:4, 38:5, 38:65, 39:4, and 41:6. Now, I haven't checked every one of them. I notice that 3 have a closing "A", which I understand does not change the meaning in the least: 2:133, 9:31, and 38:5. I don't see any way to come up with 19.

YUKSEL: The claim is false if the word had been counted with a consistent method used for the count of other specific words, such as, Raheem (Merciful=114), Yawm (Day=365), etc. There are 22 occurrences of the word "WaHd" (One) in the Quran. Lomax has missed the occurrence in verse 40:16. Here I want to congratulate Lomax for not having problem in identifying and counting a word correctly. Had he employed the same method in other cases such as in the count of *Rahim* he would not have problem with Claim 3. Evidently, Lomax deviates from this authentic method whenever it confirms the mathematical structure of the Quran, and he employs it whenever he sees that it does not bring the number 19, the number that he tries hard to escape from. In our language we call this kind of behavior as double standard. Some use double standard to increase the number of mathematical pattern, while Lomax does the same thing to blind himself to the physical facts. Where is the objectivity?

By the way, the frequency and gematrical values of divine attributes demonstrate a marvelous mathematical pattern. We have demonstrated it in "The Prime Argument" a two-round debate with Dr. Carl Sagan. That pattern would have been destroyed if the occurrence of "Wahid" were 19.

LOMAX: Yuksel is, simply speaking, making polemic where there is only caution. What I have consistently said is that Khalifa did not follow a consistent counting method. In some cases, I could not find a method of selection which would give Khalifa's results, in which case I simply reported all possible forms known to me. Yuksel now calls this the 'authentic method,' but earlier he claimed other standards, for example with counting the first word in the bismillah. I'd say the first word in the bismillah is either bi or bism. Is it not?

YUKSEL: As I stated before, the word "BSM" is the ONLY exception, and I have cited some possible reasons for that exception. Lomax, intoxicated with his agenda is not ready to reflect on it.

"Bi" is not a word. It is the pronounciation of letter "B", the second letter in Arabic Alphabet. Lomax needs to study elementary instead of advanced Arabic. Remember, that he was not able to identify Arabic letters.

LOMAX: Yuksel wrote, "In our language we call this kind of behavior as double standard. Some use double standard to increase the number of mathematical pattern, while Lomax does the same thing to blind himself to the physical facts. Where is objectivity?"

248

Indeed. What is the objective standard? That is the very question raised by the Draft FAQ, which, so far, Yuksel has not answered, instead occupying himself with this old paper. By the way, the full original paper is still available on request by e-mail from me, which is the only way it has ever been distributed, except where some persons uploaded it to AOL or Compuserve file archives. I still find the paper useful as a record of my inquiry into the 19, but it is by no means free of errors or complete.

YUKSEL: I have answered this before.

LOMAX: Yuksel wrote, "By the way, the frequency and gematrical values of divine attributes demonstrate a marvelous mathematical pattern. We will give some examples of this in the end of this first round debate, God willing. That pattern would have been destroyed if the occurrence of 'Wahid' were 19." There is no end to claims. So far, not one has produced anything more than a mild coincidence, on close examination. What Yuksel has mentioned is at the core of this hoax. If one count does not work out, another can be found which does. Since there is no practical limit to the number of ways in which the text can be analysed, there will be more and more of these "amazing facts." And some of them will be quite amazing, indeed. But as long as the probabilities are down in the millions and billions or so, it only means that some cleverness has been exercised in the search process. Now, if it were possible to verify that alif count....

YUKSEL: Whenever Lomax fears that dealing with an impressive pattern will put him in trouble, he resorts to his apparently sound general rhetoric. The pattern observed in the triple relation among the frequency of the words in Bismillah and the frequency and gematrical value of the attributes of God is enough by itself to show a "simple to understand and impossible to imitate" mathematical system in the Quran.

> 6:104 Enlightenments have come to you from your Lord. As for those who can see, they do so for their own good, and those who turn blind, do so to their own detriment. I am not your guardian"

CLAIM 39

The word "Alone" (Wahdahu), when it refers to God, occurs in (7:70, 39:45, 40:12 & 84, and 60:4); the sum of these numbers is 361, or 19x19.

LOMAX: False. The usage in 17:46 is also in reference to God, as "rbk" (your Lord). The arbitrariness of this statistic is extreme. What we have is the numbers of the suras in which the word occurs, plus the numbers of the verses. In spite of the fact that there are two occurrences in sura 40, 40 is only counted once in the sum. Why? Why do you think?

YUKSEL: The claim is true. The usage in 17:46 is NOT a reference to God. The verse is commonly mistranslated by the followers of Hadith and Sunna, since the verse clearly rejects other sources besides the Quran as the source of divine law. This verse is one of those miraculous verses that is a self-proof for its own claim. Those who are not certain about the hereafter are blocked from understanding the Quran and witnessing its miracle.

Muslim clergy, who promote mishmash collections of primitive fabrications that were compiled two centuries after Prophet Muhammad, have distorted the meaning of many Quranic verses. Lomax is referring to those mistranslations that conform to pagan medieval Arab culture disguised in Hadith books. For those who want to investigate, I will write the transliteration of the original followed by its correct translation. I will explain, God willing, why the translation of orthodox Muslims is false.

> 17:46 Wa jaalna 'ala qulubihim akinnatan an yafqahuhu wa fi azanihim waqra. Wa iza zukkirat rabbuka fil QURANI WAHDAHU wallaw 'ala adbarihim nufura.

The translation:

> 17:46 We place shields around their minds, to prevent them from understanding it, and deafness in their ears. And when you commemorate your Lord in the QURAN ALONE, they run away in aversion.

The orthodox translation of the phrase in question is "when you commemorate your LORD ALONE in the Quran..." As you can see we differ regarding where to place the adjective "wahdahu" (alone). The Arabic text uses the adjective after the Quran, not right after the word Lord. In order to make this argument clear for those who do not know Arabic, I will kick the ball to Lomax: "If you want to translate our translation to Arabic how would you say it?" Please consult those who know Arabic without reporting this debate. First ask them this question: "How can you translate this phrase to Arabic: 'when you commemorate your Lord in the Quran alone'?" Then, ask them to translate 'when you commemorate your Lord alone in the Quran."

I believe this is the right time to demonstrate Lomax' real problem: he does not accept God alone as the source of his religion. He cannot trust God when He repeatedly asserts that the Quran is detailed, clear, and is sufficient as the source of divine guidance. He chooses man-made fabrications as his second source of religion and criticizes us for not accepting them. This wrong choice is what makes him blind and deaf to the mathematical miracle of the Quran. It is the same Lomax who is an extreme skeptic in regard of mathematical system of the Quran, and the other hand is extremely gullible in accepting volumes of books full of nonsense, contradictions and lies attributed to Muhammad, the final prophet. In order to demonstrate his real problem and motives, I will quote some narration from Sunni's most authentic hadith books which Lomax has defended by any means possible on Internet.

"The Prophet never urinated in standing position" (Ibn Hanbal 6/136,192,213). "The prophet urinated in standing position" (Bukhari 4/60, 62). "A group from the Ureyneh and Uqayleh tribes came to the prophet and the prophet advised them to drink urine of camels. Later on, when they killed the prophet's shepherd, the prophet seized them, gouged out their eyes, cut their hands and legs, and left them thirsty in the desert" (Bukhari 56/152, Ibn Hanbal 3/107,163). "Moses was scared by the angel of death, thus Moses slapped him and blinded one of his eyes". "I am the most honorable messenger, on the day of the judgment only I will think of my people" (Bukhari 97/36). "Do not make any distinction among the messengers; I am not even better than Jonah" (Bukhari 65/4,5; Ibn Hanbal 1/205,242,440). "Bad luck is in the woman, the horse, and the home" (Bukhari 76/53). "If a monkey, a black dog or a woman passes in front of a praying person, his prayer is nullified." (Bukhari 8/102; Ibn Hanbal 4/86). "The prophet gave permission to kill children and women in war" (Bukhari, Jihad/146; Abu Dawud 113). "The earth is carried on a giant bull; when it shakes its head an earthquake occurs" (Ibni Kathir 2/29; 50/1). "Leaders have to be from the Quraish tribe" (Bukhari 3/129,183; 4/121; 86/31). "You shall kill all black dogs, because they are devils" (Ibn Hanbal 4/85; 5/54). "God is the time" (Muwatta 56/3). "To prove His identity, God opened his legs and showed the prophet His thigh." (Bukhari 97/24, 10/129 and the comment on the Sura 68.) "The parchment that the verse about stoning to death for adultery was written on was eaten and abrogated by a goat." (Ibni Majah 36/1944; Ibni Hanbal 3/61; 5/131,132,183; 6/269). "A tribe of monkeys arrested an adulterous monkey and stoned it to death, and I helped them" (Bukhari 63/27). "When the prophet died his armor had been pawned to a Jew for

251

several pounds of barley." (Bukhari 34/14,33,88; Hanbal 1/ 300; 6/42,160,230). "The punishment for cutting the fingers of a woman is to pay her: 10 camels for one finger, 20 camels for two fingers, 30 camels for three fingers, and 20 (twenty) camels for four fingers" (Ibn Hanbal 2/182; Muwatta 43/11). "The prophet had been bewitched by a Jew, and for several days he did not know what he was doing" (Bukhari 59/11; 76/47; Ibn Hanbal 6/57; 4/367). "Muhammad possessed the sexual power of 30 men" (Bukhari). "Do not eat and drink with your left hand, because Satan eats and drinks with the left hand" (Ibn Hanbal 2/8,33). "The prophet said:'Do not write anything from me except the Quran. Whoever wrote, must destroy it" (Muslim, Zuhd 72; Ibn Hanbal 3/12,21,39). "The prophet ordered Amr Ibn As to write everything that he speaks" (Ibn Hanbal 2/162). "Omar said: Quran is enough for us, do not write anything from the prophet" (Bukhari, Jihad 176, Gizya 6, Ilim 49, Marza 17, Magazi 83, Itisam 26; Muslim, Wasiyya 20,21,22).

Here are more from Bukhari and Muslim, two most popular hadith books:

"The intelligence and the religion of women are incomplete." "If a monkey, a black dog, or a woman passes in front of a praying person, his prayer is nullified." "To find a good woman among women is similar to finding a white crow among a hundred crows." "The marriage commitment is a kind of slavery for women." "If anybody has been required to prostrate before others beside God, the woman should prostrate before her husband." "I have been shown the dwellers of hell; the majority of them were women." "If the body of the husband is covered with pus and his wife licks it with her tongue, she still will not be able to pay her debt to him."

Yes, Lomax is rejecting the correct translation of 17:46 in order to save the hadith compilations that endorse these kinds of hearsay reports. I cannot understand how one can preach righteousness and at the same time promote these primitive lies attributed to God and His messenger.

LOMAX: I have looked at a whole series of Muslim and non-Muslim translations of this verse. Khalifa is the only one who translates it as referring to the Quran. I have been aware of his interpretation for a long time. It sounds very wrong to me, for reasons which I have explained elsewhere. But my point, as it is relevant here, is that an occurrence is omitted based on an arbitrary consideration, and, further, that the counting method is also arbitrary in other

252

ways. Normally, Yuksel admits this. But he is not doing so here, because this particular count is part of a centerpiece of Khalifite theology.

YUKSEL: The same Lomax, who proudly claims that he knows Arabic and that it is easy to read the Tashkent copy written without dots and vowels, is resorting to those translations which distort the meaning of many Quranic verses in order to make them compatible with Hadith. Why did not Lomax answer my translation by his own advanced Arabic? The verse is very short and has a very simple Arabic. Unfortunately, translations of the Quran authored by those who do not believe in the divine assertion that the Quran is complete, detailed, clear and the only source of God's religion, are full of distortions. Even non-Muslims are deceived by today's corrupt teaching. They consider hadith as the second source besides the Quran. I have written a book in Turkish on the topic: "Errors in Turkish Translations."

For those who are not scared to see common blunders of those translations, I will give our translation of a verse. As an example:

> 22:15 If anyone thinks that God cannot support him in this life and in the Hereafter, let him turn completely to heaven (God), and sever (his dependence on anyone else). He will then see that this plan eliminates anything that bothers him.

This verse is difficult to understand, unless its immediate context and Quran's overall message is studied. Now, please read "a whole series of Muslim and non-Muslim translations of this verse," and appreciate Dr. Khalifa's solitary translation. They attribute an absurd and ridiculous challenge to God Almighty. (For a collection of various translations please see the Pakistani scholar Al-Maududi's commentary of the Quran: "Tafhim-ul Quran")

For other examples, please see my response to *Signs Magazine*.

LOMAX: [trash about Hadith and other irrelevancies omitted]

YUKSEL: Wow! Lomax is trashing my quotations from his second source of guidance. Lomax is probably embarrassed to see the glaring primitive stories next to his brilliant and scientific analysis. However, they are very relevant, since they shape his mind and block his vision. I wish he had trashed them forever.

I will end this argument with one of our discussion on Internet: "Why Trash all Hadiths?" Hadith is a sneaky virus in the mind of Muslims, and it is very relevant to avoid it for the sake of our salvation.

SUMMARY OF FINDINGS

(BY DANIEL LOMAX)

Of 39 "Simple Facts," only 16, that is, numbers 2, 4, 5, 6, 13, 14, 19, 20, 21, 22, 25, 26, 27, 28, 33, and 34, are clearly true. The most impressive of the "Facts" turn out to be either unverifiable or clearly false. Of the verified facts, I cannot say that the level of signifigance is sufficient even to warrant further investigation. If there is a code verifying the accurate preservation of the Qu'ran (or disproving it, as, ultimately, Dr. Khalifa was led to claim), it has not been discovered.

A code consists of more than just a number used in its calculations. Primarily, it is a method of calculation or translation. No consistent method was followed in the analyses behind the "39 Simple Facts." Such imprecision as to analytical method destroys the signifigance of the relationships which may be discovered, as one learns in the elementary study of statistics.

Dr. Khalifa's claims, at best, fall into the category of pious fraud. At worst, he was a messenger of Satan and a conscious deceiver: quoting the title of an article in the last issue of Submitters Perspective with his name on the masthead (March 1990), "A False Messenger is a Messenger of Satan." But I do not agree with the article. Like most of his writing, it is replete with non-sequiturs and false premises.

The new, additional miracles claimed by Dr. Khalifa's followers, so far, are uniformly anecdotal. It can be expected that with continued effort, many such statistics will be discovered. In fact, eventually I would expect some truly amazing "coincidences" to be found, especially if others continue to search. Of course, if Facts 29-32 are ever demonstrated, or the like of them, the whole matter will require rexamination.

One further consideration occurs to me, which I have never seen stated elsewhere. The Quran is a clear message; it is easily recognized as the message of God by those whose hearts are free from obstruction. Had God intended the Quran to carry a code verifying its perfect preservation, he could have done it much more effectively and simply than the complex, arbitrary, and inconclusive "code" claimed by Dr. Khalifa. I would expect the code, once discovered, to be as clear as the book itself.

SUMMARY OF FINDINGS

(BY EDIP YUKSEL)

I would like to thank Lomax for articulating the view of those who reject the role of number 19 as a code of the mathematical structure of the Quran; his criticism gave us the chance to clarify the issue for those who need to hear both sides. I also congratulate him for his courage to discuss the issue publicly, which most of the Muslim scholars try hard to avoid it. It is unfortunate that his criticism is a mixture of reasonable criticism blended with blind refutation, a refutation by any means possible. We occasionally witnessed a topsy-turvy method of evaluation which reminds me the joke about a critic who finds an empty flower pot standing up-side-down on a table. The critic holds the pot and turns to his friend in astonishment: "look at this pot, it does not have opening." Then, he turns the empty pot up-side down to see the bottom. This time he exclaims in greater surprise: "wow, it does not have bottom either!" Similarly, Lomax tries to confuse the mathematical system. He tries to justify the existence of facts that he could not refute by a ridiculous and oximoronic interpretation: interesting examples of this incomplete, arbitrary, complex, anecdotal, meaningless and inconclusive mathematical structure is a plan of God (fitna) to test people who see a mathematical miracle in the Quran! Lomax is not able to see the logical problem with this interpretation.

His last paragraph, however, is the most ironic one. He claims that the Quran is clear, but the code 19 is not as clear as the book itself. It is ironic for several reasons. First, it was Lomax himself who just claimed that the verses of Chapter 74 were not clear. Second, we had a lengthy debate on the role of Hadith in Islam before this one. In that debate, Lomax repeatedly labeled many Quranic verses vague. For instance, he does not see the beginning verses of chapter 24 clear enough and accepts the pagan fabrications which inserts the punishment of "stoning to death" to Islam. Another example, despite the clear warning regarding man-made dietary prohibition (6:145-150), he accepts hadiths that fabricates numerous contradictory dietary prohibition. Now Lomax is telling us that he finds the Quranic message very clear. I hope he has changed his mind since our last argument.

After witnessing his ability to complicate the simplest facts, after witnessing his talent to cloud the clearest facts, I do not have much hope for him whether he can ever see any mathematical system in the Quran. Many smart, but extremely skeptic or fanatic people could not see the clearest miracles in the past. (6:25; 7:132; 15:14-15; 40:34). It is evident that the fuzziness and atmospherics are in his mind that complicates the simplest facts for him. It is the Quranic prophecy that those who are not happy with GOD ALONE will not be able to understand the implication of 19 after its revelation. If Lomax really wants to see the truht, he must first stop associating man-made teachings as another source besides God's word (6:114). He must start enjoying mentioning God's name alone without adding Muhammad's name in his shahada (3:18; 39:45). Then, inshallah he will be able to see one of the greatest miracles (74:35).

PS: I did not specifically responded, denied or dealt with each and every point, repetition and statement made by Lomax due to time and space constraint and to avoid redundancy. Thus, the argument should be evaluated based on what is written by both parties; without speculations on what they have not written.

***

In defense of Mr. Yuksel's admonitions

Y. Rapido
26 Jul 1995

Hello Mr. Yuksel,

Although my knowledge of Islam is very limited, I have taken time to defend your stand in a debate with one Mr. Lomax. It is certain that my limitations do not make me a powerful supporter, but since no one else came to stand up for your message, it is better than nothing, one might suppose.

Imbecility of the so-called "Muslim world" is annoying to me because, for the most part, my cause was somehow dependent of their help. Help never really came in a shape, form and strength that was required. Part of that blame goes to a cancer known as "Muslim scholars", who you have described so well! That is why a bit of frustration overcame me during this year, after watching four years of massacre of the Bosnian nation, many of them Muslims

I know it is not that good of a defense, but it is something, so I wanted to present it to you. Let me know if you got about half dozen messages that I sent you during the past week, and let me know if you got this message. Thank you...

Greetings,

> Daniel Lomax (marjan@crl.com) wrote: Indeed, and remember, dear reader, that Yuksel is the best of the: Khalifites. Few of them have troubled as much as he to verify Khalifa's: work.

Commentary: My limited "scholarly" knowledge of Islamic "schools" and teachings prevents me from entering this discussion in a "caliber" that I would like to secure for myself. However, as an educated person, and as a political activist of sorts, I am noticing some common lines of thoughts here, be that Islam or any other discipline of debate.

I am noticing that a very dedicated, pious, educated and eloquent gentleman, Mr. Yuksel, is trying to tell us all, that Islam is only a submission to God, and that God's message to human kind is only transmitted through Quran, the Book, one and the only final testament that God transmitted to mankind through a human being named Muhammad, his Prophet and Messenger. That's what Mr. Yuksel is trying to remind us of!

In order to remind us of that truth much better, and to try to convince us that that is so, Mr. Yuksel is trying to point out that Quran is a perfect book, with clear physical evidence of its perfection being readily available and visible to the humanity! So, a question comes to my mind: "Isn't it miserable that Mr. Yuksel has to spend so much energy and to endure so much public insults from one Mr. Lomax, in order to simply and emphatically try to convince so-called Muslims that Quran is the perfect Book, and the only source of Islamic true teachings?!"

Isn't it a sad, sad commentary on the state of Islam, to which many centuries of fraud, "scholarly intervention" and human greed, corruption, misery and such, has reduced this clear message... burying it under a rubble of such garbage as 60 (!) volumes of Hadith, endless mockery of God's word through Arab pre-islamic folklorism and tribalism contained in Sunna, etc. etc.

In that courageous and passionate plea to misled human species, Mr. Yuksel is referring to a life-long (interrupted by assassination, no less!) research of one dedicated Muslim, Mr. Rashad Khalifa. The "sin" that Mr. Khalifa has been accused of is that he loved Quran so much, that he wanted to show to the world its perfection, and to enter its secret codes, as he saw fit, as a Ph. D. in sciences... a searching and brilliant mind, as it is rather obvious! And in that process, which must have taken a couple of decades of enormous dedicated research, Mr. Khalifa made one human (!) error... he got carried away, and he got to believe, that his elegant and dedicated work, does merit a status of a

Messenger of sorts. If Catholic Church was dealing with those matters, it could happen that Mr. Khalifa, indeed, could have been declared a saint, for his dedicated and pious work. Monks have achieved "sainthood" in Christianity for much, much lesser deeds! There has got to be a name for a man who unearths such truths, after 1400 years of piling up of lot of quasi-religious, quasi-philosophical garbage that the "islamic scholars" have piled up on a heap that in effect almost BURRIED Quran under the rubble of man-made idiocy and heresy!

One tiny little book, written by Dr. Rashad Khalifa, and titled "Quran, Hadith and Islam" (available through Mr. Yuksel) taught me more about true Islam, just in time when I thought that there is no hope that somebody truly understood Quran. As a person educated in secular society, as a person whose mother knew Quran by heart, I have come to see that true Islam is not something that puts people off and disgusts them. Quite the contrary, true Islam is exactly what I somehow suspect -- a spiritual, religious and mental "space" in which there is no place for idiots and liars like Bukhari, Muslim, Shias, Sunnis,... and all other juggernauts that wicked human intervention created. For that single contribution Dr. Khalifa deserves a special place and title. Is that a Messenger? Perhaps not... perhaps he got carried away. So what? Does that make his message any less valid? No! Of course not! He merely made an error about which he warned everyone else. That is the proof that even the best dentist can not work on his own teeth...

Still, people who saw what he was talking about, like Mr. Yuksel, continued his research and his appealing to mankind...

And what do they get in return? NAMECALLING! Ugly, bitter, evil, demagogic, pathetic and often cheap insulting! By one Mr. Lomax, and by several others "defenders of scholarly garbage"... Does Mr. Lomax see that? It doesn't seem that way... or perhaps he does, who will ever find out...

Just one "technicality":

Does Mr. Lomax call people who turn up the light in their rooms in the evening "Edisonites", just because allegedly Thomas Edison invented a light bulb? Does Mr. Lomax call people who during industrially processed milk "Pasteurites", just because Louis Pasteur discovered bacteria under microscope and a method to fight them by boiling water, milk and such?!

It is clear that attaching person's name to something that the person discovered, reminded mankind of and or did something enlightening is a method accepted in

258

science, philosophy etc. In religion that is a bad connotation, because it suggests a cult... Well, if what Dr. Khalifa discovered is clear to me, it is that he reminded people to follow only the word of God, as written in Quran only! Perhaps the people who think that he is right, and who listen to his advice should be called "Quranites"? Or, isn't that another name that could be defined as: MUSLIMS?!

Why doesn't Mr. Lomax consider that "train of logic"? Perhaps he will replace his slanderous remarks with very true and very appropriate appellation for the ones who see the truth in Mr. Khalifa's and Mr. Yuksel's suggestions!

I only wonder will Mr. Lomax proclaim this author, Rapido, a "Yukselite" or a "Khalifite", although this author has only "met" Mr. Yuksel and Mr. Khalifa in cyberspace a month ago?!

Keep it up Mr. Yuksel. I only wish to see will you get ANY support from this numerous "Muslims" who are pounding the network with tons of such garbage that it boggles my mind...

10

Martin Gardner, a Skeptic Mathematician, Reacts

The following criticism from Martin Gardner, a mathematician, is no surprise. It is important to learn that he concedes that "**19s exceed the bounds of chance**" in Khalifa's work. That is important to hear from a mathematician. But, without Lomax, perhaps we would not have heard this confession, since Gardner, as an avowed skeptic, relying on Lomax distortions and educated ignorance, escapes from acknowledging the great prophecy. The following excerpts are from his book titled *Did Adam and Eve Have Navels?*[20]

> "What is one to make of Khalifa's numerology? It is, of course, no surprise that many 19s would show up in a book as long as the Koran, but Khalifa's 19s exceed the bounds of chance. The most plausible explanation is that he deceived himself by unconscious fudging. The best account known to me of how easily he could do this is in *Running Like Zebras*, a 1995 book edited by Edip Yuksel, the nation's top Khalifite."

Despite reading the book and learning my rejection of such a title attributed to me by my opponent, Martin prefers to call me with such title.

> "The book contains 19 x 6 = 114 pages (coincidence?) that reprint a debate on the Internet between Yuksel and Daniel (Abdul*Rahman*) Lomax, a Muslim skeptical of Khalifa's findings. He accuses the chemist of careless computer searching, of rejecting two verses of the Koran as spurious because they don't fit his calculations, and of not revealing that versions of the Koran differ in their number of words and

---

[20] *Did Adam and Eve Have Navels?*, Martin Gardner, W.W.Norton & Company, 2000, p. 260-261.

letters, and in how they divide suras into verses. Above all, he accuses Khalifa of failing to make clear what he considers a "word."

"Many Arabic words have multiple forms, and Khalifa is inconsistent in his counting rules. Sometimes he includes plural forms, sometimes not. Should a word with an affixed pronoun be called one word or two? In English the meaning of word is fairly clear because of spaces between words, but in Arabic there are no spaces. Even in English there is vagueness. Lomax's example is truck. In counting truck in a book should you include trucks, trucked, and trucking? One looks in vain for Khalifa's definition of word."

As it seems, Gardner has swallowed Lomax' deceptive example with line, hook and sinker! Since I have discussed it sufficiently in the previous chapter, I will not repeat my response to the same...

"Lomax likens the doctor to those astronomers who once fancied they could see canals on Mars. He concludes: "Dr. Khalifa's claims, at best, fall into the category of pious fraud .... Had God intended the Qur'an to carry a code verifying its perfect preservation, he could have done it much more effectively and simply than the complex, arbitrary, and inconclusive 'code' claimed by Dr. Khalifa."

"Yuksel, of course, believes he has completely demolished all of Lomax's objections. His curious book is available from his Monotheist Productions... "

Nothing new! Besides, Martin acts like a biased news reporter with nothing to contribute to the debate, with the exception of reporting one side's account with more favorable language and not reporting the opponent's criticism. Of course, I should not have expected much from him, since he des not count anything on his own to verify or falsify the claims and he lacks the desire to learn some Arabic to be able to follow the arguments deceptively raised by Lomax. However, being a mathematician, his statement about improrability of the examples of 19 in the Quran is crucial. Provided that Lomax' criticism is no more than an educated ignorance and proven to be false, such a testimony from the expert mathematician turns into a solid support.

"Now for some 19 number juggling, supplied in part by correspondent Monte Zerger. Nineteen is, of course, a prime. It is equal to 102 - 92; to 12 + 32 + 32; and to 33 - 23. The number 1,729, or 19 X 91, was involved in a famous incident between the British mathematician G. H.

Hardy and his friend Ramanujan, the Indian number-theory genius. Having taken a taxi to visit Ramanujan in a hospital, Hardy remarked that the taxicab number, 1,729, was a dull number. Ramanujan immediately replied, "No, it is an interesting number. It is the smallest number expressible as the sum of two cubes in two different ways [123 + 13 and 103 + 93J." Note that the digits of 1,729 add to 19."

"In 1989 (a multiple of 9) it was proved that every integer is the sum of no more than nineteen fourth powers. The smallest number requiring nineteen such powers is 79, the sum of four fourth powers of 2 and fifteen fourth powers of 1. The repeating decimal of 1/19 is the 2 X 9 = 18-digit number 052631578947368421. Multiply it by any number from 2 through 18 and the product has the same eighteen digits in the same cyclic order. Multiplying by 19 produces a row of 2 X 9 = 18 nines."

Delicous information from a great mathematician! When Martin puts the hat of a mathematician rather than a biased reporter his testimony regarding the probabilit and the information above becomes valuable.

Martin ends the chapter on code 19 with the following paragraphs that contain mostly trivial and disconnected observations.

"The Constitutional amendment giving women the vote was the nineteenth. The nineteenth hole in golf is the bar where golfers sink drinks like they sink puts in eighteen holes. Every nineteen years all phases of the moon fallon the same days of the week throughout the year. The Psalms is the Bible's nineteenth book. Psalm 19 opens with 'The heavens declare the glory of God, and the firmament showerth his handiwork.'"

"The numbers most often encountered in the Bible are 12 and 7. They are reflected in our calendar, with its seven days to a week and twelve moths to a year. The sum of the two numbers is 19."

"Figrue 1 reproduces a thing of strange beauty. The nineteen cells hold integers 1 through 19. Every straight row of cells adds to 38, or twice 19. It would make a wonderous amulet for the Baha'is."[21]

---

[21]   Ibid, p. 261-262

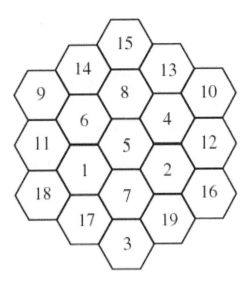

Figure 1. The only possible magic hexagon. Every straight row of cells adds to 38, or twice 19.

11

Intelligent People's Guide to Code 19

Responding Ayman's "Idiot's Guide to 19"

© Edip Yuksel (Yüksel), 19.org

May 2005

In the article "Which one do you see: Hell or Miracle?" I demonstrated the common errors in traditional translations regarding verses of Chapter 74 of the Quran, The Hidden. Those who are fluent in Arabic as well as those who have no or little knowledge of the Arabic will witness the prophetic language of the Quran in chapter 74 and its great fulfillment in our times, as long as they do not handicap themselves with prejudice and arrogance. The Secret hidden for fourteen-century is no more hidden from believers; yet it continues to be invisible to the idiot and ignorant, to the hypocrite and arrogant, as well as to many who have yet to learn about it.

I advise those who rely on traditional translations of Chapter 74 to read that article, which is available at 19.org. I had also extensive discussions on this topic throughout years and two of them are available in English as booklets, "Running Like Zebras," and "The Prime Argument." The former is an extensive discussion with AbdurRahman Lomax, a Sunni disbeliever, and the latter is a two-round short discussion with late astronomer and skeptic Dr. Carl Sagan.

Dr. Rashad Khalifa's book, "Visual Presentation of the Miracle" is a landmark book on this subject; and I highly recommend it for those who want to save time in their efforts to discover or test the prophetic mathematical-literal miracle.

Unfortunately, Rashad later was carried away by those who wanted more (74:6), and thus in later years he diluted the solid facts with speculations and arbitrary calculations, which entered the appendices of his translation of the Quran. Putting verse numbers next to chapter numbers or adding them up in arbitrary and selective way, happened to be the worst gate for abuse.

During his last year, I worked with him in Masjid Tucson, every day of the week, ten hours a day, until his last day among us. Diluting the mathematical system of the Quran with arbitrary or speculative "discoveries" was one of the main points of recurring arguments between him and me. I found myself in minority among a group of zealot believers with no concept of laws of probability. In one hand the Quran and in another a simple calculator, they were fervently trying to join the great hunt for miracles. Rashad's great care for not hurting the feelings of those who were eager to participate in the "discovery" of a miracle led to groupthink and mass addiction in abuse. Some were so ambitious; after Rashad's departure, they even went beyond the original text of the Quran and "discovered" many "mathematical miracles" in Rashad's re-re-revised English translation. A few came up with series of doomsday prophecies by arbitrarily putting verse and chapter numbers next to each other with the magical ingredient of speculations over speculations. Since their egos were inversely proportional to their knowledge, they used the same arbitrary numerical jugglery to find holy positions for their names; they came to believe their divine mission, or messengership. I have so far received personal invitation to accept their mission from about a dozen such "calculator-messengers," and ending up receiving their warnings and condemnations. It is funny until you hear that these charlatans mentally and spiritually screw up some people.

As it turned out, this abuse which was tolerated to a certain extent by Rashad, provided plenty of excuses for those who insist not to witness one of the greatest miracles; they indulge into pointing at Rashad's errors or speculations and thereby they manage to blind themselves to impressive Quranic facts that are independent of Rashad or anyone else. Perhaps, without this excuse, they would be forced to see a miracle that they did not deserve to witness. I consider this, as a divine plan to separate the discoverer from the discovery, to distinguish our human errors from the divine perfection.

God willing, I will make my Turkish book "Üzerinde 19 Var" (On it 19) available in English soon, since it is a comprehensive and delicious analysis of the subject matter; it also contains a lengthy chapter responding to the criticisms published in several languages in last twenty years. [Thank God, I was able to finish and publish **NINETEEN: God's Signature in Nature and Scripture** in 2011] Nevertheless, I believe that any sincere person who seeks the truth of the matter can witness the prophetic mathematical structure of the Quran without the aid of any other book besides the Quran. The books or articles on the subject may only help to reduce the time and energy spent to verify or falsify the counts. Besides, those articles and books may help the novice to recognize the distortion and false claims in the arguments of those who are blind to the miracle.

The "Idiot's Guide"!

Ayman, who assumes the title "Brother" as his first name, reflected his aversion and lack of appreciation of "one of the greatest" divine signs in a recent article, "*Idiot's Guide to Code 19*" published at free-minds.org. However, with the title of his article, Ayman is either trying to slyly insult those who witnessed the great sign or he is unaware of the Quranic statement that only intelligent people take heed, not idiots (3:7; 13:19; 38:29; 39:9; 40:54). Here, I will demonstrate that Ayman is not really guiding his idiots; to the contrary, he is increasing their misguidance. Ayman is an articulate and clever disbeliever of a Quranic prophecy and sign, and he is successful in playing as the devil's advocate. Unfortunately, he appears to have believed his client and became a member of one of the losing groups, prophesized in verses 74:31-37. Let's start from his first two paragraphs:

> **AYMAN**: "In the case of the so-called Code 19, the claim is made by its promoters that it is a 'mathematical miracle'. However, what is noticeable is that mathematicians who are the subject matter experts have not been flocking in droves to believe in and promote this 'mathematical miracle'. I personally, haven't heard of any mathematician or statistician who has declared his or her belief in this 'mathematical' miracle. On the other hand, it is people who are not subject matter experts in mathematics who have been making such 'miraculous mathematical' findings and promoting them. One would have expected the converse. Thus, it can be observed that there seems to be a gap between the subject matter experts and people who know little about the subject and this is what this 'idiot's guide' will try to bridge, if The God willed."

266

"Now one may argue that just because someone is an expert, this doesn't mean that they will recognize an alleged 'miracle'. However, by labeling Code 19 as 'mathematical' and 'scientific' then it must undergo the same peer review process that any mundane non-miraculous 'mathematical' and 'scientific' research undergoes, otherwise it should not be called 'mathematical' and 'scientific'."

Initially, it appears to be a strong argument. If you are his intellectually challenged "idiot" audience, then you do not need even bother reading further. Ayman's first argument against Code 19 is powerful: "Experts and smart people do not believe in it; and you as an idiot should not either." However, it does not take to be a genius or mathematician to understand the problem with his argument.

First, he claims that he has personally not heard of any mathematician or statistician who declared his belief in this system. I personally know some mathematicians and statisticians who have witnessed the extraordinary nature of this miracle, but I will not drop their names here, since I assume that they may not want to be disturbed by neither Ayman's idiots, nor fanatic terrorists who killed Rashad, the original discoverer of this miraculous code. Anyone who knows the risk of accepting this miracle, especially a public figure, such as a university professor, will understand the reason why many mathematicians are not sending their cards to brother Ayman. Not every believer may be as brave as the two magicians who declared their support of Moses despite Pharaoh's threat. Hiding one's belief to protect one's life is justified by God (3:27).

Besides, Ayman either does not know or has forgotten that the mathematical miracle of the Quran is not popular with the followers of Hadith and Sunna. Thus, more than 99 percent of Muslims and their scholars have prejudice, if not phobia against the number 19. The few mathematicians who are able to break the wall of prejudice do not come forward for fear of risking their lives. Personally, I know a few of them. As for those mathematicians who live in the West and have no knowledge of the Arabic, they are not interested in the topic for several reasons. First, they are culturally biased; they may show positive or negative interest in claims of numerical structure in the Bible, but they are not much interested in the Quran. Especially, after 9/11, there is not much good will or positive interest in the Quran by non-Muslim mathematicians to invest time in objectively studying a controversial theory, a theory that have been distorted and misrepresented by its fanatic and violent detractors. Besides, they think that they need to know Arabic to venture a scientific study on an originally Arabic text and they do not have time and much motivation to study Arabic. This also might explain why we see almost no Western mathematicians criticizing the miracle either. Richard Voss, one of the few Western mathematicians who studied the numerical structure of the Quran, is more positive than negative, as we will discuss later.

In sum, the phobia and ignorance prevalent among Muslim population, the apathy of western mathematicians to the Quran, and fear of being the target of violent Muslims, and the bias of secular paradigm against metaphysical claims are among major reasons why Ayman does not hear many mathematicians. The famous mathematician Martin Gardner demonstrated some interest in the subject matter and wrote several articles on the topic. However, his limited knowledge of the subject matter and his linguistic shortcomings necessary for a scholarly evaluation did not allow him to investigate the subject deeper. Years ago, Martin contacted me and we hade several mail and phone conversation, but he was reserved; he became neither too critical nor accepting. Putting myself in his shoes, I would perhaps do the same. He heard the loud voice of Sunni and Shiite scholars and writers pontificating that there is no such a thing in the Quran, and he even heard them asserting with straight faces that the first verse of the Quran, *Basmalah*, did not have 19 letters after all. Not knowing the Arabic language and lacking sufficient interest that would lead him to study and learn Arabic to verify or falsify the claims, though impressed and intrigued, Gardner chose to treat the subject as an agnostic.

My Experience with Mathematicians and Experts

In fact, my experience with mathematicians have taught me that mathematicians, like every one of us, are humans and they have their own walls, weaknesses, agendas, and prejudices. For instance, several years ago, I gave lectures on the mathematical structure of the Quran to the philosophy and math graduate students. Among the audience, there were also professors. I was surprised to see the difference. Philosophy graduate students were more interested in the discussion of the subject matter; but despite my challenge during and after the presentation of the mathematical structure of the Quran, at the math department of the University of Arizona, only a few mathematicians were forthcoming to discuss the issue. They followed the one-hour presentation attentively and seriously, but they did not make any negative comments regarding its mathematical integrity. They appeared to be impressed, but they did not pursue further information. I believe that they were also wondering like many other people who are consciously or subconsciously impressed by band wagons: "If there is really such a great mathematical system in the Quran, we wouldn't hear it for the first time from a lawyer and an adjunct philosophy professor; it should have been the headline in the news and we should have seen many scholarly articles about it."

Perhaps this was one of the questions spinning inside the square roots of those "experts." The few who could go beyond the convention did not feel having the necessary knowledge of Arabic to verify the presented samples.

One might wonder about the Arabic mathematicians. What about their reaction? Well, either the part of their brains that responds to the statements related to the Quran is occupied by viruses of Hadith and Sunna, or they have strong aversion against anything related to islam, or they are not willing to risk their lives by even coming close to such a dangerously controversial topic.

Towards the end of Ramazan of 2002, I was invited to participate in a popular yet prestigious weekly Turkish TV discussion panel, Ceviz Kabuğu, to discuss a best-selling book written by a young Turkish math wizard. The audience and several other experts who have been contacted by the program were also participating in the debate via phone calls. "Kuran-i Kerim Şifresi" (The Code of the Holy Quran), was alleging new mathematical discoveries and prophecies in the Quran. The young author of the book, Ömer Çelakıl, had received a first place award in a national math competition and thus his book had received top media attention. Before unveiling his "discoveries" to the media, Ömer had contacted me to get my support. I did not know him then, but he knew me from my books and media. I found some of his observations interesting yet too speculative and anecdotal. One being wizard or expert in math or sciences, as I have witnessed many times, does not guarantee that he or she is not gullible, blinded by culture or personal weaknesses. Like all of us, "experts" too are prone to committing logical fallacies, following the crowd around them. Besides, there are many smart people betraying their intelligence just to get the attention of idiots.

On one hand, I knew that most of Ömer's "discoveries" had no mathematical significance, but on the other hand, I was careful not to break his spirit; he could have been on something yet to be discovered. So, he received a mixed message from me, and he wrongly interpreted it as a support, which he later expressed it in his interviews with the media. Soon, I heard that his book was an instant best-seller and he was a celebrity. He became a regular guest of TV shows in Turkey.

The host, Hulki Cevizoğlu, had also invited Prof. Suleyman Ateş, a well-known Muslim scholar and the former head of the Turkish Religious of Affairs, (See: www.suleyman-ates.com), and Prof. Haluk Oral, a mathematician from the Bosphoros University who is considered an expert on codes. (See:

269

www.math.boun.edu.tr/instructors/oral/oral.htm). Despite the highly specialized content of our topic, our live discussion became a top-rated program by drawing millions of audience from Turkey and Europe, and lasted about three hours.

In that discussion, the top religious scholar who had written articles and books critical of the code 19, could not engage in a decent discussion with me; he lost his temper frequently and left the studio twice in the middle of the discussion. He was specially getting mad at my referring to the last prophet with his first name, as the Quran wanted me to do so (2:136). Perhaps he was using it as an excuse to avoid engaging in a face-to-face debate. Each time he ran away, he was stopped by the TV crew and brought back by the host who begged him during the breaks to stay in the panel. The host also privately asked me to be easy on him. I found myself in between of having feeling pity to an arrogant and popular scholar, or stand for the truth. The Sunni scholar was so terrified from me, he could not even engage in short face-to-face dialogues; he was like a zebra in front of a lion. He was aware of my previous TV debates with other "experts" and knew the power of my position and its performance. In one of the TV debates, the host of the show even made a public announcement that many religious scholars were declining to debate with me in public. Prof. Ateş, a critic of my books, had declined debating live with me in the past, but somehow this time he had accepted. Perhaps, the host lured him to the debate by employing his persuasive skills. In sum, towards the end of the show, I decided to be easy on him; I did not even rebuttal this so-called Islamic expert, when he made an absurd claim regarding *Basmalah*: looking in the eyes of millions of Turkish TV audience, he claimed that *Basmalah* did not have 19 letters! I let him incriminate himself in public with outrageously stupid claims. Any person with average IQ, regardless of their language, could easily see for themselves the falsity of his claim by simply checking *Basmalah* with the 28 Arabic letters.

Some of those idiots whom Ayman is hoping to guide (!) were looking for an expert in Arabic, and they found one of the top experts in Arabic language telling them that *Basmalah* did not have 19 letters. Now, they were guided! But to where? The last name of the scholar was providing a prophetic answer, which is preferred by hypocrites and disbelievers to one of the greatest miracles (74:26-37).

As for the professor of mathematics, though he was critical of Ömer like I was, he kept silent when I demonstrated the 19-coded examples of mathematical system of the Quran. I did not know him personally and I was expecting a fierce debate on statistics and probability calculations. During the debate, the mathematician preferred to be surprisingly neutral. His silence against my

270

argument for mathematical significance of examples code 19 in the Quran was later criticized by atheists. I think he found himself in a highly controversial and risky position. He had no worldly risk if he criticized the existence or statistical significance of the 19 in the Quran; to the contrary, he would be more popular and well received if he made negative remarks. However, he had a lot to risk academically and socially if he did not criticize it. Well, he did not show the courage to put his life at risk by publicly acknowledging the miraculous nature of the 19, but he took some risks by not rejecting it either.

My communication and discussion with some scientists and mathematicians, including Martin Gardner and Carl Sagan, thought me the following fact: miracles are not necessarily witnessed by those who are expert in a particular field. It requires more than that (7:146; 2:118; 27:82; 74:31)

Besides, how many "experts" would satisfy the author of the "Idiot's Guide" and his audience? If I provide a list of, say 20 mathematicians, then why should our "expert-dependent" idiots not ask the following question: "What about thousands of other mathematicians?" In other words, Ayman will not take the mathematical miracle of the Quran unless the majority of mathematicians witness it! He would not believe that the earth was round and rotating around the sun when Galileo was under inquisition. It seems that Ayman has no clue what "miracle" or "divine sign" means in the language of the Quran and according to sunnatullah (God's law).

Our brother who pretends to be "the guide of idiots," ignores the report of the holy book, which he claims to believe. The Quran narrates the presentation of many miracles by previous messengers and prophets and almost invariably, it reports the unappreciation or disbelief of the elite. Usually, it was not scholars or experts who first recognized the message. Ironically, whenever the message of a messenger became popular, then the so-called experts and scholars joined the bandwagon and used their craft to "justify" their new faith. Intelligence, objectivity and sincere search for truth (qalbin selim) is needed to witness God's signs, and experts do not necessarily have all the pre-requisites. Ayman reminds me of the people:

> 11:27 "The leaders who disbelieved among his people said, 'We see that you are no more than a human like us, and we see that the people who follow you are our lowliest and primitive in vision. We see that you do not possess any advantage over us. Indeed, we think you are liars.'"

271

Ayman mentions the case of two magicians testifying to the miraculous nature of what Moses demonstrated. He forgets that Egypt had many more magicians that they all disbelieved Moses. Besides, none of those who witnessed Moses' miracles waited or needed the approval of the two magicians; they were acknowledging the truth without seeking the approval of neither the crowd nor the elite.

Apples, Oranges and Mushrooms of the Data Set

After his appeal to experts, Ayman now is ready to attack the substance.

> **AYMAN**: "Let's start with the most common misconception. Most people who believe and promote a "mathematical" miracle think that mathematics is an objective science that can't be wrong. In keeping with the spirit of this article's emphasis on ease, I will try to use terms and examples that can be understood even by people who don't know anything about mathematics to dispel this major misconception."

After reminding us this excellent point, Ayman goes on providing a deliciously elaborate apple and orange arithmetic and concludes:

> **AYMAN**: Once we decide on a methodology then we can't change it for different apple baskets. For example, we cannot disregard the rotten apples in some baskets and count them in others. This would lead to wrong results.

> "Surprisingly, this is all we really need to understand about mathematics to evaluate Code 19. So let's recap those two simple yet important principles:

> 1. Clearly define the data set.
> 2. Use a standard methodology consistently.

> Although those well-known principles are simple and straightforward, they are often the first to be violated by phony "mathematicians". They are also the first principles to be overlooked by those who the phony "mathematicians" deceive. Thus, you will seldom find a real mathematician amongst those two groups.

I am in total agreement with Ayman regarding the consistency in methodology. However, as you will see, it is Ayman, who happens to confuse apples with oranges, and occasionally even with mushrooms and bananas.

AYMAN: As far as the so-called Code 19, defining the data set is very simple. If the data set is the Hafs version of the great reading, then only the Hafs version should be used. Hafs is one of several versions of the great reading that mainly have minor differences in the spelling of some words (see Which Quran?). One cannot go back and forth and use another version when the counts that he or she is looking for don't add up. If one uses another version, then that version must be used for all the counts and the Hafs version must be disregarded. One cannot mix and match as he or she pleases because this would be forcing a fit on the data into a preconceived result. Hence, if proponents of Code 19 decide to use the Tashkent copy for something, they must use it for all the counts and not just for fixing the count of the letter "Sin" to match a result that they want. They also can't make any modification to the data set to reach the count that they want.

Here, brother Ayman likens one ancient version of the Quran to an apple and another version to an orange. So, he hopes to prevent a comparative and critical study on multiple versions. According to him, the two manuscripts do not come from the same tree, but they belong to different trees. It seems that his apple and orange principle is peculiar; he wants to pigeonhole a critical comparative study to only one manuscript. This does not make any sense. I understand and respect Ayman's concern regarding the abuse of multiple manuscripts to concoct numerical coincidences. However, categorically rejecting the use of various manuscripts for a critical and comparative study to infer the accurate version or spelling of original Quranic verses is absurd.

Let me clarify. Dr. Rashad Khalifa discovered the 19-based system after a computerized study based on today's most common manuscripts that rely on Hafs version. If the body of mathematical evidence and some textual hints indicate a spelling error in the manuscript we are using, then why shouldn't we check other manuscripts to see whether our prediction or suspicion regarding the spelling error were reflected in them? We do not believe that any single manuscript of the Quran we inherit is error-free, since all manuscripts were duplicated by scribes who were fallible humans. Besides, some ancient manuscripts are not complete. For instance, the Tashkent manuscript is half missing. It is also a historical and archeological fact that the original manuscript written by Muhammad himself is lost. According to Sunni and Shiite sources, it was burned by Marwan bin Hakam during the Umayyad Dynasty.

Pro-Sad Party: Not Only Bad Spellers, But Also Con Artists

Ayman is referring to the correction of the letter Sin in verse 7:69. He takes one of the powerful evidences of 19, he spins and grinds it in his "clear definition" machine and it magically transforms to a defect. For those who seek the real picture of the issue here is the story and facts:

Rashad's initial counts in his early work, which were published as a book, showed the frequency of the letter Sad in three Sad-initialed chapters as 152. (*Miracle of the Quran: Significance of the Mysterious Alphabets*, Rashad Khalifa, Ph D, 1973, Islamic Productions International, Inc., St. Louis, MO.) Then, Rashad had not yet discovered the code 19, and thus there he never mentions the relationship of the 19 with the frequencies of letters, words or chapters of the Quran.

While working on his English translation of the Quran in 1969, after finishing the translation of the first chapter, he encounters the three letters initializing chapter 2, that is, A.L.M. Since he had promised himself not to skip to another verse until he understands the meaning of a verse, he decides to study their meaning or implication. Not being able to find help from classic commentaries, he decides to study the frequencies of letters in the entire Quran. In his pre-1974 books and articles, he only provides some statistical information and observations on the statistical feature of the letters. For instance, he finds that the letters initializing the 29 chapters are more frequent in chapters they initialize. He also observes that the frequency of letter Q (Qaf) is exactly equal to the number of the chapters of the Quran. He interprets this as letter Q standing for the Quran. Again, he is unaware of its relationship with the number 19 yet. He would discover it within a few months after the publication of the book.

Years later, critics discovered several errors in Rashad's computer data. One of them was the count of letter Sad; the frequency of the letter Sad should have been 153. Discovering that the frequency of the letter Sad in Sad-initialed three chapters is adding up to 153, not the expected 152, Rashad remembers the strange little note written on top of a word in verse 7:69. The superscript note is found in almost all Arabic manuscripts of the Quran currently in use. The little note has two versions. In some manuscripts there is only a small letter Sin on top of the letter Sad of the word *BasTata*. In other versions, there is a tiny calligraphy informing the reader, "yuqrau bil-sini" that is, "it is read with Sin." In other words, though the word *BasTata* is written with letter Sad, it is supposed to be pronounced with letter Sin! There is even a hadith narrating that prophet Muhammad instructed that this word to be written with Sad. Those who can read Arabic would know that there is little phonetic difference between the pronunciations of these two different letters. Though their pronunciations are close, interchanging them in the spelling of words could make big difference. For instance, the Arabic word HaSaN with letter Sin means beautiful, but with letter Sad means horse. HaSaD with Sin means jealousy, but with Sad means to harvest… Stumbling on this bizarre note, Rashad decides without hesitation that the correct spelling of the word must have been with Sin, rather than Sad.

Nevertheless, initially Rashad did not provide any historical evidence for his claim. All was he doing was to correct the spelling of a word to comply with the implication of the mathematical pattern; indeed he had good reason to be suspicious of the word *BasTata* in 7:69.

When I received this information from Rashad I was in Turkey. To me it was a major news, since I used to believe that every letter of the manuscript was protected by God. Then, my concept of the Quran was a book made of ink and paper, which was paid lip service by the majority of Muslims. Disappointed in Rashad's claim, I went to the Sulaimaniya Library to check the oldest copies of the Quran. There I found an incomplete copy of Tashkent. To have access to the manuscript I needed to use my political contacts. Finally, in year 1986 I was able to witness with my own eyes the accuracy and predictive power of Code 19. I saw that the word *BasTata* of 7:69 was written not in letter Sad, but letter Sin and the word had no tiny note on top of it. I photographed the page containing the historical confirmation of Rashad's prediction and immediately mailed it to him. I used that evidence in a Turkish magazine, 2000'e Doğru, in my response to atheists claiming that the original Quran does not exist since it was burned by Marwan bin Hakam. Later, Rashad published the same picture in his translation of the Quran as evidence for the accuracy of the mathematical structure of the Quran. It was very clear that Rashad did not have the Tahshkent copy and he had no knowledge of existence of a manuscript with the spelling predicted by code 19. Knowing Rashad and his work closely, I find Ayman's accusation regarding Rashad's intention to be false.

Now it is clear why there is a little Sin on top of the word *BasTata* of 7:69. We can easily construct the story through inferences: It was a spelling correction. But, later when people had two versions of the Quran, one with Sin, the other with Sad, they started debating. Each party claimed that their version was the accurate one. The party that promoted the erroneous spelling, rather than correcting the spelling error, added a tiny additional note "read with" next to the tiny Sin. Thus, the suggested spelling correction was transformed to justification of the spelling error with a twist. It seems that the debate on the spelling of the word became very contentious and the party defending the wrong spelling went a step further to support their claim: they fabricated a hadith narrating that the angle revealed the word *BasTata* with the letter Sad to the prophet Muhammad. Knowing that thousands of hadiths were fabricated for many reasons, including for the advertisement of a particular fruit, city or king, it is no surprise to see that a hadith was fabricated to defend a controversial spelling error.

The pro-Sad people had much bigger problem than a spelling error; they were cheaters and liars. They were the followers of hadith and Sunna. Their 19-phobic modern followers, ironically, are sharing similar problems.

In short, instead of considering the correction of the spelling in *BasTata* as a confirmation of the mathematical structure of the Quran, Ayman twists the logic and wants us use this evidence against its divinely designed purpose. Like his pro-Sad pioneers, he is defending a spelling error and accuses the pro-Sin party of fraud for correcting the spelling error. I do not think that he is even aware of the historical debate on this spelling error and his ancestors' fraudulent record; but somehow he shares with them a common problem with accepting the truth.

Standard Methodology for Hiding the Truth

Let's first read a lengthy excerpt from Ayman:

AYMAN: "Here again, proponents of Code 19 constantly violate the basic requirement for using one standard methodology. The basis of the entire Code 19 is the counting of the letters in allegedly "initialed" chapters. Here are some of the various inconsistent methodologies that they use a mixture of in order to get the result that they want:

1. (a) Count each "initial" individually in each chapter where it occurs, (b) add up the count for each initial for all chapters where it occurs and (c) divide the total by 19.

2. (a) Count the occurrences of all "initial(s)" in the chapter where they occur, (b) add up the count for the chapter and (c) divide the total by 19.

3. (a) Count only the initials in chapters with exactly the same set of initials, (b) add up the count for each set of initials for all chapters where they occur and (c) divide the total by 19.

4. (a) Count all the initials in chapters with at least one common intersecting letter, (b) add up the count for all the chapters and (c) divide the total by 19.

"The above are the basic methods described in the article 19-Fact of Fiction. In the article, brother Layth attempted to verify some of the claims of Dr. Rashad Khalifa, the original promoter of Code 19. Brother Layth concluded that 15 out of 29 chapters are confirmed as having initial counts divisible by 19 but that there is no basis for the remaining chapters or other Code 19 claims. However, brother Layth's article fell short in that it did not consider the above basic mathematical principles and hence even his approximately 50% match is likely to have been grossly overestimated. In the present article, we will apply the above two principles to get an objective estimate. In addition to the above four methods described in brother Layth's article, there are a variety of other methods that are more complex and hence more prone to disregard of the above two principles. Brother Layth did well by not treading these murky waters. Essentially, if promoters of Code 19 can't respect the above two principles for even simple counting of initials, which is the basis of their claims, then we can be sure that they won't respect them for more complicated and hence more manipulation-prone methods."

After this, brother Ayman applies all the four methodology and reaches the following conclusion:

> **AYMAN**: "As we saw, even for those basic simple counts promoters of Code 19 make extensive use of manipulation of the data set and constant shifting of the method of counting. Out of the different methods that were presented in the article 19-Fact or Fiction, method 2 gives the best result for Code 19 promoters. Using this method, 3 out of the 29 Chapters have 19 divisible counts. Of course, as we saw, given the natural probability of 1 out of every 19 numbers being divisible by 19 anyway, 3 out of 29 is within the realm of reasonable random occurrence. Certainly, there is nothing there that even remotely suggests a precise 'mathematical miracle'".

Here, I confess, Ayman is doing an impressive job. He is frame-by-frame analyzing and then hiding and twisting the facts. Before analyzing Ayman's analysis, let me give you an example from a notorious trial in the United States.

In 1991, three white Los Angeles Police Department officers beat and kicked an African-American motorist, Rodney King after stopping his car by the highway. A witness recorded the repeated beating to a camcorder. When it was broadcasted on national TV stations, it created a national outrage against police brutality. None expected those police get away with such a flagrant abuse of power. However, the lawyers for the defendant police officers had a great trick in their sleeves. They knew that they could not stop the prosecutor from showing the film to the jury members. Common sense would instruct that they should try their best to make the presentation of the film happen as fast as possible. So tat the jury would not be subjected to its emotional effect for so long. They did to the contrary. They decided to show it themselves, as a pre-emptive strike, and show it as long as possible. In fact, they would show it over and over again, frame-by-frame... The jury was made to watch the horrible beating frenzy of fifty-six baton, frame-by-frame accompanied by the distortion of articulate defense lawyers. They watched it again, and again, in normal speed or frame-by-frame.

As a consequence of this clever technique, the power of each blow was lost, the order of action and reaction was confused, the context was ignored, and the jury was desensitized to violence; thus, the hospitalized victim was transformed to a monster, and the abusive police officers were devolved to scared and panicked children. To the surprise of the public, those brutal police officers were acquitted by the jury, leading to riots in Los Angeles.

Ayman is trying to use a similar psychological technique employed by the defense lawyers in Rodney King trial. Though he focused his attention to an all-idiot jury, his technique may even sway intelligent jurors who have no prior information about the facts of the code 19. To demystify his technique, I will replay the process of the discovery of the code 19, which will put facts in their context.

Though Rashad, upon my query had explained to me the steps of his discovery of Code 19 in the beginning of year 1974, I will not share that information since I do not expect readers to rely on that hearsay testimony. In fact, it is irrelevant. I have dozens of periodical clips in several languages, early documents belonging to Rashad, and out-print books to demonstrate the major points of his discovery on a timeline. (I am planning to include them in the end of the English version of On It 19). I also know it from my own personal research experience, and I believe that the process of the discovery or verification of the mathematical structure of the Quran will almost be identical for any objective researcher who employs the scientific method. So, let me start by sharing some facts regarding the history of the discovery. (See Chapter 4)

278

A Witness from Children of Israel

Joseph Dan writes that Rabbi Judah was critical of the French and British Jews when they altered the Morning Prayer by adding a few words (Studies In Jewish Mysticism, Association for Jewish Studies, 1982, p. 88). Rabbi Judah argued that such an addition destroys the numerical structure of the prayer. Rabbi Judah's discovery is exactly similar to the one would be discovered centuries later. Here is an excerpt from Rabbi Judah:

> "The people [Jews] in France made it a custom to add [in the morning prayer] the words: " Ashrei temimei derekh [blessed are those who walk the righteous way]," and our Rabbi, the Pious, of blessed memory, wrote that they were completely and utterly wrong. It is all gross falsehood, because there are only nineteen times that the Holy Name is mentioned [in that portion of the morning prayer] ...and similarly you find the word 'Elohim nineteen times in the pericope of Ve-'elleh shemot.... Similarly, you find that Israel is called "sons" nineteen times, and there are many other examples. All these sets of nineteen are intricately intertwined, and they contain many secrets and esoteric meanings, which are contained in more than eight large volumes. Therefore, anyone who has the fear of God in him will not listen to the words of the Frenchmen who add the verse "Ashrei temimei derekh," and blessed are the righteous who walk in the paths of God's Torah, for according to their additions the Holy Name is mentioned twenty times...and this is a great mistake. "

> "Furthermore, in this section there are 152 words (152 = 19 x 8) but if you add "Ashrei temimei derekh" there are 158 words. This is nonsense, for it is a great and hidden secret why there should be 152 words...but it cannot be explained in a short treatise. ...In order to understand this religious phenomenon, we have to take the basic contention of this treatise exactly as it is stated: every addition or omission of a word, or even of a single letter, from the sacred text of the prayers destroys the religious meaning of the prayer as a whole and is to be regarded as a grave sin, a sin which could result in eternal exile for those who commit it...." (Studies In Jewish Mysticism, pp. 88-89)

The Quran prophetically promised this discovery approximately 600 years before:

> 46:10　　"Say, 'What if it is from GOD and you disbelieved in it? A witness from the Children of Israel has borne witness to a

similar phenomenon, and he has believed, while you have turned arrogant. Surely, GOD does not guide the wicked people.'"

Classic commentators of the Quran had difficulty in understanding the reference of the "a witness" since they knew that the Quran mentions more than one witness from the Children of Israel, such as Jacob, Joseph, Moses, Aaron, Jesus, etc. Whenever there was a question there was supply: a hadith fabricator came up with a Jewish name, Abdullah bin Salam, and claimed that the verse was referring to him. However, critical scholars rejected the authenticity of that hadith, arguing from chronological discrepancy. (See: Yusuf Ali's footnote in his glorious Quran). Those who found those hadith narrations to be unreliable considered the Bible as the witness, forgetting or ignoring the qualifying phrase, "from the Children of Israel." (For instance, see: Muhammad Hamidullah's translation of the Quran):

> 20:133-135 "They said, 'If he could only show us a sign/miracle (ayah) from his Lord!' Did they not receive the evidence/proof (bayyinah) that existed in the previous scriptures? Had we annihilated them before this, they would have said, 'Our Lord, had You sent a messenger to us, we would have followed Your revelations, and would have avoided this shame and humiliation.' Say, All of us are waiting, so wait; you will surely find out who are on the correct path, and who are truly guided.'"

The language of the verses quoted above is strongly related to the discovery of the 19, its function, and the reaction of many people. Note that the word BaYyeNah (evidence/proof) in verse 133 occurs exactly 19 times in the Quran. Also note that both Muhammad, his supporters and the disbelievers were told to WAIT for the fulfillment of the divine promise.

Nusemantics

Before the discovery of code 19, in 1959, Abdurrazzaq Nawfal of Egypt had discovered relationship between the meaning of the major Quranic terms/words and their frequencies in the Quran. Though I do not have proof, I heard that Rashad himself had participated in that study.

While writing his book *Al-Islamu Deenun wa Dunya* (Islam: both Religion and World) in 1959, Nawfal noticed that the frequency of the word "*Dunya*" (World) had exactly the same frequency as the word "*Akhirah*" (Hereafter), 115 times

280

each, in the Quran. In 1968, while writing another book titled "*Alemul Jinni wal Malayika*" (Universe of Jinns and Angels), he noticed that the word "*Shaytan*" (Satan) had exactly the same frequency as the word "*Malak*" (Angel), 88 times each. This interesting literary symmetry in the frequency of certain words led him to make a more comprehensive study on the numerical structure of the Quran. Nawfal published his findings in early 1980's in a book: "*al-I'jaz al-Adadi fi al-Quran al-Karim*" (Numerical Miracle Of the Holy Quran).

Some of his findings are striking. The following examples, I believe, leaves no doubt that the Quran is a mathematically designed unique book. Below are some examples from those early discoveries that left no doubt in the mind of objective people that the Quran was a Kitabun Marqum, that is, Numerically Coded Book as it is described by the Quran (83:9, 20): Before the discovery of the 19-based system, few people were aware of a symmetrical mathematical wonder in the Quran. For example:

- The word "day" (yawm) occurs 365 times.
- The word "days" (ayyam) occurs 27 times, the number of days in a sidereal month. Together with the dual plural form (yawmayn), all the plural form of the word occurs 30 times, the number of average days in a solar month.
- The word "sabt" (the seventh day, saturday) occurs 7 times.
- The word "month" (shahr) occurs 12 times in its singular form.
- The word "sana" (year) occurs 7 times in its singular form and 12 times in its plural form, totaling 19, which is the number of years in Meton cycle.
- The word "qamar" (moon) occurs 27 times, equaling the number of days in a sidereal month.
- The expression "sab'a samawat" (seven heavens) occurs 7 times.
- The words "satan" (shaytan) and "angel" (malak), each occur 88 times.
- The words "this world" (dunya) and "hereafter" (ahirah), each occur 115 times.
- The divine instruction "qul" (say) and the "qalu" (they said) each occur 332 times (excluding the "qul" of 9:129).
- The word "qist" (justice) and "zulm" (injustice), each occur 15 times.
- The word "shams" (sun) and "noor" (light), each occur 33 times.
- The word land occurs 13 times while the word sea 32 times, 13/45 and 32/45 gives exact ratio of land and sea on earth.
- The word "iman" (belief) and "kufr" (disbelief), each occur 25 times.

When we check all the derivatives of the root of these two words "*AMaNa*" (believe) and "*KaFaRa*" (disbelieve), we find and interesting detail: the frequency of all the derivatives of AMaNa is 811 and of KaFaRa is 697. They are not equal. However, the difference is exactly 114, which is equal to the number chapters of the Quran. A perfect example of nusemantics.

I learned these and many other unique and impressive Quranic facts while I was in my fourth year in Turkish prisons as a political prisoner. I checked them via the index of the Quran, al-Mujamul Mufahras. Though I found some of Nawfal's purported relationships between the words to be semantically arbitrary or speculative, many like the ones I listed above were intriguing. In late 1983 and following years, I had also discovered more examples of numerical relationships between the semantically related words, which I now call NUSEMANTICS, combination of numeric and semantics. Below are some of the nusemantic observations originally made by me and were published in my books, such as, Kuran En Büyük Mucize/Quran, the Greatest Miracle (1983-1988), *Kuran Görülen Mucize/ Quran the Visible Miracle* (1987-1988), and *Kuran'da Demirin Kimyasal Esrarı/The Chemical Secrets of Iron the Quran* (1987-1988) and various published articles. Though these Turkish books are out of print, *Kuran Hiç Tükenmeyen Mucize/Quran, the Perpetual Miracle*, contains most of my discoveries with the addition of many new ones discovered by the Quranic Research Group. You may find the English translation of the book online at www.mucizeler.com or www.quranmiracles.com. This book is so far the best and most comprehensive book on the scientific aspect of the Quran. I highly recommend it. Here are some of my observations:

The word day, besides its singular form occurring 365 times, and its plural form 30 times, all its derivatives and forms occur exactly 475 times (19x25). (Interestingly, the number 25 is the number of rotation the Sun makes around its axis while the world makes one full rotation around the Sun. In other words, in Meton cycle, the sun makes exactly 475 rotations). The plural of the word day has a detail, since Arabic has two plural forms. Unlike English Arabic has a different plural form for two things. For instance, for two days the form YaWMayn is used, while for more than three the form aYyaM is used. We see that the frequency of the dual form is 3, while the frequency of the general plural form is 27. Those who are familiar with astronomy would appreciate the detail.

I also found an interesting connection between the proper names of 27 messengers/prophets and the derivatives of the rood word "RaSaLa" (send messenger). I noticed that both the derivatives of the word RaSaLa and the names of messengers occurred 513 (19x27) times.

The Quran likens the creation of Jesus to of Adam. Interestingly, the frequencies of both names are also equal: each occurs 25 times in the Quran.

Verses 40:67 and 23:14 lists stages of human evolution and the total frequencies of the stages (dirt-17, drop of semen-12, embryo-6, bite-size fetus-3, bones-15, meat-12) and the frequency of the word "insane" (human) are the same: 65.

The chapter al-Hadid (The Iron) of the Quran is 57th chapter, and 57 is the atomic number of one of iron's isotopes. The verse 57:25 mentions the importance of iron and we include the unnumbered *Basmalah* of the chapter, it is 26th verse from the beginning. This equals to the atomic number of iron, which is 26. Furthermore, the numerical value of the word HaDYD (generic iron) is also 26 (8+4+10+4). If we add the article AL (the), that is the word AL-HaDYD (the iron), its numerical value is 57 (1+31+8+4+10+4), equaling to the number of the chapter and atomic weight of a particular iron...

In 1989, I declared a hypothesis/prediction in a speech before a Friday prayer at Masjid Tucson that the number of elements should be maximum 114, equal to the number of Chapters of the Quran. Then, the biggest element discovered in lab was 109. A chemist (Dr. Sabbahi) was present there and he was impressed by my hypothesis because he had read a journal article discussing the issue theoretically. He later gave me a journal article about superheavy elements arguing that that 114 was a "magic number" for stable elements. Years later, in 1998, when a couple of scientists discovered the element 114 in lab, I suggested them to name the element, whose temporary name is ununquadium, Quranium.

Knowing only the facts about the frequency of the word day, month is sufficient to draw the attention of intelligent and unbiased people to the mathematical structure of the scripture. I had a lengthy argument with a Sunni disbeliever who tried every possible trick to deny the fact that the word YaWM (day) occurs exactly 365 times in the Quran. After several round of a long debate, finally he had to agree with these facts. Our argument is available at 19.org and other internet sites.

Though Ayman claims that he is following the Quran alone, it appears that he has not yet purified his mind from the cultural influence of the Sunni paradigm and attitude. The ignorance and backwardness of the followers of Hadith and Sunna, unfortunately is reinforced by their arrogance. They are unable to see the evident miracles of the Quran. They only parrot their scholar's empty rhetoric that the Quran is a literary miracle without even questioning its meaning and implications; such as the subjectivity of literal taste and its lack of universal appeal. They reduce the author of the Quran to the level of their simple-minded scholars. "How could God's word have a mathematical pattern in it? Mathematics? No way!" they exclaim. They ignore the fact that all God's creation behave, interact in the language of mathematics. Because of their math-phobia or religious dogmatism, they try all the tricks and deception to reject the obvious mathematical patterns existing in the Quran.

> 74:30 "On it is Nineteen"

> 74:35-37 "This is one of the greatest (signs). A warning to the human race! For those among you who wish to advance or regress..."

Year 1974: The Prophecy is Fulfilled and The Secret is Unveiled

As a response to the doubts raised by disbelievers, the Quran refers to a secret that will be unveiled in the future.

> 10:20 "They said, 'Why hasn't a miracle come down to him from his Lord?' Say, 'Only God knows the future. Therefore, wait, and I will wait along with you'"

> 25:4-6 "Those who disbelieved said, 'This is no more than a fabrication by him, with the help of other people' Indeed, they uttered a blasphemy; a falsehood. Others said, 'Tales from the past that he wrote down; they were given to him day and night.' Say, 'This is sent down from the One who knows the secret in the heavens and the earth.' Surely, He is Forgiving, Most Merciful."

In his books and articles published before 1974, Rashad had collected data regarding the frequency of letters. He had noticed a little coincidence though: the frequency of the letter Qaf in chapters (50 and 42) where the letter Qaf is an initial alphabet letter, the total frequency of Q was 114, and it was equal to the number of Chapters in the Quran. He interpreted this numerical relationship as Q standing for Quran, like A standing for Apple. However, when he suddenly noticed the early signs of mathematical pattern he reviewed his data one more time.

Rashad did not have any knowledge that his curiosity regarding the meaning of the alphabet letters that initialize 29 chapters of the Quran would end up with the discovery of its mathematical system. His computerized study that started in 1969, the year humans picked rocks from the Moon, finally gave its fruits in 1974 by the discovery of the 14 century-old SECRET.

284

If the Code 19 was going to provide strong evidence for the authenticity of the Quran, then it is reasonable to expect that the identity of the discoverer and the time of the discovery would not be coincidental. Indeed, the events have demonstrated a prophetic design in the timing of this miraculous mathematical design.

It is an undisputed fact that the number 19 is mentioned only in a chapter called "The Hidden," the 74th chapter of the Quran. Juxtaposing these two numbers yields 1974, exactly the year in which the code was deciphered. If you multiply these two numbers, 19x74, you will end up with 1406, the exact number of lunar years between the revelation of the Quran and the discovery of the code. Adding to this prophetic mathematical design is the fact that the derivatives of the name of the discoverer, RShD (guidance), occurs in the Quran exactly 19 times. (I was the one who first discovered this fact in 1986, and Rashad learned it from me. With the conviction that God would not let this secret to be discovered by a crook, I checked the roots of the word RShD in the index. Then, I also noticed that the frequency of the word Bayyinah (proof/evidence) in the Quran was exactly 19 times.)

The Quran, as it proven by the mathematical-literal evidence, employed the numerical system of the population who first received its message. It is a historical fact that during revelation of the Quran, Arabs, like their contemporary Romans, were using letters as symbols for numbers. The system was known as Abjad. A rudimentary research on the history of mathematics will reveal that Gematria was widely used during the time of Muhammad as numbering system until two century after him. Arabs abandoned their ABJAD number system (also called Gematria) after they adopted the Indian numerals, which later would be recognized as Arabic numerals by Europeans. After being abandoned as a numbering system, the ABJAD system continued to live in public memory. Later, like many other concepts, it was abused by charlatans and psychics. Unfortunately, ignorant math-phobic clergymen and their blind followers now consider Gematria or Abjad as a Jewish innovation and a superstition unworthy of being a tool of divine communication. As a starter, I recommend Georges Ifrah's The Universal History of Numbers, a remarkable book on the subject.

The first two verses of Chapter 74 is a divine order about the revealing of the secret:

74:1. O you hidden one,
74:2. Come out and warn!

Years ago, I discovered that the Gematrical value of the 19 letters of these two verses. I found it interesting that when I considered one version of the spelling of the first word, which contains three Alifs instead of two, the number of letters in these two verses is 19. More interestingly, when we add the numerical values of each letter in these two verses the sum is a very familiar number.

Here is the value of each letter:

Index	Letters	
1	Y	10
2	A	1
3	A	1
4	Y	10
5	H	5
6	A	1
7	A	1
8	L	30
9	M	40
10	D	4
11	Th	500
12	R	200
13	Q	100
14	M	40
15	F	80
16	A	1
17	N	50
18	Z	700
19	R	200

1974

The sum is equal to 1974, according to the most popular calendar on this planet, is exactly the year when the hidden secret was discovered!

The relationship between the following numerical relations, in my opinion, is more than interesting.

- The code of the mathematical system of the Quran: 19
- The number of the chapter mentioning the code: 74
- The year of the discovery of the code: 1974

- The number of lunar years between the revelation of the Quran and the year of the discovery: 1406 (19x74)
- And the numerical value of the 19 letters of the first sentence of the chapter 74 instructing the secret to be unveiled: 1974
- The frequency of derivatives of the discoverer's name, RShD, in the Quran: 19
- Add to the above prophecies the prophetic description of the reaction of people and the continuous controversy regarding the function and meaning of the number 19:

74:30-31   "On it is nineteen. We appointed angels to be guardians of Hell, and we made their number only to be discord for disbelievers, to convince the Christians and Jews, to strengthen the faith of the faithful, to remove doubt from the hearts of Christians, Jews, as well as the believers, and to expose those who have disease in their hearts, and the disbelievers; they will say, "What did GOD mean by this allegory?" GOD thus sends astray whoever wills, and guides whoever wills. None knows the soldiers of your Lord except He. This is a reminder for the people."

The discovery of the code became possible with the use of a computer, and the Quran refers to this event centuries ago:

27:82-85 "At the right time, we will produce for them a creature, made of earthly materials, declaring that the people are not certain about our revelations/miracles. The day will come when we summon from every community some of those who did not believe in our revelations/ miracles, forcibly. When they arrive, He will say, "You have rejected My revelations/miracles, before acquiring knowledge about them. Is this not what you did?" They will incur the requital for their wickedness; they will say nothing."

74:30   "On it is nineteen!"

How to Distort and Ignore the Extraordinary Evidence

The account of Rashad's discovery, which I believe was a typical adventure of a typical researcher will shed light on what Ayman tries to shed doubt.

AYMAN: Surprisingly, the best results for Code-19 promoters were achieved not by Dr. Khalifa but by Dr. Richard Voss in his article Diamond vs. Glass. What Dr. Voss did was that he divided the alleged "initials" into "series". Using one consistent method while not tampering with the data set, Dr. Voss got 6 out of 18 "series" that are 19 divisible. He then claims that the probability of such a "pattern" is "1 in 75,000". Thus, 1 in 75,000 books would exhibit this pattern. However, the way he handled Chapter 42 seems to indicate that he tried different ways of dividing it into the series until he got the best 19 divisible totals. For example, he did not combine series 14 and series 16 into one series despite being sequential chapters. This would have given 5 out of 17 "series" that are 19 divisible, which is slightly worse than 6 out of 18.

But even using his method that gives 6 out of 18, Dr. Voss doesn't tell you why he chose this particular method out of the many thousands of possible methods. His method is even totally different to Dr. Khalifa and brother Layth's methods. It is because other methods when used consistently across all the data set actually give worse results. Hence, Dr. Khalifa and brother Layth had to use at least four different methods (or rules as Dr. Voss calls them). For example, as we saw, adding the totals for the chapters individually gives 3 out of 29 and adding the totals for the initials gives 1 out of 14. Dr. Voss's method is actually the one method that gives the best results. This cannot be a coincidence. He doesn't tell you that he tried other methods until he found the best method and chose it, but obviously he did.

If one keeps trying different methods and only presents the best method that gives a 19 divisible count, then one can find similar or better 19 or any other number divisible "pattern" in any book and the probability becomes 1 in 1 not 1 in 75,000 as Dr. Voss proposes. Moreover, since Dr. Voss didn't consider the whole great reading but just some parts, all one has to do is to show that any part of any book has a pattern using the best method and not the whole book.

For example, one can take any book by Shakespeare and keep trying different methods until he finds the best method that gives the maximum number of different groupings of some of the chapters that have a certain prime number divisible count of letters. So all that Dr. Voss proved is that he is more intelligent than Dr. Khalifa because at least he understood that to make the results appear legitimate one has to use one method and he kept trying until he found the best method that gives the highest occurrence of 19-divisibile totals. However, he failed to tell us about what would happen if one used the same approach with any book. His "1 in 75,000" probability would become closer to "1 in 1".

Here is Ayman's strategy: under the guise of conducting a statistical analysis, he disassembles the parts of an integrated system, a system that with each additional part whose statistical probability decreases manifold. It is like analyzing a video record of a fatal car collusion by replaying it in slow motion while taking it out of the context by cutting the moment of start and the impact, then declaring that there was no collusion: "Dear idiot jury members, please listen to your guide: the collusion was just an illusion! If it doesn't hit, you must acquit!" To decrease the impact of the numerical structure and to hide the intention of its designer, Ayman follows a five-folded game plan:

1. Distort the prophetic language in Chapter 74 regarding the role of the number 19 in relation to the authenticity of the Quran and accuracy of its claims;
2. Take the frequency of the initial letters out of its extraordinary numerical context involving the numbers of chapters, verses and words of the Quran;
3. Increase the chance of probability by equating the contextually most reasonable methods of counting with secondary alternatives;
4. Focus on the errors made by observers and on unfinished research, and thus ignore the amazing evidences; and
5. Impute bad motive by misrepresenting the process of the discovery.

With this method, he tries to save his audience from smashing their heads to the divine sign. In my article, "Which One Do you See: Hell or Miracle?," which is available online, I have already exposed the distortion of those who try to cover the unveiled prophecy of chapter 74 by the smoke of Hell in their imagination.

Let's now briefly see a SAMPLE picture of the context ignored by Ayman (Those with asterisks are originally discovered by me):

NINETEEN: God's Signature in Nature and Scripture. EXAMPLES:

The number of Arabic letters in the opening statement of the Quran, *BiSMi ALLaĤi AL-RaĤMaNi AL-RaĤYM* (1:1)	**19**	x 1
Every word in *Bismillah...* is found in the Quran in multiples of	**19**	
The frequency of the first word, Name (*Ism*)	**19**	x 1
The frequency of the second word, God (*Allah*)	**19**	x 142
The frequency of the third word, Gracious (*Raḣman*)	**19**	x 3
The fourth word, Compassionate (*Raḣym*)	**19**	x 6
Out of more than hundred attributes of God, only four has numerical values of multiple of	**19**	
One (*WAĤiD*)	**19**	x 1
Possessor of Great Bounties (*ŻuW AL-FaĎL AL-ÂŽYM*)	**19**	x 142
Glorious (*MaJYD*)	**19**	x 3
Summoner/Editor (*JAMeÂ*)	**19**	x 6
The frequencies of only four attributes of God are multiple of 19. Three of them used in Bismillah (see above). God's attribute Witness (ShaHYD) with its numerical value replaces the first word, Ism (name) of Bismillah, in frequency, and thus completes interlocking numerical symmetry.**	**19**	x 1
The number of different attributes of God in the Quran. (not verified yet) **	**19**	x 6
The sum of all the verse numbers where the word Allah occurs	**19**	x 6217
The number of chapters in the Quran	**19**	x 6
Despite its conspicuous absence from Chapter 9, Bismillah occurs twice in Chapter 27, making its frequency in the Quran	**19**	x 6
Number of chapters from the missing Ch. 9 to the extra in Ch. 27.	**19**	x 1
The total number of all verses in the Quran, including the 112 unnumbered Bismillah	**19**	x 334
29 Chapters of the Quran starts with 14 different combinations of 14 different letters. Their frequencies in the chapters that they initialize demonstrate an interlocking numerical pattern based on	**19**	
The number of all verses containing all those 14 letters.	**19**	x 6
Frequency of the letter Q in two chapters it initializes	**19**	x 6
Frequency of the letter Ŝ in three chapters it initializes	**19**	x 8
Frequency of the letters Y.S. in the chapter they initialize	**19**	x 15

Frequency of the letters K.H.Y.A.Ŝ in chapter they initialize	**19**	x 42
Frequency of the letters Ĥ.M. in seven chapters they initialize	**19**	x 113
The number of all different numbers mentioned in the Quran	**19**	x 2
The number of all numbers repeated in the Quran	**19**	x 16
The sum of all whole numbers mentioned in the Quran	**19**	x 8534
The number of lunar years in which the meaning of 19 remained hidden in Chapter 74, known *al-Muddathir* (The Hidden)	**19**	**x 74**
The year meaning of 19 in Ch. 74 (The Hidden) was unveiled	**19**	**74**
The number of letters in the first statement of Chapter 74	**19**	x 1
The numerical value of those first 19 letters in Chapter 74**	**19**	**74**
The frequency of all the derivatives of the root word RaShaD, the name of the discoverer, happens to be**	**19**	x 1
The numerical value of the main message of the Quran, WAĤiD	**19**	x 1
Frequency of the word Bayyina (evidence/proof)**	**19**	x 1

Let me provide a little more detail about *Basmalah*. Before presenting examples, let's summarize the gematrical (also called numerical or Abjad) value of letters (akin to chemical property of elements) and the frequency of words (akin to physical property) of Bismillah in a table. (The written letters in Arabic are shown in capital letters):

(The written letters in Arabic are shown in capital letters.):

1	BiSM	3	2, 60, 40	102
2	ALLaH	4	1, 30, 30, 5	66
3	AL-RaĤMaN	6	1, 30, 200, 8, 40, 50	329
4	AL-RaĤYM	6	1, 30, 200, 8, 10, 40	289
	Sum	**19**		**786**

In addition to the examples we have demonstrated so far, the following examples discovered by Abdullah Arik, a Turkish-American engineer whom I

291

first met at Masjid Tucson when Rashad Khalifa was among us, makes the mathematical structure of the first verse of the Quran a fascinating interlocking marvel.

PATTERN 0: The First verse of the Quran, which repeats in the beginning of every chapter except one, consists of **19** letters.

PATTERN 1: Write down the respective number of letters of each word next to their sequence number. The resulting number is a multiple of 19 as follows:

1 3 **2** 4 **3** 6 **4** 6 = 19 x 19 x 36686

PATTERN 2: Now, replace the number of letters in each word by the cumulative number of letters until that word. The result is a multiple of 19.

1 3 **2** 7 **3** 13 **4** 19 = 19 x 69858601

PATTERN 3: Replace the number of letters in each word by the gematrical value of that word. The result is a multiple of 19.

1 102 **2** 66 **3** 329 **4** 289 = 19 x 5801401752331

PATTERN 4: Now, replace the gematrical value of each word by the cumulative total of gematrical value of each word up to that word. The result is a multiple of 19.

1 102 **2** 168 **3** 497 **4** 786 = 19 x 58011412367094

PATTERN 5: Replace the cumulative gematrical value of each word up to that word above with the gematrical value of every letter in that word. The result is a multiple of 19.

1 2 60 40 **2** 1 30 30 5 **3** 1 30 200 8 40 50 **4** 1 30 200 8 10 40
= 19 x 6633695422659542210968686384316216

These are five examples observed by Abdullah, among many others.[22] Having witnessed the elegance of the mathematical pattern, I noticed that the pattern was not complete with these five examples. The first three examples constitute a

---

[22] Some of Abdullah's observations published in *Beyond Probability* could be seen as arbitrary manipulations or statistically insignificant samples. Nevertheless, his book contains many examples that are undoubtedly components of an extraordinary theomathical design, like the ones I listed here.

systematic evaluation of the numbers associated with Bismillah's words and letters. The following fourth and fifth are the next reasonable steps. Yet, the sixth step on example 5 was missing. Thus, I made a prediction and put it to the test. My prediction was verified by the system:

PATTERN 6: Consider Example 5 and replace the gematrical value of each letter by the cumulative gematrical value of each letter. The result is again a multiple of 19.

1 2 62 102 **2** 103 133 163 168 **3** 169 199 399 407 447 497 **4** 498 528 728 736 746 786

= 19 x 6642643212175349035956273652600391973549992256467035 5094

These are just a few samples out of numerous facts. A statistical analysis of these six related examples alone shows how improbable to find a sentence like Bismillah in any other men-made book. Both because of its dominating recurrence in the Quran and because its idiosyncratic structure, the Bismillah resembles a live cell. This opening formula that centers around 19 both in micro and macro dimensions remains as meaningful and effective as a genetic code.

Bismillah takes a crucial role in the mathematical system of the Quran. These, however, are too complex for Muslim scholars who cannot even count the number of letters in the Bismillah correctly. They have come up with various counts—18, 20, 21, 22, anything except 19. We do not expect the sectarian clergymen who flunk in such elementary task to appreciate this mathematical harmony.

No, they are not stupid; their brain power is mostly wasted on ridiculous and frivolous issues. By following their ancestors blindly, they have lost their chance from the start. Indeed, the Quran draws a perfect profile of them in 74:31-51. They are dogmatic and ill-fit to appreciate new ideas that may challenge their dogmas. They can split the hair into forty pieces on frivolous issues, but they are unable to properly evaluate our claims regarding the numerical structure of the Quran. Unfortunately, they have lost their chance to witness one of the greatest signs in their holy book, since in reality they do not appreciate the book itself; they give lip service in praising the book because they parrot their teachers and ancestors. The prophetic verses 74:31-51 depict their intellectual and psychological condition very clearly.

What about our atheist scientists? Well, they have their own prejudices; they are too arrogant to see God in their labs. All they do is to read the Book of Nature and find a way to utilize its principles. Generally speaking, the more a scientist reads the details the more he or she loses the big picture. The fanatic atheist shares similar problems with the orthodox religious scholar. One's right eye is blinded by goddesses of Coincidence and Chronic Skepticism, the other's left eye is condemned by gods of Blind Faith and Sectarian Dogmas. Thus, both lose their ability to see the holographic 3-D picture!

Atheists occasionally close both of their eyes to ordinary and extraordinary facts that manifest God's power in every atom and every second of the universe. Though atheism is inferior to agnosticism in terms of their philosophical defensibility, it is still attractive to some. Though materialistic atheism received a big blow with the discovery of the Big Bang, it soon overcame that shock and focused on denying God's existence through evolution. Though the atheist's twist on evolution does not address the cosmological arguments for God's existence, it somehow managed to sway people from believing in intelligent design to mutation and survival of the fittest. The dogmatic and ignorant rejection of evolution by religious people, on the other hand, increased the suspicion and contempt of intellectuals against anything related to God.[23]

This miracle is not just a personal experience; it consists of objectively verifiable facts. Independent people from different countries and background have witnessed and discovered independent of each other the same extraordinary mathematical system. Whenever we encounter denials of this extraordinary sign, our case gets an even stronger indirect support. Why? Because, the inability of dogmatic and arrogant people to witness the miraculous signs is one of the characteristics associated with such signs (6:25; 7:146; 74:31). If you are able to witness this miracle you are in a very small minority, a very lucky minority indeed.

However, witnessing the miracle does not absolve people and guarantee them eternal salvation since they must show their appreciation and pass the tests on their way. We are tested by our reaction to blessings and losess, and we are expected not to get arrogant and greedy with the first and not to get despondent and unappreciative with the latter. Verse 7:175 mentions a prototype who turned back after witnessing divine signs. Verses 2:109; 4:115; 29:38 and 47:32 provides some examples from the past.

The following verses inform us that witnessing miracles is a divine gift.

---

[23] I recommend you to read my article, *Blind Watch-Watchers or Smell the Cheese*. The article can be found at www.19.org and as an appendix at the end of the *Quran: a Reformist Translation*.

2:118 "Those who possess no knowledge say, "If only God could speak to us, or some miracles could come to us!" Others before them have uttered similar utterances; their minds are similar. We do manifest the miracles for those who have attained certainty."

83:18-21 Indeed, the book of the righteous will be in 'Elleyyeen. Do you know what 'Elleyyeen is? A numerically structured book. To be witnessed by those close to me."

74:30 "On it is nineteen!"

Chapter Starting With "Mysterious" Alphabet Letters

Knowing the facts listed above and many more, we continue to see the following facts about the letter combinations initializing the chapters. I do not include any controversial or problematic one. I distinguished some Arabic letters by using different font styles. For instance, you will see the letter Sad in bold S, while the letter Sin in regular S:

Initial Letters/Number	Sure No	Total Frequency
Q50 (Qaf)	42 50	57+57 = 114 (**19**x6)
Ŝ90 (Sad)	7 19 38	97+26+29 = 152 (**19**x8)
Y10S60 (YaSin)	36	237+48 = 285 (19x15)
K20H5Y10 Â70Ŝ90 (KefHaYaAynSad)	19	137+175+343+117+26 = 798 (**19**x42)
Ĥ8M40 (HaMim)	40 41 42 43 44 45 46	(64+380)+ (48+276)+ (53+300)+ (44+324)+ (16+150)+ (31+200)+ (36+225) = 2147 (**19**x113)
Â70S60Q100 (AynSinQaf)	42	98+54+57 = 209 (**19**x11)

I did not include many details in the table above. Ayman, as it appears, is distracted by the few errors or parameters that need more research, and thus he looses the sight of an amazing numerical structure. He reminds me a disoriented hunter who is searching the truth-value of a claim regarding the existence of a lion with the aid of a magnifier. While he is bragging about the power of his magnifier and declaring that the lion does not exist, the lion is staring at him.

Ayman is not all wrong. He levies fair criticism regarding the frequency of letter Nun in Chapter 68:

> **AYMAN**: "As an interesting side note, the promoters of Code-19 never consider that by spelling the Noun out as "Noon Waw Noon" they now have 15 and not 14 initials because of the added Waw. They are conveniently silent on what they are doing with the Waw because it messes up many of their other counts."

For the sake of the argument, I did not include NuN in the table above, since its count in chapter that it initializes (68) is 132, rather than the expected 133 (19x7). Rashad claimed that the initial letter N should have been spelled out as NuN, as an exception of the common usage. Though the ancient manuscripts supported Rashad's prediction regarding the letter Sad, we have yet to verify this prediction. The Tashkent copy is not complete and does not contain Chapter 68. A few other ancient copies I checked were too incomplete. However, the chapter itself contains some interesting clues, such as the expression of *Sahib-ul Hut* and its purpose being revealed by the *Zan-Nun* of chapter 21, and they support the expectation that the number of the letter N in the original text should have been 133, one more than the present manuscripts.

As for Ayman's question whether we should consider Waw as another member of the initials... I do not have a satisfactory answer for that. It is possible to accept the initial Nun as an exception, as the spelled out pronunciation of the letter Nun. However, I am looking forward to see a comparative study of ancient copies of the Quran. The matter is yet to be solved, and I have no doubt that the results will be in harmony with the mathematical system of the Quran. I have not tested the results yet, but the Quran Reader program provides excellent features to test the frequency of letters in 29 chapters.
(www.submission.org/quran/reader/english/)

Ayman expresses another fair criticism and justifiably pulls our attention to the errors in the count of Alifs and the initials containing them.

AYMAN: "However, this is misleading because, for example, as mentioned in the article 19-Fact of Fiction, it was impossible to verify the "Alif" counts. This is not due to any lack of effort on the part of brother Layth but even people who used Dr. Khalifa's own version of the great reading were not able to get a match of his Alif count. Moreover, inconsistent method of counting of the letter "hamza" seems to be confounded with counting errors. This automatically invalidates any results involving the 'Alif'."

"However, as we saw earlier, the count of 'Alif' has been tampered with to get the desired results and hence 'Alif Lam Ra' and 'Alif Lam Mim' results are invalid."

This is why I did not include the chapters starting with A1L30M40 in my Turkish book, *Üzerinde 19 Var* (On It 19) and in many other articles I have written on the subject. There are questions regarding their correct spellings, and modern manuscripts contain many extra Alifs. Knowing the extraordinary nature of the 19-based system, however, I have no doubt that a comparative, critical research based on the ancient manuscripts of the Quran will eliminate this problem, and we will witness that the frequencies of ALM letters in all chapters starting with those letters would give a multiple of 19. I had discussed the issue in my books and articles on this subject.

Ayman, tries hard to ignore the clear facts in the number of HM letters in seven chapters (chapters 40 to 46) starting with these letters, by reminding the reader that chapter 42, in addition to HM, has also A.S.Q. letters.

This means that they had to switch from method 3 to a variation of method 4 in the middle of the process in order to get the desired count. Thus, they violated the second basic principle.

Ayman ignores the fact that chapter 42 separates the two-letter HM (pronounced Ha Mim) from the three-letter combination of A.S.Q. (pronounced Ayn, Sin, Qaf) In other words, he wants us to ignore that HM is the first verse, A.S.Q. is the second verse of chapter 42. Besides, he does not inform the readers that the frequency of letters comprising the second verse, that is, A.S.Q. is 209, is also a multiple of 19. He might argue that the old manuscripts of the Quran did not contain verse numbers. That is correct; but old manuscripts contain colored dots to separate verses. I do not need to explain the fact that if any text contains ordered units, then that text inherently or structurally contains numbers too.

The Design in the Frequencies of Just Two Letters is Sufficient

To cut the chase, I want to share with you, one single detail in the frequency of the letter H.M. This alone should be sufficient to debunk all Ayman's criticism regarding the probability. It is the unappreciative people who are hiding the facts. Seven consequent chapters of the Quran, from 40 to 46, start with a two letter combination, H.M. (Ha, Mim). Only the 42nd chapter contains another letter combination, albeit it is separated as being in the second verse: A. S. Q. (Ayn, Sin, Qaf). We have verse by verse demonstrated the occurrence of each letter in published books.

The total frequency of "H. M." letters in the seven chapters starting with the same combinations is 2147, which is exactly 19x113.

40	64		380		444
41	48		276		324
42	53		300		353
43	44		324		368
44	16		150		166
45	31		200		231
46	36		225		261
Total	292	+	1855	=	**2147**

$2147 = 19 \times 113$

Chapter 42 stands out among other 7 chapters by containing "A. S. Q." letters in addition to "H. M.". Thus, this chapter divides the chapters with initial "H. M." into two groups. In both groups, the frequencies of "H" and "M" letters are multiple of 19. Furthermore, we find another group, comprising chapter 42 and its neighbors 41st and 43rd chapters. Now, please study the mathematical combinations created around the axis of 42nd chapter.

40	64		380		444	
41	48		276		324	
42	53		300		353	
TOTAL	161	+	956	=	**1121 (19x59)**	
						▲
43	44		324		368	
44	16		150		166	
45	31		200		231	
46	36		225		261	
TOTAL	127	+	899	=	**1026 (19x54)**	
						▲
41	48		276		324	
42	53		300		353	
43	44		324		368	
TOTAL	145	+	900	=	**1045 (19x55)**	
						▲
40	64		380		444	
44	16		150		166	
45	31		200		231	
46	36		225		261	
TOTAL	147	+	955	=	**1102 (19x58)**	
						▲

An Orchestra of Numbers and Letters

Could this mathematical system, intertwined like organic molecules, be a mere coincidence? I did consider it as such in the past. Based on probability calculations, I criticized Rashad for including this table of subgroups of HM letters to his book Visual Presentation of the Miracle. Yes, the frequency of the two letters in the main group was interesting since there was only one way of computing them according the consistent system. However, in subgroups, there were 120 possible combinations (21 of 2 chapters, 35 of 3, 35 of 4, 21 of 5, 7 of 6, and 1 of 7 chapters), and out of 120 combination, the chance of having four of them being multiple of nineteen was highly probable. Rashad could not argue against my statistical criticism. He had already published it and he hoped that one day someone might discover more patterns.

Well, Rashad did not live to see that his intuition was right. In 1992, a discovery made by Milan Sulc of Switzerland turned Rashad's insignificant observation to an extraordinary intellectual feast where numbers and letters dance together to create an impressive fusion or harmony of numbers and literature. When Milan called me from Switzerland telling me about his new observation, first I did not take him seriously. I thought it was just a mathematical property. I asked him to give me time to check his claims. Within a few minutes, I was sure that it was not a mathematical property but an intentional design, an extraordinary one. I called Milan and congratulated him for his discovery.

We have already witnessed that the sum of the frequencies of H and M letters in the seven chapters initialized by them is 113 times of 19. When we add the digits of the individual numbers for the frequency of these two letters, we again meet the number 113.

			▲									▲
40	**64**	**380**	6	+	4	+	3	+	8	+	0	=	21
41	**48**	**276**	4	+	8	+	2	+	7	+	6	=	27
42	**53**	**300**	5	+	3	+	3	+	0	+	0	=	11
43	**44**	**324**	4	+	4	+	3	+	2	+	4	=	17
44	**16**	**150**	1	+	6	+	1	+	5	+	0	=	13
45	**31**	**200**	3	+	1	+	2	+	0	+	0	=	6
46	**36**	**225**	3	+	6	+	2	+	2	+	5	=	18
TOTAL	**2147(19x113)**		26	+	32	+	16	+	24	+	15	=	**113**
			▲	D_1		D_2		D_3		D_4		D_5	▲

Now let's study each subgroup of seven chapters where the combined frequencies of ﺣﻢ ĤM are divisible by 19. In the previous pages, we learned that in each of the four subgroups the frequencies of ﺣﻢ ĤM were divisible by 19. Please note the multiplication factors of 19. For instance, note that the factor in the first group containing 1121 ﺣﻢ ĤM letters is 59, since 1121 = 19x59. The multiplication factor in the second group is 54, in the third 55, and in the fourth 58.

Now let's apply to the subgroups the same mathematical calculation we did on the seven chapters in the main group above. Result: Marvelous! (D = Digit)

Formula: $9(D_3) = (D_1 + D_4) + 2(D_2 + D_5)$

Example: $9(16) = (26 + 24) + 2(32 + 15)$

The same numerical-literal pattern you just observed between sum of the digits in the the frequencies of H and M initializing 7 chapters and the factor of 19, yes the same pattern can be seen in ALL subgroups that produce sums of H and M, which are multiple of 19.

40	64	380	6	+	4	+	3	+	8	+	0	21
41	48	276	4	+	8	+	2	+	7	+	6	27
42	53	300	5	+	3	+	3	+	0	+	0	11
TOTAL	1121 (19x59)		15	+	15	+	8	+	15	+	6	**59**
43	44	324	4	+	4	+	3	+	2	+	4	17
44	16	150	1	+	6	+	1	+	5	+	0	13
45	31	200	3	+	1	+	2	+	0	+	0	6
46	36	225	3	+	6	+	2	+	2	+	5	18
TOTAL	1026 (19x54)		11	+	17	+	8	+	9	+	9	**54**
41	48	276	4	+	8	+	2	+	7	+	6	27
42	53	300	5	+	3	+	3	+	0	+	0	11
43	44	324	4	+	4	+	3	+	2	+	4	17
TOTAL	1045 (19x55)		13	+	15	+	8	+	9	+	10	**55**
40	64	380	6	+	4	+	3	+	8	+	0	21
44	16	150	1	+	6	+	1	+	5	+	0	13
45	31	200	3	+	1	+	2	+	0	+	0	6
46	36	225	3	+	6	+	2	+	2	+	5	18
TOTAL	1102 (19x58)		13	+	17	+	8	+	15	+	5	**58**

There are so many details that would take volumes of books to demonstrate. I also did not share with you the interlocking pattern between the frequencies of God's attributes and their gematrical/numerical values and the *Basmalah*. It is by itself an extraordinary phenomenon. I have done an extensive study on the subject and have discovered new elements of the mathematical code. I will present it in my upcoming English book, God willing. Let's turn to Ayman's article:

In conclusion, very briefly and without any fancy calculations one can easily dismiss the Code-19 so-called "mathematical miracle" based on the most fundamental principles of mathematics. This is why there is nothing that the unjust people would like more than to have The God alone promoters placed in the same basket as 19ers. This would make us much easier to dismiss.

By employing the five-pronged deceptive strategy, what Ayman did was to blind himself to "one of the greatest" miracles and misguide his "idiot" audience to *SAQAR* (74:26-37). As his desire to separate himself and his audience from us whom he calls 19'ers, I welcome such a separation. As it appears that he has no philosophical depth of comprehending that without the mathematical structure of the Quran, the MODERN movement of following the Quran alone is doomed to fail. I would like to emphasize this point so that he will not miss it: If you reject Hadith and Sunna without the mathematical system of the Quran, your claim about the Quran has no foundation. If he does not get what I mean, I would be glad to engage in another debate, just to demonstrate that the unveiling of the numerical structure of the Quran is a timely divine blessing for us, since it is a necessary intellectual tool to reject Hadith, Sunna and the men-made sectarian jurisprudence, without doubting the authenticity of the Quran. None of those who promote Quran Alone can justifiably sustain their belief and cause without the prophetic mathematical evidence embedded in the Quran. Code 19 provides intellectual integrity for Muslims.

Criticism Related to the Translation of Verses
These are the signs/miracles of the Book

> **AYMAN**: "From a mathematical point of view, that is all that needed to be said to expose Code-19 promoters' mathematical prowess. Now, let's focus on how Code-19 promoters play similar tricks in the translation of the great reading. For example, here is how Dr. Khalifa translated 10:1: 'A.L.R. These (letters) are the proofs of this book of wisdom.'"

"Dr. Khalifa had to add the word 'letters' between brackets for no reason other than he was well aware of the fact that it doesn't exist in the text. To do this, he also took the word aya/signs totally out. He had to do this because the word aya is never used to mean 'letters' and in fact it is used in the beginning of chapters where there are no alleged 'initials' (For example, see 24:1 and Dr. Khalifa's own translation where he doesn't translate the same exact word aya as 'letters'). Moreover, he had to add the word 'proofs' because it doesn't exist in the original text. But this time he did it without brackets. Please read the original Arabic and verify those facts. The same thing can be said about Dr. Khalifa's translation of verses such as 12:1, 15:1, 26:2, 27:1, 28:2."

"So there is nothing in 10:1 about 'letters' being the proof of the book. In fact, we can empirically verify that the 'letters' in 10:1 are not the proof of the book since even by violating mathematical principles and using inconsistent methods and allowing changes to the data set brother Layth was unable to prove that there is a Code-19 in Chapter 10 in his article 19-Fact of Fiction."

I disagree with Ayman's criticism of Rashad's translation of 10:1; his parenthetical reference is justified by its temporal and textual context. Tough Ayman accuses Rashad of employing tricks in his translation; he is either an unqualified critic or the real trickster. He never deals with the reason why Rashad considered the reference of *AYAAT* (signs/miracles) in the verse as the INITIAL ALPHABET LETTERS. Either he is criticizing without studying the subject of his criticism or he is hiding it from his audience. Though I prefer to translate *AYAAT* as signs/evidence/miracles rather than "proofs," I am in total agreement with Rashad regarding the implication of this and similar verses in the beginning of seven other chapters. I will not translate the word *AYAAT*, so that an objective reader may reflect on its implication on their own:

> 10:1 ALR, these are the AYAAT of the Scripture of wisdom.
> 12:1 ALR, these are the *AYAAT* of the clarifying Scripture.
> 13:1 ALMR, these are the *AYAAT* of the Scripture.
> 15:1 ALR, these are the *AYAAT* of the Scripture, and a clear Quran.
> 26:1-2 TSM. These are the *AYAAT* of the Scripture of clarity.
> 27:1 TS, these are the *AYAAT* of the Quran and a clear Scripture.
> 28:1-2 TSM. These are the *AYAAT* of the clear Scripture.
> 31:1-2 ALM. These are the *AYAAT* of the Scripture of wisdom.

Rashad, in the end of his book, *Quran: Visual Presentation of the Miracle*, reminds the following fact: "The expression, "These are the miracles of this book," is found ONLY in conjunction with Quranic initials." We know that the word *aya* in its singular form is mentioned in the Quran 82 times and it is always used to mean sign/miracle. However, its plural form *AYAAT* is used for both signs/miracles the revelation contained in the scripture. Since, the singular form is never used to refer a verse in the scripture; we can infer that unless it is at least three verses, they do not necessarily count as a divine sign or miracle. For instance, the verse "Where are you going?" cannot be called *AYAAT* (signs) since it is one verse. This is very appropriate, since that expression was and is used by Arabic speaking people daily, even before the revelation of the Quran. This said, let's turn to our subject.

I concede that it is possible to understand the reference of the "these" to be the "following verses." However, I prefer the reference of "these" to be the "previous alphabet letters" for the following reasons:

- The expression "these are signs", which is repeated eight times, is used ONLY with conjunction of alphabet letters.
- If it refers to the verses of the Quran, it becomes a dull and redundant statement. A reader might ask, "Okay, I see that these are verses of the Quran. So what?" Or, if "Okay, these are signs/miracles of the Quran. How?"
- The numerous examples of the code 19 and its evident presence in some Quranic initials are sufficient to reveal the function of these letters. Though we lack a reliable data regarding the number of Alifs, based on our knowledge of the mathematical structure of the Quran we can justifiably expect that when we learn the exact frequencies of these letters in the original text will participate in the 19-based structure.

"On it is nineteen!"

(74:30)

Tested, Disturbed and Punished by Nineteen

Ayman continues to fight against God's signs:

> **AYMAN**: "Another example is Dr. Khalifa's gross mistranslation of 74:31:
>
>> 74:31 We appointed angels to be guardians of Hell, and we assigned their number (19) (1) to disturb the disbelievers, (2) to convince the Christians and Jews (that this is a divine scripture), (3) to strengthen the faith of the faithful, (4) to remove all traces of doubt from the hearts of Christians, Jews, as well as the believers, and (5) to expose those who harbor doubt in their hearts, and the disbelievers; they will say, "What did GOD mean by this allegory?" GOD thus sends astray whomever He wills, and guides whomever He wills. None knows the soldiers of your Lord except He. This is a reminder for the people.
>
> "There are at least three problems with the above translation: Firstly, the noun 'fitna' describing the count of 19 was neglected and left out of the translation and instead the verb 'disturb', which doesn't occur at all in the sentence, is inserted. "

Ayman is technically right, but substantially wrong. Yes, technically the word "fitna" is a noun and should have been translated as "disturbance" rather than as a verb. But, Rashad's translation conveys the meaning accurately. Ayman cleverly contrast the Arabic word "fitna" with the English word "disturb" and expect his audience to see a big difference. So that they will gullibly buy into his "It doesn't occur at all." A little study of the meaning of the word through browsing the verses of the Quran one would learn that the word "fitna" has the following meanings: trial; test; persecution; mischief; confusion; temptation; hardship; punishment; and disturbance. So, instead of choosing a cheap attack, Ayman could have swallowed his "at all" and suggest a technical change in the language of the expression: "as disturbance/confusion/punishment for the disbelievers."

"This is but a magic!"

But, in the following paragraphs Ayman shoots himself in the foot. It has been my experience that whenever an ardent disbeliever like Ayman attacks the prophetic miracle of the Quran God exposes them through their own words. Ayman is trying to describe the mathematical miracle as "Sihr" (magic).

A silly side issue: As you will see below, Ayman is spelling the Arabic word Sihr in a strange way as "Si7r" by using irrelevant numbers in his spelling scheme. Ayman is assuming that he is representing the Arabic letters with numbers that physically resemble them. It seems that he is oblivious to the fact that it cannot be any help for his readers who do not know Arabic. I should not expect more from someone who is fighting against a numerical sign; he will use the numbers in most ridiculous way. Interestingly, Lomax, a Sunni enemy of the miracle too was using unintelligible spelling techniques.

AYMAN: "It is ironic that the promoters of Code-19 present their 'miracle' as scientific mathematical proof. Seeing how they subtly manipulate the data set and quickly switch between methods, it didn't feel like a scientific endeavor and it felt more and more like being in the circus watching a magician do tricks. However, in this case, Code-19 promoters accepted the tricks and believed in them. This reminded me of 2:101-102:"

> 2:101-102 And when there came to them a messenger from The God, confirming what was with them, a party of the people of the book threw away the book of The God behind their backs, as if they did not know. And they followed what the devils recite regarding Solomon's kingship. Solomon did not reject, but the devils rejected by teaching people eye trickery and what was descended on the two kings Harut and Marut at Babylon. They do not teach anyone until they say: 'We are but a 'fitna'/trial/affliction, so do not reject'. So they learn from them what they separate with between the person and his mate, and they are not harming with it anyone except with The God's permission. And they learn what harms them and does not benefit them, and they have known that for who accepted it, there is no share of happiness in the ending, and miserable indeed is what they exchanged with themselves, had they knew.

"Those who teach eye trickery are described as a 'fitna'/test/affliction exactly like the count of 19 is described as a 'fitna'/trial/affliction in 74:31. It is not surprising that promoters of Code 19 rely on number 'tricks' exactly like those who teach eye trickery ('si7r'). It is also not surprising that followers of Code 19 are obsessed with it and seek to explain everything according to it exactly like those who are obsessed with eye trickery/'si7r' seek to explain everything that happens to them as either a result of evil 'magic' or good protection from evil 'magic'. Both groups seek to know the unseen future using their

307

'fitna'/test/affliction. Hence, you will find those obsessed with eye trickery/'si7r' visit 'psychics' to know about the future exactly like Code 19 followers seek to know the future, for example, about when the world will end or world war III will take place. Those obsessed with eye trickery/'si7r' use the great reading in their spells and counter-spells just like those obsessed with counting 19."

"Interestingly, magic/eye trickery ('si7r') is talked about in Chapter 74. In 74:24-25 the rejecters say this is magic that the prophet copied from someone and it is the saying of a human. If we think about it, how do we know if a saying is that of The God or not? "

Unaware of the fact that it is him who is drowning in the middle of "fitna" (confusion; disturbance; intellectual punishment; phobia), Ayman is projecting his condition to those who have witnessed "one of the greatest" miracles.

A group of scholars, including the former head of the Turkish Religious Affairs and my father, published a book critical of my book *Kuran En Büyük Mucize/Quran the Greatest Miracle*. As the title of their collective work, they picked "*The Mythology of 19*." The choice of the "mythology" or "tales from the past" was striking.

> 6:25-26 And from them are those who listen to you; and We have made covers over their hearts to prevent them from understanding it, and a deafness in their ears; and if they see every sign they will not believe; even when they come to you they argue, those who reject Say: 'This is nothing but the tales from the past!' And they are deterring others from it, and keeping away themselves; but they will only destroy themselves, yet they do not notice.

Ayman is using a different word, but again his word choice exposes him. He is no different from those who did not appreciate and did not deserve to witness the previous miracles. They confuse God's signs with magic:

> 7:132 And they said: 'No matter what you bring us of a sign to bewitch us with, we will never believe in you.'

> 54:1-2 The Hour has come closer, and the moon has split. Then they saw a miracle; but they turned away and said, 'Old magic.'

The more striking is that he has condemned himself through the Quran. His zeal to mislead himself and others from witnessing a great miracle of the Quran, which he ironically calls "great reading," has blinded him to a verse in chapter 74, just before the verse "On it is Nineteen." This is indeed a prophetic exposition of the Quran. Though he refers to the verse, he forgets to look at the mirror. God knew that the disbelievers and hypocrites would call the great mathematical sign of the Quran magic. Let's see who are those who come up with the accusation of SIHR (magic) and FABRICATION in verses leading to the number 19, and what will happen to them?

74:19   So woes to him for how he analyzed.
74:20   Then woe to him for how he analyzed.
74:21   Then he looked.
74:22   Then he frowned and scowled.
74:23   Then he turned away in arrogance.
74:24   He said, "This is nothing but an impressive magic."
74:25   "This is nothing but the words of a human."
74:26   I will cast him in the *Saqar*.
74:27   Do you know what *Saqar* is?
74:28   It does not spare nor leave anything.
74:29   Manifest to all the people. [24]
74:30   On it is nineteen.

Unfortunately, it is Ayman who is stubborn against the *AYAAT* (signs/miracles/revelation) of the Quran and he is the one who rejects them by pontificating: "it is a trick, it is a magic!" As his article is evidence, he is subjected to the *SAQAR* of 19! He will suffer from noticing examples over examples of the 19 in the scripture and nature, and he will try harder and harder

---

[24]   Traditional translations that tend to understand *Saqar* as hellfire, mistranslates the verse *Lawahatun lil bashar* as "scorches the skin." Though this meaning might be obtained by using different dialects of Arabic, the Quranic Arabic is very clear regarding the meaning of the two words making up this verse. The first word *LaWaĤa*, if considered as a noun, literally means "manifold tablets" or "manifestations" and if considered as a verb, it means "making it obvious." The second word *lil* means "for." And the third word *BaŞaR* means "human being" or "people." "For the other derivatives of the first word, *LaWaĤa*, please look at 7:145, 150, 154; 54:13; and 85:22. In all these verses the word means "tablets." For the other derivatives of the third word, *BaŞaR*, please look at the end of the verse 31 of this chapter and 36 other occurrences, such as, 3:79; 5:18; 14:10; 16:103; 19:17; 36:15, etc. The traditional translation of the verse is entirely different from the usage of the Quran. Previous generations who were not aware of the mathematical structure of the Quran perhaps had an excuse to translate it as a description of hell, but contemporary Muslims have no excuse to mistranslate this verse.

to blind himself and others to them. I hope Ayman will reflect on his position stop leading this path of un-appreciation and arrogance.

Ayman continues to distort the context and implication of verses. Let's read:

> **AYMAN**: "According to 4:82, we are told that had the reading been from someone other than The God then we would have found in it many inconsistencies. What we see in fact is that Code 19 contains many inconsistencies that had to be 'fixed' by Code 19 promoters through manipulation of the data and use of inconsistent methods. Since it contains many inconsistencies, it cannot be from The God. Certainly, The God did not give us this inconsistent Code 19 and had to wait for humans to come around and fix it. "

> "Due to the inconsistency of Code 19 and the tricks that its promoters had to do to make it appear as if it is consistent, if it is taken as part of the great reading and not a man-made falsehood then it actually inadvertently supports the rejecters' statement in 74:24-25."

A confused argument! With the same logic, we should trash ALL the translations of the Quran with their ENTIRETY, since we will find some inconsistencies within the same translation and between different translations. According to Ayman, they have nothing to do with the Quran. Instead of rejecting a particular inconsistent or erroneous observation regarding the mathematical structure of the Quran, Ayman promotes a "wholesale" attitude; he wants his "idiot" audience to reject every example of mathematical system whenever they see someone came up with an inconsistent observation or claim.

"It Is Not a Number, It is Counting; and Counting is Bad"

Now, we are coming to a novel criticism. I have read dozens of books, and hundreds of articles in several languages critical of code 19, and I have not yet seen the following argument:

> **AYMAN**: "The second problem with Dr. Khalifa's translation is that the word '3ida(t)' doesn't mean 'number', it means 'count'. "

> "Verse 33:49 makes it clear that the meaning of '3D(t)' is clearly not 'number' since the divorced women do not owe a 'number' but they owe a 'count'. They have to count three menstrual cycles. This doesn't mean that women only have three menstrual cycles or that they only get three in a year. What this means is that out of the number/3DD/عدد of

310

menstrual cycles from the time of divorce, they only owe the count/3D(t)/ عَدَّة of three."

"Verse 65:1 talks about being accurate with the count/3D(t)/ عَدَّة There is nothing accurate or inaccurate about any number. Accuracy becomes an issue only when there is measurement (i.e. counting). So once again, we see that 3D(t)/ عَدَّة is what we measure/count out of a larger population (see 9:36-37, 65:1-4) or estimate from possible alternatives (see 18:22). "

"Notice the same word used in 9:46 to indicate what we count in preparation for battle. For example, we prepare by counting supplies based on the military objective. We don't just get any absolute number of supplies. On the other hand, the Arabic word that is used in the great reading to mean 'number' is the word '3adad'/ عَدَد"

"Therefore, the 'fitna'/trial/affliction is in counting 19 and not in the number 19 itself. The Code-19 promoters are counting the letters and words of the book for no reason other than that this is the count of the 'malaika'/controllers given in 74:30-31. Hence, the other day I opened my email and found 19 new messages in my Inbox. I didn't think much of it but a Code-19 follower would have obsessively connected this to the count of the 'malaika'/controllers. This is the root cause of all their 19 counting."

First, Ayman forgets that he too could not escape from counting. To write this article, Ayman either spent hours and perhaps days counting units of the Quran, or he counted nothing. Either way, Ayman has shot himself in the foot, again.

If he wrote this article without doing any counting, then he is criticizing our counting based on ignorance and prejudice. Who knows, he might have thought that his audience might tolerate this behavior. However, if he did count, then according to his own argument he fell into the middle of the "fitna"; he joined us, the counters! Whether he was able to see a miraculous pattern in the end of his counting or not, whether he counted accurately or not, by merely counting, he has already joined our circle of "fitna."

Besides, how much counting is "fitna"? How about counting the number of chapters? Do those who count the number of chapters and reach the number 114 (19x6) fall head down into this hellish "fitna"? If yes, then what about those who do not count themselves but rely on the count of those "maftun" counters (those

311

counters who are confused; or fell into discord and mischief)? "Counting to condemn the act of counting" is a traji-comic paradox Ayman condemned himself to.

68:5   You will see, and they will see.
68:6   Which of you are confused/mischievous.
68:7   Your Lord is fully aware of those who strayed off His path, and He is fully aware of those who are guided.
68:8   Do not obey the rejecters.

Even if we concede to Ayman this point, nothing will change regarding the facts prophetically described in 74:31:

The act or result of the counting OR the number 19 will:

- Test and confuse the unappreciative people.
- To convince and remove doubt in the hearts of those who received the book, as well as believers.
- Increase the faith of believers.
- Leave the hypocrites and unappreciative people in the darkness: they will not understand the meaning and implication of the count or the number 19.

The discovery of code 19 and its miraculous system in the Quran has fulfilled and continue to fulfill ALL the four functions stated in 74:31, one by one. Ayman might fool himself and his audience by twisting the meaning of the words in the hope of creating some ambiguity in a clear language of a fulfilled prophecy, but it is Ayman and his misguided idiot followers who are the "unappreciative" people.

Let's assume for a moment that the word ADDah in verse 74:31 is referring to the ACT of counting not the count or number 19. What about the second group? I have met many people who received the book, from Jews and Christians, whose doubts were removed after they witnessed the miraculous mathematical system of the Quran. Ayman may even have met some of them on the Internet forums. If Ayman wants, I can arrange a meeting between Ayman and a group of former Jews and Christians whose doubts regarding the Quran were removed because of the number 19. Now here is my question:

Can Ayman show me a single Jew or Christian whose doubt in the Quran being removed because of witnessing "our sinister and evil act of counting"? In the past, I had similarly challenged the 19-phobic Sunnis who wished to limit the function of the number to the hell. I ask them, "Can you show me a single Jew or Christian whose doubt in the Quran being removed by the number of the guardians of hell, without its numerical function in the Quran?" They do not hear the question. Like his ditto-head Sunnis, Ayman and his followers too will most likely ignore this challenge. Well, not exactly. Ayman, as it appears, is aware of this problem and yet he is surreptitious about it. He cunningly tries to reduce the people of the book to "those who received the great reading," that is, the Quran! Any student of the Quran knows that the reference of the "people of the book" is more general in the Quran and is primarily used to describe Christians and Jews.

74:32	No, by the moon.
74:33	By the night when it passes.
74:34	By the morning when it shines.
74:35	It is one of the great ones. [25]
74:36	A warning to people.
74:37	For any among you who wishes to progress or regress.

This argument shows how desperate Ayman is. Referring to verse 33:49, Ayman wants us to believe that a divorced woman has duty of "counting", instead of her duty to fulfill the number of specified days. Perhaps he thinks a divorced woman will count One, Two, Three, Four... Perhaps a bit louder so that her ex-husband could hear. Joke aside, the Arabic equivalent of counting is "Adda-Yauddu" or "ahsa-yahussu." Let us first look at the word ADD, which means: to count, computation, number, reckon.

The Quran contains exactly 57 derivatives of the word ADDah. Guess what? I have a bad news for Ayman: this number is divisible by 19. According to Ayman's silly logic, God does not want me to count or notice the numerical

---

[25] The majority of traditional commentators incline to understand the references of this verse and the ending phrase of verse 31 as "hell fire" instead of "number nineteen." You will find translations of the Quran using parentheses to reflect this traditional exegesis. According to them, "it (the hellfire) is a reminder for the people," and "this (the hellfire) is one of the greatest (troubles)." Their only reason for jumping over two textually closest candidates, that is, "their number" or "nineteen," was their lack of knowledge about Code 19. They did not understand how a number could be a "reminder" or "one of the greatest miracles." Indeed their understanding was in conflict with the obvious context; the topic of the previous verse 31 is the number nineteen, not the hellfire. Finally, the traditional understanding contradicts verse 49. Those who know Arabic may reflect on the word *ZKR* in the ending phrase of verse 31 with verse 49 and reflect on the fact that hellfire is not something the Quran wants us to accept and enjoy!

313

harmony in His book, but somehow He has filled his book with such numerical harmony! He even provides us with another example of 19 in the very word that means COUNTING and NUMBER, the word that Ayman wants so desperately to distance from the number 19!

Not for Ayman's intellectually challenged audience, but for my intelligent audience, I am going to quote the verses that contain the eight occurrences of the very word ADDah, a word that Ayman is trying to hurl into abyss of ambiguity. I will also quote the verses where this word is attached to pronouns. I will highlight the translation of the words so that you will reflect on its meaning in their context. To test how fitting is Ayman's claim, you may substitute our Number/Count with his suggested Counting/Act of Counting:

2:184   Numbered days. Whoever of you is ill or traveling, then the same NUMBER from different days; and as for those who can do so but with difficulty, they may redeem by feeding the needy. And whoever does good voluntarily, then it is better for him. And if you fast it is better for you if you knew.

2:185   The month of Ramazan, in which the Quran was sent down as a guide to the people and a clarification of the guidance and the criterion. Therefore, whoever of you can observe the month, let him fast therein. And whoever is ill or traveling, then the same NUMBER from different days. God wants to bring you ease and not to bring you hardship; and so that you may complete the COUNT, and glorify God for what He has guided you, that you may be thankful.

9:37   The NUMBER of the months with God is twelve months in God's record the day He created the heavens and the Earth; four of them are restricted. This is the correct system; so do not wrong yourselves in them; and fight those who set up partners collectively as they fight you collectively. And know that God is with the righteous.

9:38   Know that accelerating the intercalary is an addition in rejection, to misguide those who have rejected by it. They make it lawful one calendar year, and they forbid it one calendar year, so as to circumvent the COUNT that God has made restricted; thus they make lawful what God made forbidden! Their evil works have been adorned for them, and God does not guide the rejecting people.

18:22   Some would say, 'They were three; their dog being the fourth,' while others would say, 'Five; the sixth being their dog,' as

314

they guessed. Others said, 'Seven,' and the eighth was their dog. Say, 'My Lord is the best knower of their NUMBER.' Only a few know them. Therefore, do not argue with them; just go along with them. You need not consult anyone about this.

33:49 O you who believe, if you marry the believing women, then divorced them before having intercourse with them, then there is no INTERIM required of them. You shall compensate them, and let them go in an amicable manner." ()

65:1 O you prophet, when you people divorce the women, you shall ensure that a divorce INTERIM is fulfilled. You shall count (aHSuW) such an INTERIM precisely."

74:30-31 On it is nineteen. We appointed angels to be guardians of Hell, and we made their NUMBER to be only a trial/punishment for the unappreciative, to convince those who received the book, to strengthen the faith of the faithful, to remove doubts from the hearts of those who received the book, as well as the believers, and to expose those who harbor doubt in their hearts, and the disbelievers; they will say, "What did GOD mean by this allegory?" GOD thus sends astray whoever wills, and guides whoever wills. None knows the soldiers of your Lord except He. This is a reminder for the people.

Now let's finish this section with a verse that contains both the act of counting (HaSaYa) and the number (ADaD):

72:28 This is to ascertain that they have delivered their Lord's messages. He is fully aware of what they have. He has COUNTED the all things in NUMBERS.

Galileo was so right when he declared that the language of the universe is mathematics. If the designer of the universe also the designer of the Quran, then we should expect to find His mathematical signature embedded in the His word too. **But, Ayman's god is a god whose only argument against a disbeliever who claims that the Quran is manmade is to burn that person in hell-fire!** Ayman sees hell, fire and smoke in chapter 74, while we see a great prophecy with an intellectual and spiritual blessing and an intellectual punishment for the unappreciative and hypocrites.

315

Choosing hell over miracle, choosing regression over progress

The following twist, however, demonstrates how cunning and tricky people become just to protect their darkness from the light, their hell from the blessings, and their false imaginations from the truth:

> **AYMAN**: "The third problem is that the structure of the sentence was changed so that now 'disturb the believers' is one of the objectives instead of being part of what accomplishes the objectives that follow."
>
> "The list of objectives is not related just to 'assigning their number' as Dr. Khalifa claimed, it is related to 'making their count as a 'fitna'/trial/affliction for those who have rejected.'"
>
> "Thus, it is not the 'number' that accomplishes the following purposes but the 'count being a 'fitna'/trial/affliction for those who have rejected' that accomplishes the following purposes:
>
> a) Result in certainty and elimination of doubt of those who were given the book and those who had faith and increase in faith those who already had faith.
>
> b) Result in those in whose hearts there is sickness (they are the hypocrites as per 2:10) and the rejecters asking: 'What did The God want with this example?'"

Since my answer to this claim might take too much space, and already this response far exceeded the length initially I was expecting, I will refer you to my article, "*Which One do you See: Hell or Miracle?*" which is available at 19.org and some other internet sites. The article is a conclusive refutation against this and similar distortions. As I discussed in detail, the language of verses in chapter 74, miraculously has double meaning. If someone chooses to see hell and discord there, they are left with some excuses to get what they wish for (3:7; 17:82; 18:29; 74:31)

But, those who are open for God's blessing and are not arrogant to witness miracles, they will witness a great prophecy of a fascinating miracle and good news. Many verses of the Quran emphasize this divine design.

He is also distorting the implication of the question, "What did God want with this example?" He wants his audience to believe that the verse is criticizing

316

those who are curious about the meaning of the Quranic verses, such as "On it 19." In fact, the question depicts the ignorance of disbelievers and hypocrites who are not willing to understand the examples given by God. The Quran provides examples for people to think. Thinking starts by questioning: "What does it mean?" Without even asking that question, by God's will, Rashad and some believers witnessed the meaning of the example of the number 19. But, it is Ayman and his ilks are still in darkness regarding this example. They are the ones who have that question. They may not utter it publicly, but it is in their mind.

God does not cite examples in order to stop us from thinking and reflecting on them, as Ayman wants his audience believe. To the contrary, God cites examples for us so that we may take heed with them. Let's reflect on the relationship between the following three verses. I indicated the root of key words in our discussion:

> 14:25-27 It produces its crop every season, as designed by its Lord. GOD thus cites the examples (MaThaL) for the people, that they may take heed (ZaKaRa). The example of the bad word is that of a bad tree chopped at the soil level; it has no roots to keep it standing. God strengthens those who believe with the proven word, in this life and in the Hereafter. And God sends the transgressors astray. Everything is in accordance with God's will.
>
> 3:7   Only those who possess intelligence will take heed (ZaKaRa).
>
> 38:29   A Scripture that We have sent down to you, that is blessed, so that they may reflect upon its verses, and so that those with intelligence will take heed.
>
> 74:36-37 A warning/warning (ZaKaRa) to the human race. For those among you who wish to advance, or regress!"

In brief, the Quran does not condemn those who show intellectual curiosity to learn the meaning of its verses. The Quran does not reprimand the curious believer who asks, "What is the meaning of this?" To the contrary, it reprimands the arrogant and ignorant person who closes his ears, eyes, and mind to God's signs and then wonders in darkness, "What is the meaning of this?"

Though Ayman and his ilk are trying to hide that they are the ones with that question burning their minds; they have come up with an answer; a false one. In fact, it is evident from Ayman's article that Ayman has asked the very question,

"What God mean by this" (74:31) and came up with only this answer: The number that will misguide those who witness a miraculous numerical structure in the Quran, such as the number of chapters and letters of Bismillah! So, by trying to find an alternative answer to the meaning of the number 19, Ayman himself must have asked the very question. Since, he handicapped himself to come up with the right answer he is still in the category of people who are expressing their lack of understanding by this question.

Interestingly, none of those who witnessed the mathematical miracle of the Quran witnessed it by wondering about the 19. Even the discoverer never asked such a question. The history of discovery shows that Rashad wondered about the meaning of alphabet letters in the beginning of chapters. He asked his question, followed a scientific method to find the answer, and ultimately he was blessed with an marvelous answer. After the discovery of the pattern, he remembered the number 19 of chapter 74 and at that moment, the connection was established. As for Ayman and his ilk, despite the fulfillment of the prophecy, they are still wrestling with the question. Once a while they see a part of the miracle, but it is as short as lightening. The moment lightning stops, they are left in their darkness suffering from the hell fire in their imagination.

How much of the Quran can/should we understand?

AYMAN: "We see that those who ask: 'What did The God want with this example?' actually pursue the 'fitna'/trial/affliction and get answers that they claim provide the book's original true interpretation ('ta'awil') as described in 3:7. Based on their imagined (since none knows it except The God) original true interpretation ('ta'awil') they confidently make prophecies and give exact dates of future events and remove parts of the great reading that they are sure 'didn't originally belong in the book'. "

"We see that those who pursue the 'fitna'/trial/affliction of counting 19 actually reject parts of the great reading and we see the hypocrites pursue the allegorical signs while wrongly believing to be reformers (2:10-12)."

"We can see that the majority of those who follow Code 19 actually still rely on Hadith despite denying it (for example to reject 9:128-129) and are likely to be found next to the Sunnis and Shias spinning seven times around their pagan black cube idol and performing other similar mindless rituals such as stoning some stone pillars."

318

"When those who were given the book and those who had faith actually see the application of 3:7 as demonstrated in front of our eyes through 'making the count of 19 as a 'fitna'/trial/affliction for those who have rejected', this results in their realization of the truth of the book as described in 3:7, elimination of doubt and increase in faith. The sign in 3:7 is crucial to how we properly approach the book. Thus, realizing it through a clear demonstration of its application has profound implications on increasing our faith and eliminating any doubts. It is amazing how The God accomplishes a positive objective from something negative. We see this pattern in all of His creation."

Before discussing how the meaning of verse 3:7 has been distorted by Ayman and Sunni scholars, I would like to briefly address his criticism regarding several points: I do not reject 9:128-129 based on Hadith; I reject them based on the testimony of a numerically coded book, that is, the Quran. Does Ayman have any evidence, besides the number of people believing in them? Let me repeat in other words so that Ayman will understand my challenge: since you reject the testimony of the Quran based on numbers, then what is your evidence that this or that verse belongs to God? Excluding the number of the heads in a particular religious crowd? As for Ayman's innuendo of "black cube idol," it sounds a good slogan, yet I am not the right address to receive it; I have personally never revered such an idol! Thirdly, I do not believe in throwing pebbles to Satan residing somewhere in the dessert.

Ayman is establishing his argument or accusation on the distorted meaning of 3:7. This is one of the worst and most common distortions in the Quran, since it discourages us from studying and learning the Quran. Ignorant people have abused the verse to promote ignorance, exactly what Ayman is trying to do. I will quote one of my previously written articles on this very important issue.

Is the Quran meant for everyone to understand? Or are some parts inaccessible to mere human beings?

Verse 3:7 is one of the most commonly mistranslated verses; it is extremely important, since it deals with these very questions. Different interpretations of this verse can lead to two totally different conceptions of Islam!

The traditional rendering suggests that some Quranic verses can never be understood fully. Yusuf Ali's translation of this critical passage represents the orthodox understanding:

319

> 3:7   "He it is Who has sent down to thee the Book: in it are verses
> basic or fundamental (of established meaning); they are the
> foundation of the Book: others are allegorical. But those in
> whose hearts is perversity follow the part thereof that is
> allegorical, seeking discord, and searching for its hidden
> meanings, but no one knows its hidden meaning except God.
> And those who are firmly endowed by knowledge say: 'we
> believe in the Book; the whole of it is from our Lord:' and none
> will grasp the Message except men of understanding."

Here is our translation of the verse:

> 3:7   "He revealed to you this scripture, consisting of straightforward
> verses, which are the essence of the scripture, as well as verses
> with multiple meanings. Those who harbor doubt in their hearts
> will pursue the multiple-meaning verses to create confusion, and
> to seek their meanings. No one knows their meaning except God
> and those who are deeply rooted in knowledge; they say, 'We
> believe in this; it all comes from our Lord.' Only those who
> possess intelligence will take heed."

I will share an excerpt from the Introduction of *The Quran: a Reformist
Translation*, where I defended our translation to the Orthodox distortion on 3:7.
Since it is written by me, I will not use quotation nor give indentation.

The Arabic word we have translated as "multiple meanings" is "*mutashabih*at."
The word comes from "shabaha" ("to became similar").

The word can be confusing for a novice. Verse 39:23, for instance, uses
"*mutashabih*at" for the entire Quran, referring to its overall similarity -- in other
words, its consistency. In a narrower sense, however, "*mutashabih*at" refers to
all verses which can be understood in more than one way. The various meanings
or implications require some special qualities from the person listening to or
reading the Quran: an attentive mind, a positive attitude, contextual perspective,
the patience necessary for research, and so forth.

It is one of the intriguing features of the Quran that the verse about *mutashabih*
verses of the Quran is itself *mutashabih* -- that is, possessing multiple meanings.
The word in question, for instance, can mean "similar," as we have seen; it can
mean "possessing multiple meanings"; it can also mean "allegorical" (where one
single, clearly identifiable element represents another single, clearly identifiable
element).

As you may have noticed, interpretation of the last part of 3:7 depends on how one punctuates the verse. (There is no punctuation in the original Arabic text.)

If one stops after the word "God," then one will assume, as centuries of Sunni and Shiite scholars have, that even those who possess deep levels of knowledge will never be able to understand the *mutashabih* verses. However, if the sentence does not stop there, the meaning will change to the opposite: Those who possess knowledge will be able to understand the meaning of allegorical or multiple-meaning verses.

Here are five reasons we prefer the second understanding of this verse.

REASON ONE: The passage clearly emphasizes the unhealthy intentions of those who fail to understand multiple-meaning verses. With the disease of doubt in their hearts, they try to confuse others by focusing on their own faulty interpretations of these verses. We believe the passage emphasizes this point because the Quran tells us elsewhere that only sincere believers possess the qualities necessary to understand the Quran (as emphasized in 17:46; 18:57; and 54:17).

REASON TWO: The Quran tells us repeatedly that it is easy to understand. (It does so at many points, including 5:15; 11:1; 26:195; 54:17; and 55:1-2.) If one punctuates this verse in the traditional way, there is an apparent contradiction – the Quran is, at least in some places, impossible for any human being to understand -- and Muslims maintain that the Quran does not contradict itself.

We believe the Quran broadcasts a very clear, coherent message. However, there is sometimes a problem with our receiver. If our receiver does not hear the broadcast or cannot understand it well, then something is wrong with our receiver and we have to check it. If the signal is weak, we need to recharge our batteries, or reset our antennas. If we do not receive a clear message, we need to set our tuning to the right station in order to get rid of the noises and interference from other sources. We may, of course, ask for some help from knowledgeable people or experts for this task. If the receiver does not work at all, then we have to make a sincere effort to fix the broken parts. However, if we believe that the problem is in the broadcast, then nobody can help us.

REASON THREE: It is beyond dispute that the Quran encourages believers to study its words with patience. It advises us not to rush into understanding without sufficient knowledge (20:114). Nevertheless, it claims to be easy to understand (see REASON TWO, above). This, however, is not a contradictory position!

321

Experience with the book suggests that both of these statements are accurate. Although it can be explored for a lifetime without conquering all of its subtleties, the Quran, as a whole, is in fact quite easy to understand, revolving as it does around three basic ideas:

There is only one God.
This life is a test.
There will be an accounting for each individual after death.

We believe the Quran really is comprehensible and worthy of sustained, careful study, just as it promises. We believe that whoever opens his/her mind and heart as a monotheist and takes the time to study it, will understand it, and that this understanding will be enough for salvation. We believe that such people will also be inspired to explore it deeply, and will find ample rewards for doing so.

REASON FOUR: In order to believe in all the verses of the Quran, one does not need to be deeply rooted in knowledge. To be a "believer" is a sufficient condition to believe all the verses. However, one needs to have deep knowledge of the Quran in order to understand *mutashabih* (multiple meaning) verses accurately. Therefore, 3:7 mentions a narrow category (those who deeply rooted in knowledge) in relation to those multiple meaning verses.

REASON FIVE: If we follow the orthodox punctuation and translation of 3:7, then, we must, by logical extension, establish a clear definition of what the *mutashabih* verses are ... in order to avoid trying vainly to understand them or teaching others based on them. We thus need a definitive list of the *mutashabih* verses in order to avoid being among those who are condemned in this verse. There is a problem, however: No one has ever been able to compile such a definitive list! What could the criteria for the list possibly be? Surely one person's lack of understanding of a verse should not make a verse "taboo" for all other people. If that were the case, the lowest degree of understanding would be the common denominator for understanding and interpreting the Quran! In this Alice-in-Wonderland school of Islam, there would be a perpetual race towards ignorance!

Unless one is committed to determining the truth by majority vote, then one may want to reflect upon the five reasons listed here for interpreting the verse as we have.

A side note: There are a few Sunni commentators who support our understanding of this verse. For instance, the classic commentary of al-Baydawi prefers this understanding. Please note that Yusuf Ali also acknowledges this fact in the footnote of 3:7: "One reading, rejected by the majority of Commentators, but accepted by Mujahid and others, would not make a break at the point marked Waqfa Lazim, but would run the two sentences together. In that case the construction would run: 'No one knows its hidden meanings except God and those who are firm in knowledge. They say', etc."

The Art of Mutating Alphabet Letters to Words

In order to educate his "idiots", Ayman resorts to one of the old yet failed attempts to explain the combination of letters initializing 29 chapters, 19 of them as independent verses:

> **AYMAN**: "According to 16:103, the great reading is in a clarifying Arabic language ("lisan 3arabiy mubin"), not in some inconsistent esoteric 'code'. To refute the traditionalist and 19ers theories about the alleged initials, all that had to be done is prove one set of the alleged initials to be a word that has meaning and fits in the context. In the article Language Barrier, not just one but eleven of the sets of initials were proven to be Arabic words that have meanings that fit in the context."

> "For some of those who are too deep into the 'fitna'/trial/affliction of counting 19 it might be too late. For others, it may not be too late. My only advice to all is to seek the mercy of The God. At the end, we are not saved by anything except His mercy. On the God we rely, our Lord, do not make us a "fitna"/trial for the unjust people."

Ayman's claim regarding the meaning of letters is interesting. Many scholars who knew Arabic much better than him tried their best to do exactly what he claims to have done. A few even tried to invent some words to explain the initial letters. Some claimed that they are abbreviations, each letter standing for a word, and came up with arbitrary abbreviation schemes. For instance, some claimed that Nun is another word for fish. Commentators like Razi and Suyuti claimed that they were letters to pull the attention of the audience. Zamahshari thought that they were musical sounds to prepare the reader to the Quran. Reportedly, Ibn Abbas argued that ALM stood for *Ana Allahu Ara*, that is, "I am God and I see". He even derived the divine attribute Al-*Rahman* from the combination of three different initials: ALR, HM and N. Taberi contains many hadith narration regarding the meaning of these letters.

Al-Qamus al-Muhit does not even list the two letters, TaHa, as an Arabic word, yet some hadith narrations claim that TaHa is another name for Muhammad.

The closest word to TH I found in Lisan al-Arab meant many things, including "meat," not "O Man." Nevertheless, a hadith reported in the name of Ibn Abbas claims that TaHa means "O Man." Taberi narrates this hadith among many other speculations. To see the extent of the zeal to come up with a meaning, any meaning, let's read the following narration:

> "And it has been narrated from Ibn `Abbas and others that Taha means `O Man'. It has been narrated from some Sunni scholars (`ammah) that Ta refers to the purity (taharah) of the Holy Prophet's heart and Ha refers to its receiving guidance (hidayah) towards God. And it has been said that Ta means the delight (tarab) felt by the inhabitants of paradise and Ha signifies the disgrace and degradation (hawan) felt by the inmates of hell."

Since, there were too many people trying to fabricate new meanings without evidence, each claim was as good as the other. Thus, all failed. Ayman is now joining the crowd and serving us the same spoiled food concocted and cooked by his Sunni and Shiite ancestors. Ironically, he blames Sunnis for not coming up with a meaning for these letters while it is no other than him who is "delving into" or "sinking deep into" the least reliable hadiths narrations found in Sunni sources. Instead, if he delved into the Quran, he would witness one of the greatest miracles in the frequencies of its units, including letters.

No, I will not be satisfied by just one example as Ayman sneakily wants us to accept. He should follow the same standards he was expecting from the numerical code. After ignoring many examples of numerical code in the design of its chapters, verses and words, he even ignored many 19-based numerical patterns in the frequency of letters: he wanted more. He even ignored the fact that he himself did not have a conclusive count of some letters, such as Alif, and did not acknowledge the fact that we needed more research. Now, he wants us to believe his theory, just by one example! An example based on a fabricated hadith! If he is not a used car sales person, he should try that as his career. He has dug numerous hadith books and the footnotes of the dictionaries to come up with a meaning for the letters, and he has failed miserably.

Worse than that, he fell into his imaginary hole he dug in verse 3:7. Now what would be his answer to those Sunnies who would accuse him of trying to understand the meaning of *mutashabih*at (multiple meaning verses) by asking, "What does God mean by this?" If he is claming to know the meaning of alphabet letters, according to his own misunderstanding of 3:7 he could easily be labeled a hypocrite!

324

I checked Ayman's article titled "Language Barrier" at free-minds.org. According to him, the letters initializing the 29 chapters are meaningful words. One of his arguments is the separation of HM and ASQ.

> **AYMAN**: "Also, from Chapter 42 it is clear that the words at the beginning are words and not initials and we can't haphazardly separate them as initials because of the deliberate separation between 'حم' and 'عسق' when they could have been joined and written as: حمسعق"

As you read above, the numerical structure of the Quran has already explained the reason for this separation. Ayman is good at hiding the facts, and thereby introducing his speculation as the only answer. However, the following argument breaks Ayman's record in absurdity:

> **AYMAN**: "Even chapters that start with single letters should be read out as words and not spelled as initials. Spelling them out would be like spelling the single letter ' و at the beginning of Chapter 103 as 'waw', for example, instead of reading it as 'wa'. The God gave us the great reading not the great spelling."

I appreciate Ayman's punchy and slick slogan in the end; it could be a good line in a TV commercial but it is misrepresentation of the facts. Quran is not just an ordinary "reading book;" it is a unique book. God gave us the great reading together with a great evidence/miracle authenticating the extraordinary claim of that reading. To Ayman, the great reading does not have even a tiny evidence for its authenticity. When we query the nature of his praise of the Quran, we will find it empty, a mere lip service repeated by a billion Sunni and Shiite Muslims. Ayman's backward mentality is not capable of appreciating this extraordinary and unique feature, so he wants to belittle those letters with the word "spelling."

None considers the letter W (Waw) of 103 as a mere alphabet letter, since it is clear from its context that it is a part of the first statement. Confusing the Arabic letter W that is used to denote the importance of a word, with Q, N, or S, is just absurd. It is like confusing the English indefinite article A with letter Z. Let me translate the irrelevancy of his thesis to English by listing some statements starting with combination of letters:

KT. Honk if you love peace and quiet.
EE the early bird may get the worm, but the second mouse gets the cheese.
EE. give me ambiguity or give me something else.

325

ET there is no such a thing as "idiot-proof," since someone will always produce a better idiot.

YD. Lottery: a tax on people who are mathematically challenged.

DSK. read this book.

EERSC. be more or less specific.

TSST. Understatement is always best.

EDSE. One-word sentences? Eliminate.

SM always pick on the correct idiom.

SE. Always remember you're unique, just like everyone else

KT. Look at the end of the first and the end of the last.

F unappreciatif people do not witness divine signs and miracles.

A clear conscience is usually the sign of a bad memory.

The letter A in the beginning of the last statement in the list above is obviously not similar to the previous statements. By pulling our attention to the function of the indefinite article "A" initializing the last statement, Ayman wants us to believe that all the initial letters are like the letter A. He claims that DSK means Desk. He then goes further and combines the letter F with the following word "unappreciative" and dividing it into two pieces and come up with "Fun appreciative people do not witness divine signs and miracles." As for the rest, he wants his "idiot" followers to use their own imagination and come up with something.

Ayman screams, "Eureka!" and pontificates, "Since the A in the beginning of the last sentence is a word and F stands for "fun," all the letters initializing other sentences must be meaningful letters and words too!" This is almost exactly what is Ayman doing. If Ayman wants us to consider all other letters like A, wouldn't we ask him, "Are you okay?" (Here is a little challenge for the curious: What about finding out the relationship between the letters and the statements listed above? Clue: ponder on the third sentence starting with **KT**. A word with deliberate spelling error too might help:)

Here is what Ayman suggests for the letters TS (Ta, Sin):

> **AYMAN**: "Another example is in chapter 27 where the word 'tas'/ طس appears. According to Classical Arabic dictionaries, the word means, 'delve/sink deep into something', amongst other meanings. Thus, 27:1 could be translated as follows: 'Sink deep into these signs of the reading and a clarifying book.' (27:1)."

Ayman is not sinking deep into the signs, but he is sinking deep into his imagination. The verse does not have "Fiy" (in, into) after the letters T.S. If they were a verb meaning "tis" (sink deep) then a preposition "fi" (into) should have followed it. None of Ayman's suggested words exists in the Quran. Yes, NONE. I checked Lisan'ul Arab and I could not find TaS there either. However, I found it in Qamus al Muhit listing it under TaSA, indicating that it has also a different spelling without Alif in the end. It provides many meanings including what one Ayman suggests, however, with a little twist: it means to dive into water and it must be used with preposition "fiy" (in, into). Al-Mufradat fi Gharib al-Quran of Asfahani dismisses the "TaS = verb" theory as nonsense. Taberi reports another speculation for TS; it stands for God's attributes Sami (Hearer) and Latif (Kind).

I cannot spend too much time on his fabrications that ignore the grammatical structure of Arabic language, common sense, and most importantly ignores the most reasonable explanation: their frequencies in the very chapters they initialize!

None of the words suggested for initial letters by Ayman and his predecessors exist in the Quran. Neither their derivatives exist in the Quran. None. Zilch. Zero. Out of 29 initial letter combinations, he is not able to show a single one that is used by the Quran other than being initials. Why? Because they are fabricated by hadith books. Some of the words he is suggesting could not even make entrance into the footnotes of the most comprehensive Arabic dictionaries, such as Lisan ul-Arab. Then, why in the heavens, ALM initializes six chapters, if they are functionless letters or meaningless words? Should we wait for Ayman until he is able to make up a word out of them? Or, should we trust Ayman's productivity since he dug up some fabricated meanings for a few others? If HM is a word as Ayman suggesting, then why the Quran repeats this word SEVEN times in the beginning of chapters, and NOT A SINGLE time in the Quran? Ayman might accuse me of asking the dangerous question: "what does God mean by this?" He can ask the question secretly and invent absurd answers for it, but when we question his claims then we become guilty of asking dangerous questions!

41:53   "We will show them our proofs in the horizons, and within themselves, until they realize that this is the truth. Is your Lord not sufficient as a witness of all things?"

12

Intelligent People's Guide to Code 19 (2nd Round)

Connecting the dots of the "letters that are dotted"

In his quick answer, Ayman ignored many points made in my defense of the mathematical structure of the Quran. Those who read my article will notice that Ayman have ignored numerous crucial points made against his position. Now let's deal with his second round attack. If Ayman does not produce new arguments, then his criticism soon would dwindle into a few trivial knit-picking or face-saving gobbledygooks. Let's see:

> **AYMAN**: "1. Mistranslation of the great reading: A new meaning for 'marqum' as 'numerically structured' is invented without providing any argument as to why the clear Arabic meaning was not used. The meaning has no basis in classical Arabic dictionaries. Here is what the dictionaries say about the word: ...."

> "Clearly, Arabic dictionaries say that 'marqum' means 'written/recorded' or 'whose letters are dotted'. There is nothing in there about 'numerically structured'."

This type of silent semantical inventions insults the intelligence of the reader because it doesn't even attempt to provide a justification and hence assumes that the reader is not intelligent enough to make up their own mind about whether the invented meaning is justified or not. It also assumes that the he or she will not verify the meaning against other translations or classical Arabic dictionaries and discover the distortion.

Ayman, the author of the article titled "Idiot's Guide..." has somehow become too defensive regarding his intellect. If I explained why I did not repeat the errors of traditional translations for every correct translation I presented, my lengthy response would be much longer. (However, we will discuss the difference in our translations in the upcoming *Reformist Translation* of the Quran, God willing).[26] It does not take much intelligence to expect that Ayman or another zealot opponent of the mathematical miracle would notice our translation if it differed from traditional one. However, I expected that inquisitive people would learn the reason for our differing from the traditional translations on their own; it seems Ayman is not one of them.

The Arabic word KiTaB (book) comes from the verb KaTaBa (write). So, the word KiTaB means "writing, letter, book" Another form, MaKTuB means "written thing, and letter." If Ayman wants to collapse the meaning of MaRQuM to of MaKTuB, then, the phrase in the verse, "*Kitabun Marqum*" means "a written writing." This superfluous expression is what Ayman suggests to fit "the great reading" in his mind! I argue that the correct translation is "a numbered writing," or "numerically-structured writing."

In the forum of 19.org, Ayman states, "Although 'raqam' means 'number' in modern Arabic, it doesn't mean 'number' in Classical Arabic." What an interesting coincidence! By mere chance, modern Arabic has turned the Quran to a numerical book! Ayman should save his "great reading" from the pollution of the numbers! Well, had he little reflected on the dictionaries he is quoting he would know that the modern Arabic use of the word comes from ancient times.

Ayman is able to use some classic reference books. That is a great plus. However, he is not able to connect the dots, to make sound inferences, or think out of his math-phobic religious box. That is a greater minus. I do not think that Ayman is incapable of making sound inferences; but his dogmatic, arrogant and reflexive reaction to the code 19, unfortunately deprives him from using his intellectual skills properly. Ayman should reflect on his innermost intentions. The following verse describes a dogmatic and ignorant believer who would not be satisfied with any miracle despite his claim of being objective:

> **6:109** They swore by **God** using their strongest oaths; that if a sign came to them they would acknowledge it. Say, "The signs are from **God**; and how do you know that once it comes, that they would not reject?"
>
> **6:110** We divert their hearts and eyesight, as they did not acknowledge it the first time; and We leave them wandering in their transgression.*

---

[26] *The Quran:a Reformist Translation* has been published by Brainbow Press and it is available at Amazon.com, Barnsandnoble.com and other online stores.

6:111 If We had sent down to them the controllers, and the dead spoke to them, and We had gathered before them everything, they still would not acknowledge except if **God** wills. But most of them are ignorant.

Ayman, with his dogmatic and fanatic reaction to the manifest examples of mathematical pattern in the Quran, fits the profile described above. Here are the dots that Ayman failed to connect: in his claim that "marqum means whose letters are dotted." Had Ayman educated himself regarding the history of mathematics, and especially Abjad and Arabic numerals, he could easily connect the dots in that definition. In old manuscripts belonging to the 7th and 8th centuries, we see that dots are rarely used. There are extensive scholarly studies about the emergence of dots in Arabic letters, and archeological evidences show that dots were in use centuries before the revelation of the Quran. Though, Arabic alphabet had 28 letters, and words differed from each other based on those 28 letters, many early scribes did not bother to put dots on letters; they could differentiate letters and words within their immediate and proximate context. They preferred simplicity and economy in writing.

I have studied Quranic manuscripts written without dots. Though I am not a native Arabic speaker, I can read those un-dotted texts. The reading speed increases with more practice. For some words, one needs to solve an intricate literary puzzle. Imagine that the letters B, T, TH, Y, N are all written the same, with no dots! To read an un-dotted book, one needs to be proficient in Arabic, understand the text, remember the context, and be intelligent enough to solve the puzzles quickly. In other words, the early manuscripts were automatically eliminating the illiterate, the context-ignorant, and the idiot from the pool of potential readers. Thus, reading the early manuscripts of the Quran implied comprehension of its literery meaning, its contextual integrity and the intelligence of the reader. With the use of dots and diatrical marks, now even the parrots and idiots pretend to read the Book; albeit, with no or little understanding.

Let's turn back to the definition of *maRQuM*, which according to Ayman "whose letters are dotted." Why? Archeological evidence shows that the letters of the early manuscripts were mostly un-dotted. Then, why the Quran should describe itself with dotted letters? The answer is in Abjad, or Gematria. We know that during the revelation of the Quran, like their contemporary Romans, Arabs were using Arabic alphabet letters in sequence of ABJD to represent numbers. When they used letters text they usually did not need to dot the letters since the peculiar combination and the context would narrow down the alternatives dramatically. But to represent numbers they HAD TO DOT THE LETTERS. Otherwise, none could differentiate the number 2 from 10, 50, 400, or 500. None could differentiate 8 from 600, 70 from 1000, etc. Thus, dotting the letters was not essential for prose or poetry, but essential for representing numbers. Thus, the word RaQaM means digits, and maRQuM means DIGITIZED, NUMBERED, or NUMERICALLY STRUCTURED.

There is another verse where a derivative of the word RaQaMa is used. Verse 18:9 is about the young monotheists who escaped from he oppression of their people and were put in sleep in a cave for 300+9 years. The verse describes them with expression, "*ashab ul-kahfi wal-raqym*" (people of the cave and numbers.) The following verses inform us about a debate regarding numbers. The allegorical language is similar to of those in chapter 74 and it implies a wondrous mathematical sign yet to be discovered.

18:9	Did you reckon that the dwellers of the cave and the numerals were of Our wondrous signs?*
18:10	When the youths hid in the cave, and they said, "Our Lord, bring us a mercy from Yourself, and prepare for us a right direction in our affair!"
18:11	So We sealed upon their ears in the cave for a number of years.
18:12	Then We roused them to know which of the two groups would be best at calculating the duration of their stay.
18:13	We narrate to you their news with truth. They were youths who acknowledged their Lord, and We increased them in guidance.
18:14	We made firm their hearts when they stood and said, "Our Lord, the Lord of heavens and earth, we will not call besides Him any god. Had We said so then it would be far astray."
18:15	"Here are our people; they have taken gods besides Him, while they do not come with any clear authority. Who then is more wicked than one who invents lies about **God**?"
18:16	So when you withdraw from them and what they serve besides **God**, seek refuge in the cave, and your Lord will distribute His mercy upon you and prepare for your problem a solution.*
18:17	You see the sun when it rises, visiting their cave from the right, and when it sets, it touches them from the left, while they are in its spacious hallow. That is from **God**'s signs. Whomever **God** guides is the guided one, and whomever He misguides, you will not find for him any guiding friend.

18:18 You would reckon they are awake while they are asleep. We turn them on the right-side and on the left-side, and their dog has his legs outstretched by the entrance. If you looked upon them you would have run away from them and you would have been filled with fear of them!

18:19 Thus We roused them so they would ask themselves. A speaker from amongst them said, "How long have you stayed?" They said, "We stayed a day or part of a day." He said, "Your Lord is surely aware how long you stayed, so send one of you with these money of yours to the city, and let him see which is the tastiest food, and let him come with a provision of it. Let him be courteous and let no one take notice of you."

18:20 "If they discover you, they will stone you or return you to their creed. Then you will never be successful."

18:21 Thus We let them be discovered so that they would know that **God**'s promise is true and that there is no doubt regarding the moment. They argued amongst themselves regarding them, so they said, "Erect a monument for them!" Their Lord is fully aware of them. Those who managed to win the argument said, "We will construct a temple over them."*

18:22 They will say, "Three, the fourth is their dog." They say, "Five, the sixth is their dog," guessing at what they do not know. They say, "Seven, and the eighth is their dog." Say, "My Lord is fully aware of their number, none know them except for a few." So, do not debate in them except with proof, and do not seek information regarding them from anyone.*

18:23 Do not say of anything: "I will do this tomorrow;"

18:24 "Except if **God** wills." Remember your Lord if you forget and say, "Perhaps my Lord will guide me closer than this to the truth."

18:25 They remained in their cave for three hundred years, and increased nine.*

18:26 Say, "**God** is fully aware how long they remained, to Him is the unseen of heavens and earth, He sees and hears. They do not have besides Him any ally, nor does He share in His judgment with anyone."

18:27 Recite what has been inspired to you from your Lord's book, there is no changing His words, and you will not find besides Him any refuge.

Under the subtitle "More mistranslation," Ayman reacts to the neutral word PHENOMENON in our translation of 46:10. He does not suggest any other word as reference of "similar."

> **AYMAN**: "Of course, the reader can guess what Code-19 promoters claim this 'phenomenon' is. Why do they always have to work so hard at twisting the meanings, adding words out of the blue and inventing meanings? The answer is always that they are trying hard to make the great reading fit their preconceived 'Code 19 miracle'. "

Phenomenon does not mean numerical sign or code 19. However, in the context it could refer to it. Ayman has problem with any implication of divine sign that is extraordinary, that is great, that is phenomenal. He wants to reduce the Quran to another reading book, with an empty word "great" attached to it. A mere lip-service! Whenever, a great feature of that book is mentioned Ayman will be there to fight against it; by hook or crook. Let me present the verse in discussion without the word PHENOMENON. The following argument is prophetic, since it happened in the past, present, and will happen in the future.

46:7 When Our clear signs are recited to them, those who rejected said of the truth that came to them: "This is evidently magic!"

46:8 Or do they say, "He fabricated this!" Say, "If I fabricated this, then you cannot protect me at all from **God**. He is fully aware of what you say. He suffices as a witness between me and you. He is the Forgiver, the Compassionate."

46:9 Say, "I am no different from the other messengers, nor do I know what will happen to me or to you. I only follow what is inspired to me. I am no more than a clear warner"

Rabi Juda: A Witness from the Children of Israel

46:10 Say, "Do you see that if it were from **God**, and you rejected it, and a witness from the Children of Israel testified to its similarity, and he has acknowledged, while you have turned arrogant? Surely, **God** does not guide the wicked people."*

46:11 Those who had rejected said regarding those who had acknowledged: "If it were any good, they would not have beaten us to it." When they are not able to be guided by it, they will say, "This is an old fabrication!"*

46:12 Before this was the book of Moses, as a role model and a mercy. This is an authenticating book in an Arabic tongue, so that you may warn those who have transgressed, and to give good news to the righteous.

He is the one who is trying hard to distance the Quran from its prophecy of "On it is Nineteen." Since he cannot accuse Rabbi Judah of conspiring with us, he prefers to stay silent regarding the discovery of code 19 in the Bible by someone from Children of Israel. Like all the extraordinary evidences, he perhaps considers this another coincidence. (Here is an idea for Ayman and his target audience: you can always accuse us for conspiring with the Jewish Rabbi. Sunnis and Shiites would jump on to this allegation with joy!)

> **AYMAN**: "On a positive note, Edip finally admits that the word *aya* means 'signs' and not 'proofs' and hence indirectly admits that this meaning was invented by 19ers to distort 10:1, 12:1, 13:1, 15:1, 26:1-2, 27:1, 28:1-2, 31:1-2. Unfortunately, he fails to grasp the implication of such admission and instead resorts to strange arguments to continue to justify that *aya* refers to so-called 'initials'. Here is an example:

> "Edip gives up the invented meaning of 'proofs' only to invent another new meaning of 'verses'. It is an indisputable fact that the word 'verses' in Arabic means 'abyat' NOT *aya*. Hence, we hear about 'verses of poetry' as 'abyat shi3r' NOT 'ayat shi3r'".

You can usually learn how desperate is the critic by looking how many times he or she is dancing on splits of hair. Here, Ayman is harping on an inexact choice of a word that is very closely related to the original one. Signs, in the context of the Quran are intellectual, numerical, natural, and spiritual evidences that guide a person to recognize God's message and reform his mind and action accordingly. Thus, Quranic signs prove the authenticity of the divine message. Since, the verses of the Quran contain intellectual, numerical, natural, and spiritual evidence they prove themselves; they are unique, they are self-testifying evidences. That's why the verses of the Quran are referred by the Quran with the plural *AYAAT*. The Quran never uses the word ABYAT for the verses or numbered sentences of the Quran. It is interesting to see Ayman trying so hard not to use the word *AYAAT* or VERSE in his references to the verses or *Ayaat* of the Quran. He just uses numbers to refer to the NAMELESS UNITS of the Quran. Though, this idiosyncratic treatment emanates from Ayman's imaginary linguistic theories. I like it. Numbers are delicious! Hopefully, Ayman one day will appreciate the numerically designed message.

> **AYMAN**: "It can also be logically seen that verses and signs are not equivalent. The purpose of a sign/'aya(t)' is to provide GUIDANCE. Half a sentence or a 'verse' that is meaningless without what came before it doesn't provide guidance and cannot logically be a

335

sign/'aya(t)'. For example, we hear at the end of the story of Moses in 26:10-67: 'in this is an *aya*/sign (singular)'. Surely, The God doesn't mean that the 'aya(t)' is described in just 26:67. Instead it is what is described in all the story from the beginning that provides guidance and hence is a sign/'aya(t)' (singular) despite encompassing several 'verses'. In fact, 26:67 is meaningless on its own and hence doesn't provide any guidance. On the other hand we have 2:164 that contains several 'aya(t)'/signs despite it being a single so-called 'verse'. "

Ayman, as it seems, have not read my response carefully. I argued that since the singular word*aya* (sign/miracle) was never used as a reference to the literal statements of the Quran, one verse or statement cannot be considered a sign or miraculous. I even provided an example for this claim. Ayman repeats my position for an unrelated argument. He conveniently collapses two different words, sign and guidance, into one, while a little study of the Quran will inform the reader that*aya*, in singular form, is used consistently to denote miracles. Yes, miracles are obviously different then guidance, since not all of those who are given miracles is guided by those miracles.

> **AYMAN**: "I agree with Edip that it is dull and redundant to anyone who mistakenly thinks that 'ayat/signs' means 'verses'. The fact is that right at the beginning of Chapter 24, 24:1 points to 'ayat/signs' being in this chapter despite the chapter having no initials. Since 24:1 cannot possibly be pointing to anything other than what is coming AFTERWARDS, now Edip will claim that 24:1 is 'dull and redundant'."

Ayman hides the difference in sentence structure of 24:1 with those that come with initials. Let's read the translation of both verses.

> 10:1   "A.L.R., these are the *AYAAT* (signs/miracles) of the wise book."
>
> 24:1   "A chapter which We have sent down and decreed as law, and We have sent down in it clear *AYAAT* (signs, revelations) that you may remember."

The 24:1 is not just a mere reference, but a descriptive reference. There is a difference between saying "these are statements" and "these are unambiguous statements so that you take heed." The first one is dull and redundant, but the second one is fine and necessary.

336

Moreover, his other point about "these" only pointing to the so-called "initials" is completely negated by 2:1-2 where the same type of pointing device comes right after "ALM" and is pointing to the book. Clearly, the ALM is not the book.

Another distortion or misrepresentation! In my previous response, I did not argue for "these" but for "these are signs." Yes, let me repeat again so that Ayman will not try to distort the facts and my words in order to keep himself and his audience blind to the mathematical miracle: The expression "Tilka *ayaat*" (these are signs) occurs in 8 verses, and in all of them with conjunction of initial letters. The irony is, from his first attack on, Ayman accused us of employing tricks and twisting the meaning of words, while in reality it is him who has made it as his method of rejection of God's signs.

Blinded by the Smoke of Hell in Chapter 74

> **AYMAN**: "2. 'Doctors who smoke' syndrome. At any rate, Edip goes into another lengthy discussion about how the world is not fair to the few closet 19ers scientists and mathematicians. Firstly, let me say that there may be a few 'scientists and mathematicians' here and there that believed in Code 19 much as there are a few doctors who smoke. Yet, no doctor that smokes will ever publish a paper in a peer reviewed scientific journal claiming that 'smoking is good for your health'. Similarly, no scientist or mathematician will ever publish a paper in any peer reviewed scientific journal endorsing Code 19. This is not because the editors of scientific journals such as Nature are bearded Sunni fanatics or are apathetic to the great reading, but it is because they are apathetic to false science."

I already listed the reasons why at this time mathematicians do not show interest in dealing with code 19. Without refuting my reasons, Ayman repeats his demand, this time with more elaboration. He pretends to engage in a debate, yet he is entertaining himself with his monologues.

Would Ayman believe in code 19 if it was published in a peer-reviewed journal? Which journal would be satisfactory? How many journals? Without providing these specifics, whatever happens, Ayman will find another excuse to reject one of the greatest divine signs. If someone finds so many excuses to reject the prophecy of chapter 74, the Hidden, and its amazing fulfillment, that person is able to blind himself to journal articles, even to angles coming down from the sky.

Furthermore, Ayman confuses undesired addiction to intellectual conviction. How in the world, one can liken a mathematician who is intellectually convinced about the existence of a mathematical pattern in a book to the doctor who smokes? It can only be possible if one is blinded by the smoke of hell, a hell that is guarded by 19 angels!

Another Twist

To my response against his criticism regarding the letter SAD, Ayman retorts with the following:

> **AYMAN:** "3. The Code-19 Archeologist. Edip Yuksel wrote: "I understand and respect Ayman's concern regarding the abuse of multiple manuscripts to concoct numerical coincidences. However, categorically rejecting the use of various manuscripts for a critical and comparative study to infer the accurate version or spelling of original Quranic verses is absurd."
>
> "I would like to know, as I am sure many readers, what critical and comparative study did Dr. Khalifa, Edip, or any Code-19 promoter conduct "to infer the accurate version or spelling of original Quranic verses" for words that have no bearing on their counts? "
>
> "The silence is so deafening that you can hear a pin drop."
>
> "Of course 19ers only conduct 'critical and comparative study to infer the accurate version or spelling of original Quranic verses... that fixes their counts'. As for words that don't factor into their counts, they are not worthy of 'critical and comparative study to infer the accurate version or spelling of original Quranic verses."
>
> "It is clear that the "Code-19 archeologist" objective is not to uncover the truth about what really happened. It is to fix a count so that it adds up to 19."

My statement quoted above was a response to Ayman's accusation that we change the spelling of the Quran to fit the count. Ayman was giving the example of letter SAD. When Ayman learned my refutation regarding his claims about the letter SAD, instead of showing the courage and wisdom by accepting the falsity of his accusation, now he is repeating the same accusation with a twist. Instead of talking on specifics, now he wants us to show him the results of our other studies! Ayman has lost the specific argument that he started and now without even referring to it, he is switching to general accusation. He is smart enough to know that general and unsubstantiated accusations do not take many calories nor evidence to come up with. My silence on his new general accusation might be deafening, but his silence on my refutation of his specific accusation is screaming! Almost his target audience can hear.

The "Standard Methodology" Mantra

> **AYMAN**: "4. Evading the Standard Methodology Issue. Instead of addressing the issue, Edip evades it and again hides behind a lengthy nostalgic story recounting of the history of the discovery of the 'miracle'. He seems to be doing this in the hope that the reader will forget about the inconvenience of the requirement for a "standard methodology" because those two words demolish Code-19."

I gave numerous examples in my response and they speak up the methodology. If Ayman found any inconsistency then he should be able to point at them. But, now again he is escaping from specifics to seek a refuge behind esoteric and abstract demands. Similar trick was tried by Lomax, his Sunni ditto-head in our internet debate which was first published in limited edition of Running Like Zebras in 1995. He cried repeatedly about METHODOLOGY and DEFINITONS. Knowing that all was part of his trick to avoid discussing specifics, I did not respond that demand initially. Finally, I provided him with our standard of counting words and letters. Nevertheless, he continued his demand for more and more definitions and methodology. Below is my final answer to Lomax and his ilk:

I do not have any motivation to come up with a coherent statement of exactly what the miracle is for someone who is not able to see the simplest facts. You cannot help someone who stubbornly closes his eyes to the light and complain of not being able to see it. What if that person demands a coherent statement of exactly what the light is? Personally, I would not bother to bring a technical and comprehensive definition of the light for someone who hates the light and demonstrates all kind of blindness in the past.

Ingrate!

> **AYMAN**: "5. Appealing to 'Nusemantics'. Again, instead of addressing the issues, Edip goes on a tangent that has nothing to do with Code-19. He gives a dozen or so instances where the count of words may add up to certain numbers. Of course, none of those counts add up to 19 so even assuming that the different forms of a word were

counted objectively, this has nothing to do with proving that there is a Code-19 in the great reading."

The examples I provided was to demonstrate that the Quran indeed have an extraordinary mathematical structure. Thinking that Ayman might have deep phobia against the number 19, I hoped that sharing with him the examples of nusemantics would allow him to see that the Quran is not just a literary book. Instead of acknowledging and appreciating those examples, he now complains that they were not multiple of 19. (In fact, some demonstrated interesting connection with 19). As for those examples that were directly related to code 19, Ayman just ignored them.

AYMAN: "6. Disregarding the rules of mathematics. Edip then tries to impress the reader by providing table after table of calculations while implicitly admitting that "just two letters" out of all the so-called initials give some kind of a pattern. However, even for just those two letters he does not provide the most basic info needed to evaluate the results in those tables:"

1. An objective measure of the statistical significance of the results.
2. A clearly defined method of how he got the results and how that method was constantly applied.
3. A clear definition of the data set to which the method was applied.
4. A clear explanation of why this particular method was selected.
5. A clearly documented audit trail of BOTH successes and FAILURES. For example:
    - Trials where this method did not give a 19-divisible number.
    - Other methods that were tried and did not give a 19-divisible number.

"Surely, Edip does a fine job of presentation but without the above information then one can make equally impressive presentations about any document. Similarly in his section entitled HOW CAN ONE DISTINGUISH, he didn't provide any of the above info."

Ayman belittles the pattern in H.M. by "just two letters out of the so-called initials." First, these two letters separately are 2 out of 14 alphabet letters used in initials; they are one seventh of all the letters. The combination of H.M., on the other hand, initializes SEVEN chapters. In other words, 7 out of 29 chapters… It involves approximately, one fourth of all chapters with initials. Besides, there is only one more initial that could possibly show a similar pattern: A.L.M. I expect that we should see a similar pattern in the subgroups of ALM-initialed chapters when we obtain the exact frequency of the letter A in this initial.

Though Ayman was presented with the clear and interlocking examples regarding the frequencies of H.M letters and the mathematical relationship between the digits of those frequencies, he wants us to provide him with extensive report. Imagine a skeptic who challenges you that the green hose on the floor is not connected to running water. You turn the faucet on and water the flowers, but the skeptic is still not convinced. He asks you to prove it by providing a document from the Water Company. Well, to prove him the existence of running water you turn the hose towards him and soak him with water. He is now wet from head to toe, yet he still insists that there is no water and challenges you to bring the official proof or scientific explanation how the water runs throug the hose!

I present Ayman with an extraordinary mathematically example from the Quran. Instead of looking at it and appreciating its great design, he is asking for bureaucratic mumbo jumbo. Ayman is soaked with the extraordinary pattern in the H.M initials. Instead of cleaning himself with it, he asks about elaborate reports of his situation! Here is my challenge to him: take it to mathematicians and ask their opinion, sure without telling them the source of the pattern. Or, to pull their attention, tell them that you found it in one of books of Babbage, Euler, Leibniz, Pascal, Newton, or Charles Dickens.

Ayman can always reject each of God's signs by shrugging his shoulder "this is just one example!" He has already received dozens of "just one examples" of extraordinary numerical system. Like the jury of the Rodney King, he will watch one bat at a time, and ignore the magnitude of individual bats. Since we are still discussing initials of the Quran, I would like to share with you a new discovery just posted at 19.org/forum by brother Asad. I did not yet verify or falsify it; but if it is correct it perfectly fits the 19-based system.

The Quran contains 14 initial letters without repetition:

A L R Ĥ M K H Y A Ŝ T S Q N

There are only 114 verses in Total Quran that all initial letters exists in them:

2:140; 2:187; 2:19; 2:213; 2:217; 2:220; 2:228; 2:237; 2:260; 2:275; 2:282; 2:285; 2:286; 2:61; 3:120; 3:154; 3:93; 4:102; 4:129; 4:34; 4:4; 4:43; 4:46; 4:47; 4:69; 4:90; 4:92; 5:110; 5:13; 5:2; 5:3; 5:33; 5:5; 5:53; 5:6; 5:75; 5:89; 5:95; 5:96; 6:119; 6:139; 6:151; 6:152; 6:71; 7:131; 7:160; 7:168; 8:1; 9:121; 9:29;

341

9:34; 9:42; 9:71; 10:24; 10:27; 10:4; 10:93; 11:81; 11:88; 12:101; 12:36; 12:51; 12:80; 12:9; 13:31; 13:4; 13:41; 14:22; 16:76; 17:33; 18:18; 18:28; 18:45; 18:63; 20:130; 20:132; 20:135; 20:71; 22:11; 22:19; 22:54; 23:27; 24:31; 24:41; 24:56; 24:58; 24:61; 26:49; 27:59; 28:10; 28:19; 28:82; 28:23; 28:38; 30:48; 33:71; 35:11; 38:22; 38:24; 39:6; 40:64; 40:78; 42:52; 46:30; 47:15; 47:32; 48:29; 49:7; 57:25; 58:4; 61:14; 65:1; 72:28; 73:20;

As we know that the number 114 is not only 6 times 19, it is also the number of chapters of the Quran, the frequency of Bismillah, and the number of attributes of God, and much more.

> **AYMAN:** "To use Edip's words one could say: 'I want to share with you, one single detail in the frequency of words in Edip's article. This alone should be sufficient to debunk all criticism regarding the probability. It is the unappreciative people who are hiding the facts. The total frequency of the word Ayman in Edip's article is actually 114 (6x19)!'"

> "Yes folks, this is true. Is it a miracle? Perhaps some 19ers would think so. On the other hand, I would say that it is a sign that Edip was focusing too much on me as opposed to on the issue. "

This has nothing to do with the example I presented to be sufficient. The reason is obvious: the example I presented contained several mathematically intertwined remarkable patterns. Besides, Ayman later corrected his count from 114 to 118. Nevertheless, it is nice to see him to connect the dots between a number representing the frequency of his name and its implication. It seems one example was enough for him to notice the implication. This shows that he has no excuse regarding his intellectual capacity to witness numerous extraordinary patterns in the Quran. He needs to check his intentions and get over his math-phobia in relation to the Quran.

They insist to see smoke and fire rather than a miracle!

> **AYMAN:** "7. Ignoring the difference between '3ida(t)/count' and '3adad/number'

> "Had Edip read carefully 74:31 then he would have seen that not just any "count" is the "fitna/trial/affliction" it is the count of the guardians of hell. When I count anything and I get the result as 19 or 19 divisible, I don't link what I am counting to the count of the guardians of hell in

342

74:30. On the other hand, 19ers do. That is the difference between obsessive and normal behavior."

"There is nothing ambiguous about what I am saying. I am clearly saying that 'ADDah (3ida(t))' consistently means 'count'. If I were 'trying to hurl the word 'ADDah' into the abyss of ambiguity then I would haphazardly interchange its meaning between 'number' and 'count' as Edip is doing below: "

Ayman is exaggerating the difference between the word "count" and "number" so much so that novice people might think that it is a crucial point in our discussion. Ayman hopes to "demolish" the patterns of 19 by dancing on a split-of-hair. Even if we replace the word COUNT with the word NUMBER in the translation of 74:31, NOTHING will change regarding the prophetic function of 19 and its emergence in the number of letters, words, verses, their sequences and gematrical values. Ayman is over-exaggerating a trivial word choice, since he has no substantial argument left against code 19. Both in Arabic and English, the words are very close to each other. To translate this issue to English let's look at the Webster's dictionary regarding the word "Count" as a noun:

Count-n. the act of counting; enumeration; reckoning; calculation. The number obtained by counting; the total. An accounting. ...

It is a fact that the number 19 in the Quran, or the act of counting up to 19 in the Quran, has become a fitna/test for Ayman. He is tormented by it and it will only grow with time. Just see a guy named Zlatan, who is now suffering from what I call "chronic nineteen syndrome."

My faith has increased by the number or by the act of counting. Some Christians and Jews who have witnessed the mathematical system based on this number or the result of the counting became convinced regarding the authenticity of the Quran. Hypocrites and unappreciative disbelievers are still confused regarding the implication of this number or count. All these support the miraculous nature of the number 19 in 74:31.

Patently confused about the number of months

I did not know that Ayman had specialty in distinguishing the Count from Numbers. He seems to have dedicated a considerable time to support his thesis.

AYMAN: "Moreover, saying that the number of months is twelve is patently false. If what is meant is the absolute number from the time of

creation of the heavens and the earth, then this is in the billions and not twelve. If what is meant is the number of constant periods in a year then that is never twelve either in any calendar. It is 12.3 in the lunar calendar, solar calendars are forced to have inconsistent months, while luni-solar calendars are forced to have years with 13 months. Therefore what is meant is not the number but what we should count. For more information, please see:
http://www.free-minds.org/articles/science/timing.htm "

Ayman is patently confused. Let's look at The American Heritage Dictionary of Science:

> **AYMAN**: "Lunar month, Astronomy. The period of one complete revolution of the moon around the earth; the interval between one new moon and the next: The proper lunar month, which is called they synodical month, is the period between one new moon and the next, an average of 29 days, 12 hours, 44 minutes, and 2.8 seconds."

> "Lunar year, Astronomy. A period of 12 lunar months, about 354 1/3 days: The Greeks had begun to compensate for the defect of the lunar year, by the occasional addition of an intercalary month."

The Quran informs us that the count/number of months is twelve. Period. That means, after 12 months (12 is a number by the way), we have to start counting from the beginning. Well, that makes it a lunar year! Ayman is not able to comprehend such a simple idea.

> **AYMAN**: "Again, talking about ambiguity and inconsistency, Edip fluctuates between "count", "number" and "interim" for no apparent reason when "count" would fit perfectly in all occurrences."

Of course it does; since the meaning of the word COUNT includes the word NUMBER. Just, look at any good size English dictionary. Thus, depending on the context, the word count could be more precisely translated as NUMBER. There is no inconsistency, ambiguity, or confusion there.

> **AYMAN**: "1. Instead of the action 'to disturb' which doesn't occur in the sentence, now the translation correctly describes the count of the guardians of hell as the 'fitna'/trial/affliction."

> "2. Now 'the number' is not the reason why 'the faith of the faithful is strengthened'. Instead, the reason why 'the faith of the faithful is

344

strengthened' is that they see that 'their count is only a trial/punishment for the unappreciative'."

"Now that Edip largely corrected Dr. Khalifa's mistranslation, he should take the next step and see the clear implication of his correction."

That is a wrong reading of my translation. However, there is a reason why some people like Ayman are allowed to read it such away that they will be blocked to witness the visible miracle of the Quran.

> 6:104 "Visible evidences have come to you from your Lord; so whoever can see, does so for self; and whoever is blinded, does for self. I am not a watcher over you."

Wrong reading, again!

AYMAN: "8. Acknowledgement of the inconsistency in Code-19. As I said at the beginning, there are many positive developments in Edip's response. Here he implicitly acknowledges the inconsistencies in Code-19."

"Edip Yuksel wrote: A confused argument. With the same logic, we should trash ALL the translations of the Quran with their ENTIRETY, since we will find some inconsistencies within the same translation and between different translations."

"Indeed, we should reject inconsistencies in translations. The whole premise of Code-19 is that it is a 'precise mathematical miracle'. Therefore, inconsistencies destroy that basic premise. It is no longer a 'precise mathematical miracle' but is a 'miracle in the eye of the beholder'. Kind of like the golden calf that the descendents of Israel built. Remember, the calf didn't speak back to them or guide them much like Code-19 doesn't guide anyone."

Ayman either does not read carefully the statements he is criticizing or has problem in comprehension. Though I typed the key word in capital letters, he still missed my point. As for his likening Code 19 to the calf, it shows his ignorance. What is analogous to calf is not one of the greatest miracles, but it is Ayman's ego that arrogates him against God's will.

6:24	"See how they lied to themselves; and that which they invented deserted them."
6:25	"And from them are those who listen to you; and We have made covers over their hearts to prevent them from understanding it, and a deafness in their ears; and if they see every sign they will not believe; even when they come to you they argue, those who reject Say: 'This is nothing but the tales from the past!'"
6:26	"And they are deterring others from it, and keeping away themselves; but they will only destroy themselves, yet they do not notice."
6:27	"And if you could see when they are standing over the Fire, they Say: 'If only we could be sent back, we would not deny the signs/revelations of our Lord, and we would be believers!'"

It is no coincidence that among all the 114 attributes of God, only one attribute has numerical value of 19: WAHeD (6+1+8+4=19), which means ONE. Oneness of God is the main theme of the Quran. It is also no coincidence that among all the 57 attributes of the Quran only one has a numerical value of 19: HuDaY (5+4+10=19), that is, GUIDE. The main purpose of the Quran is to guide us to the straight path. (I will inshallah document these and many more numerical facts in the English version of "On it 19".)[27]

Audacity of Ignorance

> **AYMAN**: "9. The misunderstanding of 3:7. The parallel between 74:31 and 3:7 and how they are tied by 2:26 is very compelling. If anything people should be very careful about saying that they know the original meaning of the clearly allegorical sequence leading to and including 74:30."

Ayman ignores all my criticism regarding his mistranslation of 3:7 and repeats himself by associating some verses together. The verses tied together has nothing to do with point of argument, since the only verse talking about the nature of *Mutashabih* (multi-meaning) verses is 3:7 and there the reading must give possibility of understanding those verses, otherwise, the problems I listed would emerge.

---

[27] The book is published under the title NINETEEN: God's Signature in Nature and Scripture, Brainbow Press, 624 pages, 2011.

Ayman wants us to accept the traditional theory of "holy ignorance" regarding the "allegorical sequence leading to and including 74:30" His ignorance leads him to Hellfire in those verses, while our knowledge leads us to witnessing a great miracle. This is audacity of ignorance!

Triple inverse imagination

Though I refuted his fabricated suggestions for initials, Ayman continues his sales-pitch without a blink. To his silly W example, he is now adding another one:

> **AYMAN**: "10. The myth of initials. Firstly, I gave meaning for 11 out of the 29 sets based on classical Arabic dictionaries not based on any Hadith. Moreover, all the meanings I gave fit in the context. So this is not just about giving one example."

> "Here is another example of a word that appears ONLY at the beginning of a chapter and NEVER appears elsewhere. Yet no Code-19 promoter takes this word as initials. For example, Chapter 78 starts with the two letters '3in' and 'mim'. This word '3m' never occurs in any other place in the great reading and it is written as a connected word exactly like '7m', 'alm', 'alr', etc. How come no one ever claims that the word '3m' at the beginning of Chapter 78 is actually initials?"

> "We can even see that Chapter 78 can be read as they read the chapters with alleged initials: "

> 3in Mim. They are asking one another about the great news...

> "Now if we take 3in Mim as a word and not initials: "

> 3M they are asking one another about the great news...

> "Edip will claim that the above is ungrammatical. 2

None other Ayman has considered the first word AMma (contraction of An and Ma, meaning "from what?"), as initial. I have not seen any commentator of the Quran suggesting or even hinting at that. Of course, he could have discovered something that all others have not, but he needs to substantiate, provide evidence for why we should choose his way. However, Ayman lives in a fantasy world. He is so desperate to come up with a meaning, any meaning, but a numerical one! Therefore, he imagines that people could accept the first word of chapter 78 to be an initial. Then, he imagines that he would come up with the meaning

347

"from what?" Then he imagines again that I would claim it to be ungrammatical! I have never seen a fictional scenario this short with so many complications. It is a triple imagination that inverses the facts thrice to reject many more facts!

27: 92 "'And that I recite the Quran.' He who is guided is guided for himself, and to he who is misguided, Say: 'I am but one of the warners.'

27: 93 "Say: 'Praise be to God, He will show you His signs and you will know them. And your Lord is not unaware of what you do.'"

74:30 "On it is Nineteen!".

12

Intelligent People's Guide to Code 19 (3rd Round)

AYMAN: "Edip, it is clear to those who read your article that nostalgic stories about the history of the 'precise mathematical (sigh) miracle' have sentimental value to you and hence you spent a lot of space on them. However, as an author you should keep your audience in mind when writing an article. You should not expect the reader to share your fondness for such 'crucial' stories. I did you a favor by focusing the discussion on ten issues instead of 'crucial' nostalgia and personal attacks. This is evident in that your response to the ten issues was prompt, more to the point, and with less name-calling (you can thank me later). In this response, I will also keep to the ten points."

I am engaging in this debate not only for you, but also for present and future generations who might be curious to know about the details that you call 'nostalgic.' You have indeed ignored to address many substantial points, which is fine. The truth-seeking readers will notice those points and they will learn that the darkness has no way to win against the light; the fire of hell has no power to burn the divine signs; the arrogance and ignorance has no power to cover the clear evidences.

You are the one who started this debate and you were the first one who called us names, such as TRICKY, MAGICIANS, HYPOCRITES, etc. One thing you should learn that you are debating with someone who is different from your targeted audience whose characteristics were defined by you in the title of your article, in accordance to God's will. (47:29; 10:95-102). You might now regret for the title of your 19-phobic article, but it is God's plan to expose people and incriminate them with their own utterances.

AYMAN: "1. Mistranslation of the great reading. Where did I compare your translation to a traditional translation such as that of Yusuf Ali? I never did. I only pointed out the invention of words to make the great reading fit a preconception of Code 19.

"Ayman has misunderstanding. I was referring to his accusation; he accused me of deliberately hiding my untraditional translation from the reader. "

"I don't think that 'rarely' is an accurate description. There are even examples of very early dotting dating to the 4th century and earlier on mundane inscriptions. Please see:"

> http://www.islamic-awareness.org/History/Islam/Inscriptions/
> jramm.html
> http://www.islamic-awareness.org/History/Islam/Inscriptions/
> raqush.html

"Moreover, if 'dotting' was used on such mundane average everyday documents as business letters early in the 7th century (for example see: http://www.islamic-awareness.org/History/Islam/Papyri/PERF55 8.html) then we can be certain that it was widely used:"

"More importantly, as Edip himself pointed out there are early manuscripts of the great reading that are fully dotted."

I was referring to the earliest available manuscripts of the Quran. I recommend you and others to see *The Quran: A British Library Exhibition World of Islamic Festival*, 1976, Martin Lings and Yasin Hamid Safadi. I have studied some of the oldest available manuscripts, some were originals and some were photocopies of the originals. The point of my argument was simple: To read a text, during the revelation era, native Arabs did not need to dot the letters. However, they had to dot the letters when they used letters as digits that add up to numbers. Thus, the description of the Quran as MaRQuM, that is dotted or digitized, is a reference to the Quran's numerical structure. A structure that was hidden from everyone until 1974, and will be hidden from the people who are unappreciative and have disease in their hearts, until the end of the world. (See: The Hidden, 74:31)."

AYMAN: "Yes I agree. However, Edip conveniently forgets to mention that in none of those archeological evidences the dotted Arabic letters are used to denote numerals as he will claim later."

Well, you just proved your lack of knowledge on the subject matter. I recommend you to study the books and the history of mathematics or numerals. In the first round, I referred to Georges Ifrah's The Universal History of Numbers, a remarkable book on the subject. But, it seems that you are scared to find out that Arabs during Muhammad times were using their alphabet as digits/numerals. Closing your eyes to this fact may preserve your 19-phobia, but it is evident in archeological evidence and massive scholarly research.

AYMAN: "There are examples of very early quranic manuscripts that are fully dotted. So Edip's "mostly" is meaningless because his argument is completely demolished by even one early manuscript that is dotted. Also, the great reading itself tells us that it is dotted as he admits below."

How early? Which century? I am referring to the earliest available manuscripts. You are trying to hide behind the ambiguity of the word "early." Besides, this was not my main point. It seems that unable to rebut the main points, you are going after secondary supportive arguments. The main point of my argument was this. I cannot believe that I have to repeat myself so many times to make your ear hear it. I am going to put in bold, so that you have no excuse of not getting it.

During the era of the revelation of the Quran, Arabs could read texts without dotting them or without adding vowels to them, since they could recognize the ambiguous letters of the words from its proximate context. To be able to read undotted text one needs to have knowledge of language and a certain level of intelligence. The earliest available Quran manuscripts testify to this fact. During the era of revelation, Arabs did not use the so-called Arabic numerals; they used their Alphabet, which was in a different order than the modern Arabic Alphabet, to represent numbers. (The order of Modern Arabic alphabet is a late innovation by those who abandoned the Quran and followed Hadith and Sunna). To use numbers as digits/numerals, they had to DOT them. Thus, the Quran describing itself as MarQuM, that is dotted or digitized, has implication to its numerical structure.

AYMAN: "As a translator, Edip should try to put the closest meaning from Classical Arabic dictionaries, which he admits is "dotted" and let the reader reach his own conclusion about why 'dotted' is used based

351

on the evidence instead of answering the question with his own wild speculations based on preconceptions. A translator should not be "advertising" a so-called miracle and instead should be closely reporting what is being said. The answer to Edip's question is because we do have examples of early manuscripts that are fully dotted. Also, as I showed above, even mundane everyday texts such as business letters were fully dotted to reduce ambiguity."

Ayman's ancestors changed the order of Arabic alphabet and its numerical function about two centuries AFTER the revelation of the Quran. So, the meaning and implication of the DOTTED/DIGITIZED is lost for their children. Had Arabs still used the original SEQUENCE of their alphabet as digits or numerals, then Ayman and his ilk would not have declared his ignorance, and I would not need to bother to connect the dots. Ayman is an educated person who feels the confidence to take a debate on this issue, and yet he does not know this simple historical fact. In sum, the word MaRQuM means numbered, as it is evidenced from historical facts and its use among living Arabs today. Ayman ignores both the past and present to distort the meaning of a word.

Arabs Used Alphabetical Numerals during Revelation

AYMAN: "The Hebrew Abjad (Aleph, Bet, Gimmel, Dalet) was used to represent numerals but there is zero evidence that Arabic was used in this manner. Hence, what Edip are saying is nothing more than baseless speculation. In fact, the proper name "Gematria" that he uses is the name for Hebrew Numerology which Jewish Rabbis practiced extensively. They too considered the number 19 sacred. As truly stated in 74:31, the count of 19 has ONLY been a 'fitna'/trial/affliction. This is true not just for modern followers of Dr. Khalifa but throughout history. For example, the Babis were obsessed with the number 19 long before Edip was born and so were the Hebrew Rabbis. Even Christians are obsessed with it and they prove through similar Gematria numerology as the one Edip uses that 19 symbolizes their idol Virgin Mary!"

This is another patently false claim. Any reader, including your pupils can learn the facts about the use of Gematria or Abjad by Arabs during the era of Prophet Muhammad. Ayman is again in the business of turning things backward; he is trying to use an evidence supporting Code 19 against it. If the book given to Moses was by God, then what is more natural to find similar code in that book too? In fact, that would be a reasonable expectation. If we did not know about the existence of Code 19 in the original text of the Old Testament, then Ayman

352

and his ilk would justifiably question that by saying, "God uses common attributes for Torah and the Quran; then why there is not even a hint of presence of such a code in Torah?" Like all other chronic disbelievers, no evidence and sign will make him see. What Ayman is doing is taking advantage of the negative connotation of the word Jew in the minds of his audience and thus appeal to their tribal emotions.

It is true that Bab and Bahaullah were intrigued by the numbers 9 and 19. That is it. Their interest has nothing to do with the function of the number 19 in the Quran as it is prophetically described in Chapter The Hidden. As for Christians, I was not aware of their "obsession" with the number 19, and its association with Virgin Mary. Even my search via google among millions of websites did not provide such an "obsession." In order to topsy-turvy the facts, in order to appeal the xenophobic emotions of his audience, Ayman is creating new obsessions for Christians!

> **AYMAN**: "Moreover, Edip is demonstrating his ignorance of the fact that the Roman numerals do not use an Alphabetic numeral system like the Hebrew inspired Code-19 system. The Roman numerals use mixed tallies and alphabets. An example of a tally system is: I for one, II for two, and III for three."

So what? There are different ideas among archeologists regarding the origins of the Roman numerals, but as you also acknowledged, they used letters for numerals, if not for all numerals. In his remarkable book, *The Universal History of Numbers*, Georges Ifrah has dedicated the Chapter 17 to "Letters and Numbers," Chapter 18 to "The Invention of Alphabetic Numerals" and Chapter 19 to the "Other Alphabetic Number-System". There the athor provides extensive information supported by archaeological evidence, their pictures and graphics, on the Hebrew, Armenian, Phoenician, Greek, Syriac, Arabic, and Ethiopian Alphabetic Numerals. So, you may try to hide yourself behind the semi-tally nature of Roman numerals, you cannot ignore all the evidences regarding many ancient Alphabetic Numerals, and especially the Arabic ones.

> **AYMAN**: "The fact is that we never see pre-quranic inscriptions that use Arabic alphabets as numerals. On the other hand, we see plenty of pre-quranic inscriptions that show that pre-quranic Arabs used a type of tally system. Their system was clearly a base-10 system because one can see that while the symbols for 1 to 9 are mostly simple vertical tally lines, the one for ten changes to a curve or a horizontal line with a dot. This is significant in light of the fact that we see a same base-10 system

in the great reading where "completing the count" can be clearly seen as counting ten."

This is the continuation of Ayman's sinking into his denial of facts. Enough said. Anyone can easily investigate the question for himself or herself. I highly recommend Ifrah's book since it is the most comprehensive work on the subject and is highly acclaimed by scholars.

When they used letters text they usually did not need to dot the letters since the peculiar combination and the context would narrow down the alternatives dramatically. But to represent numbers they HAD TO DOT THE LETTERS. Otherwise, none could differentiate the number 2 from 10, 50, 400, or 500. None could differentiate 8 from 600, 70 from 1000, etc. Thus, dotting the letters was not essential for prose or poetry, but essential for representing numbers. Thus, the word RaQaM means digits, and maRQuM means DIGITIZED, NUMBERED, or NUMERICALLY STRUCTURED.

> **AYMAN**: "The above is pure speculation for which Edip has zero evidence. There are no pre-quranic inscriptions showing dotted letters used as numbers. On the contrary, as Edip admitted earlier and as the evidence I provided shows, we can see mundane pre-quranic inscriptions where dots are used. Moreover, we see that pre-quranic inscriptions use a base-10 tally system for numbers and not a letter-based system like that of Hebrew. In fact, the Classical Arabic dictionary meanings match with both the archeological evidence and modern Arabic much better than Edip's speculation. As presented earlier, archeological evidence shows that pre-quranic Arabs used base-10 tallies that consisted mostly of inscribed parallel vertical lines or horizontal lines with dots. Thus, the meanings of 'marqum' as 'recorded/inscribed', 'lined' and 'dotted' converge and explain how in modern Arabic the word evolved from the three older meanings to have something to do with numbers. On the other hand, Edip's forced meaning doesn't explain how the other meanings relate and is only based on speculation about 'dotted'."

Again, these claims about the early Arabic numerals is no more than claiming that the earth was flat during the time of Prophet Muhammad. Anyone can learn the facts about the history of Arabic numerals by visiting the closest library or asking professor google.

> **AYMAN**: "Edip, firstly the statement "dwellers of the cave and the digits/numbers" is pure nonsense. How can one dwell in the

354

digits/numbers? You are obviously rushing to link 18:9 to your Code 19 without properly thinking about whether things make sense or not."

"Moreover, if one reads the sign in 18:9-26, one can clearly see that OTHER PEOPLE at the time of the prophet were discussing the matter and speculating about the story. How did those other people find out about those 'companions of the cave'? They must have known about it from a record that tells their story. We also hear in 18:21 that there was some kind of shrine built to commemorate them. As usual for this kind of shrine, in all likelihood it had an inscription talking about the story. So here 'raqim' talks about a neatly lined inscription. 'The people of the cave and the inscription' makes a lot more sense than 'the people of the cave and the numbers', fits in the context and is consistent with the clear Classical Arabic meaning. The only reason why anyone would need to doubt the Classical Arabic dictionary meaning is if something doesn't make sense or doesn't fit in the context."

Here, for the first time Ayman caught an error in my response. Congratulations. Yes, the translation of the word "Ashab" as "dwellers" may not fit both the cave and numbers. (Though, when the prophetic implication of the verse is fulfilled in the future, our understanding might dramatically change, like the ones in chapter 74). Therefore, a more accurate translation for us would be the "the people of the cave and the numbers." Ayman is demonstrating his number-phobia by finding the association of the "people of the cave" to "inscription" to be okay, but "number" not to be okay. He claims that his choice of "inscription" fits in the context. Ironically, he is the one who ignores the context where 13 verses after the word RaQYM there is a detailed debate about NUMBERS in relation to the people of the cave. His ignorance and arrogance make him blind to that intriguing contextual relationship. The number of letters from the beginning of the Chapter to the phrase "three hundred and increased by nine (300+9)" in verse 25, being exactly 309 and few other details, is a hint regarding another great numerical sign in relation to RaQYM.

> **AYMAN**: "My objection was to Code-19 promoters using 46:10 as proof of their 'precise mathematical miracle'. It is another one of their endless circular arguments. 46:10 talks about the 'phenomenon' of Code-19 because there is a Code-19 miracle in the great reading and the proof that Code-19 is a miracle is that 46:10 talks about it."

No, it is not a circular argument. The verse refers to A WITNESS among the Children of Israel, that is the Jews, who you wish your audience to dismiss just because of their race, AND we learn that indeed A WITNESS from Children of Israel testified to the same numerical structure in the original text of the Old Testament. This cannot be a coincidence.

I am not sure what you mean here. The word *aya*/signs is ALREADY in 46:7. You didn't do us any favors and "add" anything.

It seems that you will never get it. A study on the meaning of the singular form *aya* and its plural *AYAAT* will inform the reader what I meant.

> **AYMAN**: "Actually, even the Christians discovered a Code 19 in the Bible that justifies taking the Virgin Mary as a sacred idol. The fact is that the Bible that Edip has is as false as the book of Hadiths. Moreover, Edip doesn't tell you how much guidance and extra understanding Jewish Rabbis and Christians got as a result of their 'extraordinary evidences'. Why doesn't he? Because they got exactly the same guidance and extra understanding that 19ers got as a result of their extraordinary 'evidences' (or should we say 'claims'): A big extraordinary ZERO."

If we apply your logic, we should dismiss Jesus as God's messenger, just because Christians idolize him! However, I cannot hide that that I really loved your last sentence. It is such an extraordinary uppercut; sure to the number ZERO.

Numbers Everywhere!

> **AYMAN**: "What splitting hairs? The words *aya*/signs and 'abyat'/verses are two completely different words. Firstly, unlike poetry the great reading is not in 'verse' form. This is indisputable. As for 'numbered sentences', since you looked as early quranic manuscript you should know very well that those numbers were not part of the original text but were added long after the fact for the purpose of ease of reference. So this is also indisputable."

As if I claimed or even suggested that the word *aya* and ABYAT are similar words, Ayman is reminding me that they are "two completely different words." As for Ayman's argument regarding the number of verses: From the earliest available manuscripts, it is evident that verses of the Quran were separated by big dots. I use the word "verse" because of the convention. A better word for the statements in the Quran would be the plural word "signs." I should perhaps start using that.

If there is order of units, in Ayman's argument, such as letters, words, and sentences, then the numbers are embedded in Ayman's paragraph quoted above. For instance, Ayman's 46th word in the paragraph above, which is in 4th sentence (4:46) is about numbers! If numbers are made up arbitrarily by my imagination, then Ayman does not have any 2:15 in his paragraph! (Should he check the Quran instead of his paragraph, he would find the consequence of his actions.)

> **AYMAN**: "Yes, the numbers refer to arbitrary "nameless units" of the great reading. You can't say that they refer to "abyat/verses" because the great reading is not poetry. You also can't say that they refer to "numbered sentences" because the numbering is a modern device and many are not even sentences but can be several sentences or can even be meaningless half sentences. This is in line with the fact that the number schema was done purely for ease of reference and took mainly aesthetic reasons into account."

Numbers are everywhere. In the computer screen, in the genes, in the periodic table, in the apple pie or the circle's pi, everywhere there are numbers. There are numbers in the snail's shell, in your nose, and the lenses of your eyes. And it is in the Quran too, as one of the greatest miracles! A prophetic miracle that will not be witnessed by backward people (74:37)

> **AYMAN**: "Ayman, as it seems, have not read my response carefully. I argued that since the singular word *aya*(sign/miracle) was never used as a reference to the literal statements of the Quran, one verse or statement cannot be considered a sign or miraculous. I even provided an example for this claim. Ayman, repeats my position for an unrelated argument. He conveniently collapses two different words, sign and guidance, into one, while a little study of the Quran will inform the reader that *aya* is used consistently to denote miracles. Yes, miracles are obviously different then guidance, since not all those who are given miracles are guided by those miracles."

> "Had Edip "studied a little" of the great reading, he would not have made such clearly false statements. For example, 16:12 talks about the night, the day, the sun, the moon, and the stars as being *aya* (plural). Of course all those cosmic phenomena are signs NOT miracles. The problem is that Edip is trying to force the meaning of miracles (in Arabic: "mu3jizat") on the word signs/*aya*. The fact that "mu3jizat" and

357

not *aya* is the Arabic term that closely means "miracles" is indisputable."

This is the second time Ayman is catching my error. But, if he was fair, he would have easily noticed that it was an unintended simple error. In the beginning of the paragraph, I defined singular form *AYA* in parenthesis with two words: sign and miracle. In fact, I had indicated previously that the singular form *aya* in the Quran means sign, miracle, evidence, or lesson. However, while writing in haste, in one of the references to Ayat, I forgot to mention all its meanings. Ayman is ready to forget its context and all the other statements, just to attribute a false idea to me. Furthermore, the singular form *AYA*, contrary to what Ayman claims, is frequently used to describe many miracles, such as the parting of the red sea and reviving the dead bird.

> **AYMAN**: "Clearly, when taking this approach of considering all the relevant information, then the word *aya* in 10:1 cannot be pointing to 'alr' but is pointing to the book in our possession."

Ayman is beating the horse or his dead donkey. I have conclusively discredited his fabrications regarding the meaning of initial letters. If he checks the first volume of Zarkash'is Al-Burhan Fi-Ulumil Quran, he would learn that many scholars tried hard to assign meanings to those letters, but none received recognition, since all were arbitrary and subjective fabrications. Now, despite the 8 occurrences of the expression "tilka ayat" (these are signs) in conjunction with the initial letters, he is still trying to sell his silly fabrications. Besides, he is unable to see the signs/evidences anywhere in the entire Quran.

He Wants "Suhufun Munasharah" (Published Pages)

> **AYMAN**: "2. 'Doctors who smoke' syndrome. Yes, if it is published in a scientific journal, then I would believe your labeling of Code 19 as 'scientific/mathematical' and not as mere 'numerology' tricks.

Any distinguished scientific journal, preferably dealing with the subjects of mathematics and statistics. Here are some examples:

Advances in Applied Mathematics
Advances in Mathematics
Annals of Mathematical Logic
Applied Mathematical Modeling
Applied Mathematics and Computation
Applied Mathematics Letters

Applied Numerical Mathematics
Bulletin des Sciences Mathématiques (Sister Marie would like this one)
Computers & Mathematics with Applications
Discrete Applied Mathematics
European Journal of Combinatorics
etc..."

Thanks for the list. I will ask a mathematician who witnessed the code 19 to be the co-author of an article for those journals. I cannot do it this year, since I have deadlines to beat. I know how fussy Journal editors are; usually on technical and procedural details, rather than the substance.

> **AYMAN**: "I am sometimes called upon to referee papers for a few journals in the area of mathematics (not the ones that I listed though). If I get a paper on Code 19 I will ask the same questions about methodology that I asked you here and I am sure most of my peers will. So it is better that you answer here and save yourself a polite rejection letter from the editorial staff."

Are you bluffing or the journals you are called upon to referee were not worth mentioning their names? Your identity is hidden from public and you can claim whatever you wish. You might be an astronaut too. If your peers are as blind and arrogant as you are, I would not be surprised to get rejection letters from them. To the divine signs, peers like you do not have ears.

I have no doubt that Ayman will not witness the miracle of the Quran because it is published in the pages of journals. As the prophetic language of 74:52 describes and criticizes the demand of modern disbelievers regarding the code 19 reminder (ZiKR; ZiKRa; taZKiRah): "suhufun munasharah" that is "published pages."

> **AYMAN**: "3. The Code-19 Archeologist. Edip, you claimed to use various manuscripts 'for a critical and comparative study to infer the accurate version or spelling of original Quranic verses', when in fact it is clear that you are only using them to 'fix your counts so that they add up to a 19 divisible'. Hence, I am not surprised that you still haven't answered my question. Perhaps you are buying some time so that you can conduct some real unbiased studies."

You accused us about fixing the letter Sad in Chapter 7 of the Quran, and when I presented the facts, you suddenly forgot about your specific accusation. Instead of apologizing and accepting the facts, you prefer to run from the number 19.

AYMAN: "Edip, do you seriously think that asking you about 'critical and comparative studies Code 19 promoters conducted to infer the accurate version or spelling of original Quranic verses... that have no bearing on your counts' is an accusation? What is it an accusation of? I really want to know."

Here is what you wrote in your first article: "Hence, if proponents of Code 19 decide to use the Tashkent copy for something, they must use it for all the counts and not just for fixing the count of the letter 'Sin' to match a result that they want." If this is not an accusation, then what is an accusation?

Stereograms and Cyclopes; Lions and Zebras

AYMAN: "4. Evading the Standard Methodology Issue. How is it that METHODOLOGY and DEFINITION is avoiding discussion of specifics? Aren't METHODOLOGY and DEFINITION specifics? I really would like to know how you rationalize this."

"How do you expect to ever get your work published in a scientific journal without being specific on METHODOLOGY and DEFINITION?"

"I apologize if I bored you with dull and mundane concepts such as METHODOLOGY and DEFINITION that are not as exciting and flashy as MIRACLE, but they are a necessary part of scientific validation."

I can easily write down the methodology and definition for the code 19 phenomenon in the Quran. This might be a necessary piece of information for those who are not familiar with the Quran. However, people like you who claim expertise in the Quran should be able to infer the methodology and definition based on the examples I provided. If you are not able to infer the methodology and definition from the examples I provided, then you have no clue about the subject you are arguing. Perhaps you want to sidetrack the issue and bog down the argument on the procedure rather than the substance.

Did I make myself clear now why I do not respond to your demand? I know your modus operandi, and I know how you are trying so desperately to find a way to bore and distract the audience with lengthy procedural details so that they would not see the obvious facts and would not leave you alone in your darkness.

360

The closest analogy to the miracle of 19, to its both subjective and objective nature, is perhaps stereograms. Here is a brief description of Random-dot stereograms:

"Random-dot stereograms take advantage of the fact that our brain's visual processing apparatus pays particularly close attention to elements that are identical in our left and right visual fields. By embedding identical but displaced dot-pattern images in the left and right sides of a field covered with randomly placed dots, random-dot stereograms create patterns that look like visual noise in two dimensions, until you look at them the right way. The most amazing thing about the random-dot stereograms is that they make it possible to see startling 3-D illusions without the help of special glasses or head-mounted displays or optical devices of any kind." (*Stereogram, Cadence Books*, San Fransisco, 1994).

I know that now Ayman will pick the word "ILLUSION" from the description and jump into the air with excitement hoping that he is justified to use the same word for the Code 19. That's fine; he is at liberty to fool himself. I will not go in details regarding the differences between the Code 19 and the analogy of Stereogram. He either learns something from this analogy or confuses himself by the 'noise.' Interestingly, the Quran contains some literary random-dots to create noise for those who approach the numerical pattern with arrogance and prejudice; they will never pass the noise and see the intentionally embedded patterns and pictures! Interestingly, the random-dot noise starts with the first letter of the Quran. Those with prejudice, at that point start losing the focus and terminally blind themselves to the marvelous patterns that follow. The Quran guides those who seek guidance and increases the misery and confusion of hypocrites and disbelievers (2:20; 3:7; 5:64; 5:68; 17:41; 17:82; 74:31).

I have a stereogram hanging on the wall of my office. To the eye of the uninformed, it is just a bunch of colored random dots. No recognizable pattern is apparent. However, if you stay in front of it in a certain distance, focus your eyes and take little time to practice, you will find that the hidden pictures will emerge. A mother lion with a baby lion! The figures exist there and they are perceived by your mind. The fact that people without communicating with each other can see EXACTLY the same pictures, demonstrates that the hidden pattern is not an illusion, but an objective fact. An objective fact that is seen by some minds and is hidden to others! The 3-D picture in the brain of he witness, is made by the interaction of intentionally designed objective data AND an open-minded brain that gave the possibility for the presence of such a hidden picture among random-dots. The human language is like random-dots in comparison to

361

mathematics, and finding mathematical pattern in it is akin to the design of stereogram.

I invite Ayman to my office and ask him to look at the stereogram hanging on the wall. Some of my friends who are present there testify to the fact that the picture contains two lions. In fact, one of Ayman's friends who had no prior knowledge of the details about the lions in the picture is asked to look at the picture to find out the details about the lions. After staring at the picture for a while, his friend declares his discovery of two lions among the random dots! He discovers exactly the same details as others have done. One is big and the other is small. The bigger one is walking little ahead of the baby which is on the right side. There are some trees nearby, etc. Ayman, with a smirk on his face, instead of focusing on the picture in a certain distance, claims that it is all an illusion, a fitna for us. I show him the written testimony of a former owner of the picture, who happened to be a Rabbi lived long time ago. He too claimed to have seen the pictures of two lions among the random dots. He mocks with the nose of the Rabbi and then he accuses me of tricking people. He starts asking me about the methodology of designing and the definition of the stereogram. "What about publishing it in a reputable magazine?" he innocently demands, "If the editors of magazines could see the picture, then I would too would see it!" Perhaps I would provide him with an answer. But, if I see him turning his head to the opposite direction and on top of that stubbornly closing one of his eyes, from fear of falling into fitna, then I would not take his demand serious! Would anyone waste his time on these kinds of discussions with a lion-phobic guest who is unable to look at the artistically hidden picture in front of him!

No, I will not waste my time to help a blind in his mission to justify and glorify blindness. Ayman has no intention to check objectively whether there is indeed a picture among the apparently disordered dots! I am writing this article for the sincere people, who would open both of their eyes AND their minds, and spend a few minutes or hours in proximity to the picture to see the hidden figures in the conspicuously titled picture, The Hidden. In fact, under the title there is a subtitle which reads, "Only people with two open eyes and minds will see the lions!" A picture in which many people from different countries, ages, and background could see exactly the same details. Obvious for people. Precise. Universal. A picture that changed and keep changing the paradigm of many people. A picture that is hidden from the arrogant and paranoid enemies of the lions. Well, though Aymand did not witness the lion in the picture, I did not know that he was even scared from hearing the word L-I-O-N.

362

Let me be kind to my lion-phobic friend, and explain the issue of methodology and definition with a simple example. A few pages above, to demonstrate that each text containing ordered units inherently contains numbers, I used one of Ayman's paragraphs that started with his question, "What splitting hairs?". Did I need to write the methodology of my count of words and sentences? Did I need to make the definition of word and sentence? Of course not! A person with average IQ could easily infer that I was counting the expression *aya*/signs as two different words. If you count them with your MS Word program, you will find out that it counts the pair as one single word. Obviously, for a reason, I did not follow the counting method of MS Word in its entirety. So, I did not need to provide my non-idiot readers with fancy-sounding METHODOLOGY and DEFINITIONS. They would infer it on their own, in a few seconds.

Now, what Ayman wish is to drag me into an argument about MS Word's definition of words versus Edip's definition of words, and thereby let the main point—the numbers being implicit in the structure of ordered units—get lost.

> **AYMAN**: "I think I speak for everyone at Free-minds on both sides of the debate when I say that they are not impressed by this sort of answer. Coherent statements are all that we can go on. I am sure that you are a very nice person and if we met we would get along just fine. This debate is not about how nice or charismatic each person is but it is about seeking the truth. The truth can only be served with "coherent statements". Also, please don't compare anyone at Free-minds with Mr. Lomax. He was inhibited in his debate by his illogical sectarian Sunni baggage. I have no 'holy' prophets to idolize, 'holy' stones to spin around, or 'holy' books to extract an esoteric code from. I only holy The God."

Yes, I too believe that we could get along just fine. In fact, let me invite you to a face-to-face debate so that we could discus this issue even finer. If you prefer to do it at the University of Arizona, you will be my host. I will arrange it. Like Ali Sina, an evangelical opponent of the Quran and its signs whom I debated on Internet, you too may decline this offer. Well, if you are scared to show your face in public, then we could arrange a PalTalk discussion. Please give me a good excuse, such as that you are a mute mathematician, you are unable to think fast, or you are a beast with a one eye in middle of your fore-head ☺

God uses many of His attributes for His books, including the Quran. Therefore, your attempt to belittle God's infallible words under the pretext of exalting God, tells me a lot about your faith in the "great reading". Among the 57 attributes of

363

the Quran, there are the attributes of "MaJeeD" (Glorious), "HaKeeM" (Wise), KaRYM (Honorable), HuDY (Guidence), AJaB (Interesting/Wondrous), and MaRQuM (digitized/numberically structured). Your last statement may sound poetic to the ears of your audience but it is an empty statement. One could use exactly the same empty rhetoric to reject the entire Quran: "I have no holy books to extract guidance from. I only holy The God." Or, one could say, "I have no great reading to get signs from. I only accept the greatness of The God." By the way, which Arabic word are you using for your Holy?

The 0.3rd of the Moon and 0.0007th of a Year

> **AYMAN**: "5. Appealing to 'Nusemantics' What meaning or message is there in those Nusemantics? Is the message that 'dunya' and 'akhira' are equivalent or that 'qist' and 'zulm' are equivalent because they occur the same number of times? Clearly not. If 'shams' and 'noor' occur the same exact number of times, does that mean that the 'shams'/sun is the only source of light. We hear in the great reading about The God's light/'noor', so should we idolize the sun because it is equivalent to 'noor' according to your Nusemantics? What is the message in 'ayyam' occurring 27 times? We can empirically verify that the number of days is not 27 in any month (lunar is 29.5 days and the sidereal month is 27.3 days). Moreover, the frequency of the different words is more or less subjective because sometimes people count the different forms in Arabic haphazardly to get at the result. Nusemantics is largely based on selective counting and then speculation about a meaning for the result. I prefer to focus on reading and understanding."

Your questions demonstrate your confusion. Sun is the major of source of light. If you ask people on the street to associate a word to light, most likely would get the Sun, unless, they have phobia against finding a numerical structure in God's book. Symmetric occurrences of the semantically associated words do not mean that we should treat them the same, and none of us has claimed such. I noticed that you frequently employ a trick: you attribute to us some lousy arguments of your own fabrication and declare your victory against them. This is a well-known logical fallacy called "straw man". You even confused a fan of yours at the forum; she was mislead by you about the word AMma (from what?). She thought that it was indeed our thesis; while in fact, it was your own fabrication.

If you could show me how the word "ayyam" may occur 27.3 times in the Quran, then your objection would be valid. Right now, I looked at the frequency of the word QaMaR (Moon) in the Quran and I found it too to be exactly 27. You can check it via Fuad Abdulbaqi's concordance. You may exclaim,

"Beware, it is a fitna, it is another coincidence!" Somehow, there is harmony and design in the things you call fitna! Yes, indeed it is the fitna in your impaired mind. You will now ask me about the 0.3 of the word Moon! Well, it is left for you to hang your disbelief on, so that you can blind yourself to God's signs in the design of the Sun and Moon, both in the nature and the scripture. You share the same sickness with the Turkish scholar who could not object the fact that the word YaWM, in its singular form, indeed occurred exactly 365 times in the Quran; he came up with the following objection: "What about the 6 hours?" I told him that God left him six hours to deny the 365 days! He had only 0.0007th of a year as his portion!

You have no idea how magnificent is God's system and test. You have fallen into the fitna of your own making, your head down. You have become an unappreciative of one of the greatest signs, and like all your ancestors, you think that you are guided! You say that you focus on reading and understanding, but you have demonstrated that you are unable to understand the Quran and many of its signs.

Letter A, the Last Resort for the Innumerate Blind

> AYMAN: "For the HM, please give the following and then we can see if it is 'little' or not. This is not just for me but the information that you provide will help you to present your findings to the scientific community:
>
> 1. An objective measure of the statistical significance of the results.
> 2. A clearly defined method of how he got the results and how that method was constantly applied.
> 3. A clear definition of the data set to which the method was applied.
> 4. A clear explanation of why this particular method was selected.
> 5. A clearly documented audit trail of BOTH successes and FAILURES. For example:
>   - Trials where this method did not give a 19-divisible number.
>   - Other methods that were tried and did not give a 19-divisible number. "

If you check the www.mucizeler.com you will find a statistical analysis of the pattern in frequencies of these letters. But, you are not really looking to test the facts. All you are doing is to evade facing a numerical pattern shining in the Quran. Since you are so knowledgeable about the Quran, about the METHODS and DEFINITIONS, why don't you try to trash the numerical pattern in the frequencies of HM letters?

> **AYMAN**: "By the way, you could obtain the exact frequency of the letter A needed to achieve a similar result very easily using a computer program that tries to solve the Hm formula for the Alm through some simple bounded trial and error or more efficient algorithms. I am surprised that you or other 19 counters haven't tried it yet. This will at least tell you if there is a possible combination or not. However, I guarantee that you will NEVER be able to obtain a frequency of the letter A that adds up to 19 and is true because the counting of 19 that you are doing is nothing more than a "fitna". If there was a true 19 pattern that is part of the great reading then The God would have prevented it from getting corrupted."

You wanted and you got a problem: HM letters. These two letters alone has dazzled your eyes. Unable to deal with that EXTRAORDINARY pattern, now you are seeking refuge in another letter, the letter A. You are hoping that we will never be able to substantiate a true count of the letter A, and you are staking your disbelief in the dark cave of that letter. Congratulations, you got a letter and some time! And, enjoy your imaginary hell-fire in the Hidden.

> **AYMAN**: "Had Mr. Asad provided you with the info that I requested above you would have been able to verify it much more effectively. Also, before you add a section to your first response with 4 Ayman, the frequency of my name in your article was 114 so as you can see there is nothing special about this number."

> "If someone had as much time as Code 19 miracle hunters, they would be able to produce a remarkable intertwined pattern from your article as well as hidden messages about the end of the world."

Obviously, there is something special about that number, and thus it pulled your attention. As I said before, the mockery of yours will incriminate you in the Day of Judgment. You were able to see numerical relationship and association between numbers and their implication. However, you ignored them when God, the greatest mathematician, showed them. I wonder whether you will reflect on your position after reading the following verses:

> 6:4   Whenever a sign came to them from their Lord, they turned away from it.

366

6:5   They have denied the truth when it came to them. The news will ultimately come to them of what they were mocking.

26:4   If We wish, We could send down for them from the heavens a sign, to which they would bend their necks in humility.

26:5   Not a new reminder comes to them from the Gracious, except that they turn away from it.

26:6   They have denied, thus the news will come to them of what they used to ridicule.

38:1   S90, and the Quran that contains the Reminder.*

38:2   Indeed, those who have rejected are in false pride and defiance.

38:3   How many a generation have We destroyed before them. They called out when it was far too late.

38:4   They were surprised that a warner has come to them from among themselves. The ingrates said, "This is a magician, a liar."

38:5   "Has he made the gods into One god? This is indeed a strange thing!"

38:6   The leaders among them went out: "Walk away, and remain patient to your gods. This thing can be turned back."

38:7   "We never heard of this from the people before us. This is but an innovation."

38:8   "Has the remembrance been sent down to him, from between all of us!" Indeed, they are doubtful of My reminder. They have not yet tasted My retribution.*

74:36   A warning to people.

74:37   For any among you who wishes to progress or regress.

29:1   A1L30M40*

29:2   Did the people think that they will be left to say, "We acknowledge" without being put to the test?

29:3   While We had tested those before them, so that God would know those who are truthful and so that He would know the liars.

29:4   Or did those who sinned think that they would be ahead of Us? Miserable indeed is their judg-ment!

78:27   They did not expect a reckoning/computation.

78:28   They denied Our signs greatly.

78:29   Everything We have counted in a record.

74:56   None will take heed except if God wills. He is the source of righteousness and the source of forgiveness.

I hope Ayman reflects on the verses quoted above and reforms himself.

Major Arabic Dictionaries Refutes Ayman's Distortion Regarding the word ADdah

> **AYMAN:** "7. Ignoring the difference between '3ida(t)/count' and '3adad/number' "
>
> "Isn't understanding 74:31 crucial to this discussion and to you? Isn't the word 'count' in 74:31?"
>
> "If you think that nothing will change, then in your translation replace '3idat' with the more correct 'count'."
>
> "The so-called Code 19 was born from mistranslations of 74:31 and not surprisingly, it dies with the proper translation of 74:31. How come suddenly a central word in the most important 'verse' for 19ers is so trivial? "
>
> "Yes, unlike number, count involves an act of counting."

First, it is your hallucination that the Code 19 is born from the mistranslation of 74:31 and it is based on a single word, a word of your choice. You hope that if you first declare, in our name, that a particular word is the only base for our argument and if you manage to confuse your audience, then you will be able to cover the numerous numerical signs in the Quran! What a delusion!

But let me grant your wish! I am going to take your claim in the last statement above more seriously. Based on your insistence on this point, it is safe to assume that you think that translating ADdaH of 74:31 as "act of counting" rather than "number" is your biggest argument against the Code 19. The space you allocated on this point and the tone of your language supports this assumption.

First let me summarize my position. I accepted that the word ADdaH could be translated with the word COUNT and thereby it could mean both number or the act of counting. In either case, the implication does not change. Since EITHER the count of 19, or to please you, our count of the 19, OR the number 19, has become *fitna* for those who are unappreciative. The verse does not say that unappreciative people will count 19. The act of counting or the result of counting will create a negative reaction among unappreciative people!

Now, I will show that you have no basis regarding your uncompromising claim that our translation of the word as NUMBER is wrong. To please you, I had agreed to replace the word Number with the word COUNT, with its double meaning. If the word count does not mean NUMBER, then it cannot be the exact translation of the word ADdah.

There is no consensus regarding the Arabic word ADdaH. Some linguists consider the word as one of the plurals of ADaD (number) similar to ADaD (numbers). Other linguists consider ADdaH as an infinitive. Lisan-ul Arab, the most extensive Arabic dictionary, informs us about the difference among linguists regarding the word ADdaH:

والعَدَدُ مقدار ما يُعَدُّ ولفَّظُهُ، والـجمع أعداد وكذلك العِدَّةُ؛ وقيل: العِدَّةُ مصدر كالعَدَّ، والعِدّة أيضاً: الـجماعة، قَلْتُ أو كَثُرَتْ؛ تقول: رأيت عِدَّةَ رجالٍ وعِدَّةَ نساءٍ، أَنْفَذْتُ عِدَّةَ كُتُبٍ أي جماعة كتب

As it clear now, that you are wrong in rejecting NUMBER as a correct translation of the word ADdaH in 74:31! Since, you will seek your "fitna" by picking one of the multiple meaning of COUNT, I will translate the word in the verse as NUMBERS/COUNT.

I am not the one counting 19 everywhere. Zlatan is not counting 19 everywhere either.

The Quran does not condemn the COUNTERS, that is, the ADdyN. But, it condemns those who do not appreciate (Kafaru) the act of counting or the numbers of the guardian of Hell, which is 19. Read this sentence again to sink in your brain, my innumerate friend. Glorifying innumeracy is not an act the Quran promotes.

To my statement, "My faith has increased by the number or by the act of counting," Ayman gives the following answer:

369

This is not a function described in 74:31!

> AYMAN: According to your corrected translation of 74:31, "the act of counting" is ONLY a "fitna" for the rejecters. It is not the reason why "the faith of the faithful is strengthened". Instead, the reason why "the faith of the faithful is strengthened" is that they see that 'the count of the guardians of hell is ONLY a trial/punishment for the unappreciative.'"

> "I am afraid that the way to increased faith is not that easy. The God has surrounded people and they will certainly fill hell. Read 17:60 carefully and see an example of 'fitna'."

Ayman should go back read my response above. The verse you are reminding me is about those who took Tree and Fire literally, got confused, and fell in *fitna*. Since, I know that the description of Hell and Heaven in the Quran are allegorical, I do not have such confusion. But, you are obviously confused about the number Nineteen. Each evidence of the importance of the 19 in the Quran, from the number of its chapters to the frequency of its words and letters, you will be forced to close your eyes and ears tighter and tighter. There is a perfect description of your situation in the Quran in verse 2:15-20. That is your fitna. That is what you are struggling for. In the Day of Judgment, when you end up with guardian of Hell, you will see 19 all around you, and you won't be able escape from that number! Even the letter A, your last refuge, will not provide you shade there.

To my statement, "Some Christians and Jews who have witnessed the mathematical system based on this number or the result of the counting became convinced regarding the authenticity of the Quran. Hypocrites and unappreciative disbelievers are still confused regarding the implication of this number or count. All these support the miraculous nature of the number 19 in 74:31," Ayman gave the following answer:

> AYMAN: "This is not a function of the count. It is clear from your corrected translation that 'the count of the guardians of hell is ONLY a trial/punishment for the unappreciative'. Everything else results from this function."

Ayman has not seen a SINGLE Christian or Jew whose faith increased because of NOT witnessing the number 19 in the Quran. Not a single Christian or Jew has accepted the divine authenticity of the Quran either because they witnessed our wrong calculations, OR because some people are failing test/fitna because of not appreciating our correct counting. Either way, Ayman is not able to substantiate his claim.

Edip's Hat Confuses Ayman

> **AYMAN**: "Edip, you clearly ignore the fact that according to 17:12 the year in the great reading is clearly solar not lunar.
>
>> "17.12. We have made the night and the day as two signs: the sign of the night We have obscured, while the sign of the day we have made visible so that you may seek favors from your Lord, and so that you may know the number of years and the calculation, and We have explained all things in detail."
>
> "Given that night and day is certainly determined by the sun and not the moon, the sign in 17:12 leaves absolutely no doubt that the year is solar. More confirmation is in 12:47-49 that talk about the "year" definitely as a solar year because of reference to agricultural cycles and planning of crops that would have to be done according to seasons that don't change from one year to the next."

To the contrary, what Ayman says, the Quran accepts both lunar and solar year. Each has their use. Ayman reads too much into the verse 17:12. The verse refers to both Moon and Sun and describes their luminosity. It nowhere restricts the number of years or calculation to Sun alone.

To the contrary what Ayman speculates, verse 55:5 reminds us that both Moon and Sun are following a pattern that could be calculated. Verse 2:189, on the other hand, inform us that the function of the phases of the Moon is to determine the time. Furthermore, verse 18:25 gives the number of years in a peculiar expression, "three hundred plus nine," thereby giving both solar and lunar number of years the youth slept in the cave. It is no coincidence that 300 solar years are equal to 309 lunar years.

> **AYMAN**: "The word number is just informational and doesn't involve any 'act'. On the other hand, the word count involves an act of counting. This is clear from the dictionary definition that you provided."

If the dictionary definition for the word COUNT, "total number reached by counting," includes the act of counting, not just number, then translating the word ADdaH in verse 74:31 with the word Count is wrong, since according to

371

the Arabic dictionaries the word is primarily a plural of ADaD, that is number. Lisan-ul Arab gives the infinitive "to count" as the secondary meaning. In any case, Ayman is not justified in eliminating the word NUMBER or NUMBERS in connection to the number 19.

> **AYMAN**: "You correctly describe the count of the guardians of hell as ONLY a 'fitna'/trial/affliction."
>
> "The reason why 'those given the book are convinced and the faith of the faithful is strengthened' is that they see that 'the count of the guardians of hell is ONLY a trial/punishment for the unappreciative'. The word 'yaqin' and its derivatives are used in the great reading in connection with empirical observation."

The word ILLA (only) includes all the reactions described by the verse. Let's reflect on the following English statement:

Putting the hat on her head was ONLY to show her my love, to declare our friendship, and to adorn her head, to protect her from sunshine, and to increase Ayman's confusion; he will say 'what did Edip mean by this'?

Ayman will argue that, "Edip put the hat on his daughter's head, only to show her his love.[28] Because Edip loved her, her head was adorned by the hat. Similarly, because the sun and her head witnessed that Edip loved her they cooperated to protect her from the sunshine, and witnessing that Edip was capable of loving, Ayman will get confused." This is the translation of Ayman logic on 74:31!

> **AYMAN**: "8. Acknowledgement of the inconsistency in Code-19. Your point is irrelevant. Except for Dr. Khalifa's translation, translations do not claim to be divinely authorized or free of inconsistencies and errors. On the other hand, 19ers claim that Code 19 is a 'precise mathematical miracle' despite the fact that it is full of inconsistencies. This is indisputable and hence you had no difficulty admitting it. If you change the description of Code-19 to an 'inconsistent miracle' then I will agree with you."

---

[28] I do not have a daughter. It is an imaginary daughter, like the imaginary hat to wake up Ayman from his imaginary senarios to justify his denial of a manifest sign.

"Remember that 'consistency' is the only criteria we need to determine if something that claims to be part of the great reading is not from The God (see 4:82)."

Again, Ayman did not address my question. If had done so, he would not continue repeating this silly argument. Just replace the word "miracle" with the word "meaning" and you will get what I mean. There are millions of inconsistent understanding or meaning of the Quran in people's mind. Now, based on this inconsistency, can one reject understanding or any ascribed meaning to the words of the Quran? Ayman should go back and read my response again and reflect a little before engaging in knee-jerk reaction.

> **AYMAN**: "9. The misunderstanding of 3:7. I was hoping that you would carefully read 3:7, 2:26, and 74:31 together before responding."

I read carefully, but you did not address any points critical of your misunderstanding of 3:7. How can you expect to understand the relationship of three verses, while you are unable to understand neither of them?

> **AYMAN**: "10. The myth of initials. None of the commentators suggested that Alm, Alr, etc. are there for a Code 19 either. The question that you need to ask yourself is why hasn't any commentator suggested that 3m is a set of initials and not a word?"

> "It is because it has a meaning and the meaning fits in the context. This is the same for the other 11 instances out of the 29 where the words had meaning and the meaning fits in the context. 11 out of 29 is far better than the 3 out of 29 we get with Dr. Khalifa's methods or even Dr. Voss's best method."

> "I have already shown that 'pre-quranic' Arabic makes extensive use of abbreviations. So the remaining 18 sets are in all likelihood abbreviations too."

> "Instead of picking on sister Marie, try to explain how the words are initials despite clearly being written as words in the Arabic script."

> "Peace on who followed the guidance, Ayman"

I discussed this issue enough and I will rest my case here. As for Marie, she was misled by you since she thought that we were considering the first word of Chapter 78. Ayman did not correct her by saying that it was him who fabricated such a claim in our name. Now, he has the audacity to criticize me for correcting her against his false accusation.

74:49 Why did they turn away from this reminder?*
74:50 Like fleeing zebras,
74:51 Running from the lion?
74:52 Alas, every one of them wants to be given separate manuscripts.
74:53 No, they do not fear the Hereafter.
74:54 No, it is a reminder.
74:55 Whosoever wishes will take heed.
74:56 None will take heed except if **God** wills. He is the source of righteousness and the source of forgiveness.

Peace, Edip

13

Intelligent People's Guide to Code 19 (4rd Round)

I have to skip another round of arguments which was partially redundant and will share final correspondence:

AYMAN: Edip, I am sorry that you see this debate as a threatening war. While you see threat to your dogma, I see opportunity for you and me together to expose falsehood and discover truth. Only your self can defeat you, while in this debate everybody wins because it gets us closer to the truth.

I would like to sincerely thank you for all your responses and for taking time from your busy schedule to engage in this good debate. It helped me to learn a lot, for example:

1. The proper understanding of 74:31 based on 3:7 and 2:26.
2. The pagan origins of Gematria.
3. Like all mystic numerology, Code 19 is indefensible when cornered on methodology and definitions.

I also see that this debate continues to have many positive effects on you, for example:

1. You corrected Dr. Khalifa's mistranslation of 74:31 and provided a very good translation for it. As I said before, this is the most important result of this debate and if this is the only result achieved then we have both won. For now, I can only hope that one day you will see the implication of your OWN correct translation of 74:31.

2. You seem to be gradually realizing that "mathematical miracle" is a misnomer of Code 19. Hence, you use the terminology of "mathematical miracle" less liberally than before. You used "miracle" 103 times in your first article, 16 times in your second response, and 7 times in your third response. You used "mathematic/mathematics/mathematical" 113 times in your first article, while using it 15 times in the second essay, and only 6 times in your third response. Interestingly, at the same time you started to increase your use of the word Gematria (current name for Hebrew Numerology). So, at least in your mind, you seem to be gradually repositioning Code 19 from a "mathematical miracle" to something akin to "Hebrew numerology".

I also would like to apologize if anything that I said personally offended you, this was not my intention. This debate has nothing to do with you or me personally but it has to do with exposing falsehood and getting closer to the truth. I hope that we are both after this same objective. I look forward to our continued discussions.

EDIP: Ayman: I am leaving the decision to our "idiot" and "intelligent" readers. The intelligent ones will find the books about the history of numbers rather than relying on my or your selective scanning. I am not going to post the pages of the book here, for the following reason:

My targeted audience are not idiots; they will find the book on their own and they will see the chapters I referred to. They will also check other books to find the truth of the matter. I provided them with some information and gave them the name of the most comprehensive and authoratative book on the subject. But, if one day I am going to publish our debate as a book, then I might add some excerpt from the book to show the falsity of your claims.

As for your supporters, they will be glad to see what YOU CHOSE to show them, and they will be blind to what you are hiding from them. Ayman conveniently shows a diagram from the beginning of the book, and somehow ignores to look at the most relevant chapters: 17, 18, and the chapter 19. What he is sharing with is us is ONE THOUSAND year before the revelation of the Quran. He is obviously scared to come closer to the era of revelation since he will face the ALPHABET LETTERS ALSO FUNCTIONING AS NUMERALS. He might wants us to excuse him for him not having time to go that further in the book, but even in the introduction, the author of the book informs the reader about this issue. Here is a paragraph from page xxi of the introduction:

"Given their alphabets, the Greeks, the Jews, the Arabs and many other peoples thought of writing numbers by using letters. The system consists of attributing numerical values from 1 to 9, then in tens from 10 to 90, then in hundreds, etc., to the letters in their original Phoenician order (an order which has remained remarkably stable over the millennia)."

As I told, I do not have time to scan chapters of an excellent book that is available at libraries. You may declare your victory if you wish. I should not care for someone who is so stubbornly insisting to turn a miracle to hell-fire! What you choos is what you will deserve.

If I find time in the future, I will, for the las time, refute your charges, expose your fallacious arguments and red herring, God willing. Who knows, I might find more time to scan the pages from the chapters I mentioned.

Anyone who studies Aramaic, Hebrew and Arabic will find out the similarities of the language and their alphabet. However, by invoking the tribal emotions of his audience against Jews, Ayman wants to disconnect historical relationship between the two languages, their alphabet, and their numbering system.

It is an interesting coincidence that Ifrah dedicated Chapter 19 of his book to the ABJAD numbering system. Ayman is rejecting the undisputed fact that Arabs adopted Hindu numerals long after revelation of the Quran. He is however, unable to suggest an alternative: What were Arabs then using as numerals during the time of revelation of the Quran, not a thousand year earlier? Perhaps, he wishes us to believe that they were drawing a line for each number! For forty, forty lines, or a mixture of lines and symbols! Or, he would hope that we would just forget about numerals, and think that they did not have any numerals. That they were using the names of the numbers. Nothing else! According to Ayman, during the era of revelation, Arabs who were trading on three continents had no clue about numeral systems, which were used by many nations around them!

For those who wish to learn that Arabs were using their original alphabet order ABJAD before they adopted the Hindu numerals in 760 AC, please check the following links before getting hold of Ifrah's book:

http://en.wikipedia.org/wiki/Arabic_alphabet
http://en.wikipedia.org/wiki/Abjad
http://www-gap.dcs.st-and.ac.uk/%7Ehistory/HistTopics/Arabic_numerals.html
http://alargam.com/english/ragm364.htm
http://www.counton.org/museum/gallery2/gal2p3.html

Didn't I say that NOW just after we cut off their ONLY Rope connecting them to their Hell and Fitna, they will be fabricating other ropes? It is what exactly you did. You pretended to answer my posting mainly devoted to your ONLY argument and you managed to act deaf and blind to the numerous verses we listed as refutation of your argument. Wow! That requires great athletic skills; an Olympic record in somersault.

In the past, Zlatan, Ayman, many other 19-phobic regressive people, waved the ONLY FITNA rope, though they occasionally associated MAGIC and HELL to the number 19. Recently, you also hanged yourself with the same rope to your imaginary hell and fitna. When you are challenged by numerous verses debunking your ONLY, you do not even address the issue. You write a lengthy "response" without addressing the main point of that article, perhaps pretending that if you make it long enough people will not remember the main point of your opponent. How convenient.

The unappreciative enemies of God's signs are exposed to their skeletons. However, you are too young and there is hope that one day you will realize the debt of your misguidance and arrogance.

I am not calling names. I am stating the Quranic facts. Either we are liars or you are. Either we are hypocrites or you are. Either we are the unappreciative of God's signs, or you are. Either we are the progressive ones or you are. Either we are condemned to hell fire or you are. So, I will not be playing politics and caress your ego by hiding the Quranic facts.

Consider me as your friend, since friends tell the truth; they do not deceive their friends, especially in a matter that involve eternal salvation.

NOOR: Those sincere ones would only say: (11:28-29). Brother Edip has his point of view and brother Ayman has his own. Ayman is rejecting others' works, but he is neither introducing any that seems fruitful nor wit! Ayman is opposing other's interpretations, but he is not offering his! Ayman is debating others, but he is not clarifying his stand!

Brother Ayman & Edip reminded me with the old and the young man story. Here is the story. I will refer to the old man as Edip, and the young man as Ayman. (There is no intention of offence, though you may feel as such):

Once upon a time there was a young man by the name of Ayman and an old man named Edip. Ayman, unfortunately amid the hectic days and speed events, lost his home keys on streets. So he stopped and started looking for them. Few moments, Edip showed up and offered his assistance to Ayman, saying: *"whether can I be of any support to you."* Ayman (busy and exhausted while searching for the key) said: *"Well, I lost my keys and I have spent half the evening looking for them"*. Edip asked:" *where have you lost them and when it came up to your mind about the lost keys."*Ayman told Edip that he could not figure out the exact time he lost his keys, but pretty sure he lost them, **there, and pointed to the other direction opposite to where he stands.** Edip with exclamation mark on his face:" *If you lost your keys there, why are you looking for them in here"*. Ayman said – *"well, here there is light, but there are no lights over there. "*Edip looked at him, then said:" *why do not you move the lights to the other side where you lost your keys, you might be able to find your keys!!*

That is exactly what is going on with Ayman and Edip as I view the outcomes!

Ayman is now sticking to the books of classical Arabic and few dictionaries of Lisan Alarab, Almuhit, Almanthur, and is referring to the books of different authors in a **sincere approach** to make it up with the lost past days of ignorance,same as our previous days,and our search for making it up for the lost past. (4:82) & (18:54)! But as I feel, he has nothing to offer except few points that were made up in order to dismantle others their remarkable achievements.

In this post, **Edip** is shouting at Ayman, trying by all means to pull him by force to look on the other side, on the Quran, where he may find his quest. But, Ayman had preferred to tie himself to the trunk, so that Edip would not be able to move him. Added to that, Ayman has closed his ears and eyes, so the shouting and the *whole scene of nagging* will not influence him.

Dear Ayman, why do not you bring the lights to the book of the Lord (4:82) & (5:101) instead of looking at the books of other creatures (6:114) & (6:116)?

I know you were eagerly attempting to learn the book of God, and your start was a sincere one. But be advised that you are not the only one to have the keys for the truth.

74:52 "Or else you are looking forward to have your own manuscript!"

Let us **suppose** you have them,though I doubt, It doesn't mean to mock others' belief and their interpretation and what most **ironic** is that you are inventing idioms and ideas that are out of the context (40:35) and you are attempting to **nullify** the work of R.K. and claiming other as "not faithful"(4:94)!

379

I am fully aware that some of R.K views are not final, but need a deep study (*by the way, I am not a khalifite*). But I am pretty assured that R.K. was the one to trigger you to search for the truth inside the Quran. But unfortunately, though you are targeting the quran but mistakenly heading to other books than the Quran.

You may feel happy with it. I am only advising you, and feel free to look wherever you wish, but you should take a deep look at the following verses:

> 17:89   In this Quran We have cited every example for the people; but most of the people refuse to be anything but ingrates!
>
> 20:2   We did not send down to you the Quran so you may suffer.
>
> 20:114   Then High above all is God, the King, and the Truth. Do not be hasty with the Quran before its inspiration is completed to you, and say, "My Lord, increase my knowledge."

If you deem that you are on the right side, you would have said:

> 11:28   He said, "My people, do you not see that I have proof from my Lord and He gave me mercy from Himself, but you are blinded to it? Shall we compel you to it while you are averse to it?"
>
> 11:29   "My people, I do not ask you for money, my reward is with God. Nor will I turn away those who acknowledge, for they will meet their Lord. But I see that you are a people who are ignorant."

I know you are fluent in **English and Arabic**, but you should take care & be aware that though the Quran is in Arabic language (41:44). It does not mean an Arabic speaker is the guided one.

I hope this helps. Or should I expect!

> 18:42   So his fruits were ruined, and he began turning his hands for what was destroyed upon its foundations though he has spent on it. He said, "I wish I did not make any partner with my Lord!"

EDIP: I thank brother Noor for his insightful comments. Though I have been hard on Ayman's position and his methodology, I appreciate his cunning intelligence, his speed, agility, ambition to defend his position by any means possible, and his patience in conducting this debate with me. Who knows, I might be entirely wrong about the level of his arrogance, prejudice, fanaticism

380

and 19-phobia; one day he might be able to break the walls of his prejudice and fears, and may witness the great signs of the Quran. Then, he would be a good witness, by God's will.

To an ingrate who repeats the same number-phobic objection:

Appreciating God's words will leads a person to appreciate the implication, meaning of his words, and its prophecies, such as the one about 19, and the structure of His words, such as 114 chapters, 19-lettered Basmalah.

I have spent more than enough time and words to share with the unappreciative people why some people and I have witnessed a great divine sign/evidence mathematically embedded in the Quran, without an atom's weight of doubt. I also know that witnessing this miracle is not for everyone. God has forbidden the unappreciative and hypocrites to witness the miracle. The unappreciative and hypocrites will find many excuses and false arguments to blind themselves to it.

The hell-fire is for those who are eager to see it in the Hidden, the signs are for those who are appreciative of God's blessings. It is up to you to which group you will join.

Below I am posting some of the many verses describing the unappreciative disbelievers and hypocrites. See, where you will find yourself. I do not need to remind you that the test is serious and our eternity is at stake!

After reading the following verses, ask yourself whether you really understand them and whether you really appreciate them. Then, if you wish answer my question in the end:

> 1:1   In the name of God, the Gracious, the Compassionate
> 2:17   Their example is like one who lights a fire, so when it illuminates what is around him, God takes away his light and leaves him in the darkness not seeing.
> 2:18   Deaf, dumb, and blind, they will not revert.
> 2:118   Those who do not know said, "If only God would speak to us, or a sign would come to us!" The people before them have said similar things; their hearts are so similar! We have clarified the signs for a people who have conviction.*
> 3:7   He is the One who sent down to you the book, from which there are definite signs; they are the essence of the book; and others, which are multiple-meaning. As for those who have disease in their hearts, eager to cause confusion and eager to derive their

interpretation, they will follow what is multiple-meaning from it. But none knows their meaning except God and those who are well founded in knowledge; they say, "We acknowledge it, all is from our Lord." None will remember except the people of intellect.

6:4 Whenever a sign came to them from their Lord, they turned away from it.

6:5 They have denied the truth when it came to them. The news will ultimately come to them of what they were mocking.

6:25 Among them are those who listen to you; and We have made covers over their hearts to prevent them from understanding it, and deafness in their ears; and if they see every sign they will not acknowledge; even when they come to you they argue, those who reject say, "This is nothing but the tales from the past!"

6:26 They are deterring others from it, and keeping away themselves; but they will only destroy themselves, yet they do not notice.

6:104 "Visible proofs have come to you from your Lord; so whoever can see does so for himself, and whoever is blinded, does the same. I am not a guardian over you."

6:158 Do they wait until the controllers will come to them, or your Lord comes, or some signs from your Lord? The day some signs come from your Lord, it will do no good for any person to acknowledge if s/he did not acknowledge before, or s/he gained good through his/her acknowledgement. Say, "Wait, for we too are waiting."

7:132 They said, "No matter what you bring us of a sign to bewitch us with, we will never acknowledge you."

7:146 I will divert from My signs those who are arrogant on earth unjustly, and if they see every sign they do not acknowledge it, and if they see the path of guidance they do not take it as a path; and if they see the path of straying, they take it as a path. That is because they have denied Our signs and were heedless of them.

14:25 It bears its fruit every so often by its Lord's leave. God cites the examples for the people, perhaps they will remember.

14:26 The example of a bad word is like a tree which has been uprooted from the surface of the earth, it has nowhere to settle.

20:133 They said, "If only he would bring us a sign from his Lord!" Did not proof come to them from what is in the previous book?*

20:134 If We had destroyed them with retribution before this, they would have said, "Our Lord, if only You had sent us a

382

messenger so we could follow Your signs before we are humiliated and shamed!"

20:135 Say, "All are waiting, so wait, and you will come to know who the people upon the balanced path are and who are guided."

25:4 Those who rejected said, "This is but a falsehood that he invented and other people have helped him with it; for they have come with what is wrong and fabricated."

25:5 They said, "Mythologies of the ancient people; he wrote them down while they were being dictated to him morning and evening."*

25:73 Those who when they are reminded of their Lord's signs, they do not fall on them deaf and blind.

27:82 When the punishment has been deserved by them, We will bring out for them a creature made of earthly material, it will speak to them that the people have been unaware regarding Our signs.*

27:83 The day We gather from every nation a party that denied Our signs, then they will be driven.

27:84 Until they have come, He will say, "Have you denied My signs while you had no explicit knowledge of them? What were you doing?"*

27:85 The punishment was deserved by them for what they transgressed, for they did not speak.

27:92 "That I recite the Quran." He who is guided is guided for himself, and to he who is misguided, say, "I am but one of the warners."

27:93 Say, "Praise be to God, He will show you His signs and you will know them. Your Lord is not unaware of what you do."*

29:1 A1L30M40

29:2 Did the people think that they will be left to say, "We acknowledge" without being put to the test?

29:3 While We had tested those before them, so that God would know those who are truthful and so that He would know the liars.

29:4 Or did those who sinned think that they would be ahead of Us? Miserable indeed is their judgment!

29:54 They hasten you for the retribution; while hell surrounds the ingrates. (Please read from 29:48)

38:1 S90, and the Quran that contains the Reminder.*

38:2 Indeed, those who have rejected are in false pride and defiance.

38:3 How many a generation have We destroyed before them. They called out when it was far too late.

38:4   They were surprised that a warner has come to them from among themselves. The ingrates said, "This is a magician, a liar."

38:5   "Has he made the gods into One god? This is indeed a strange thing!"

38:6   The leaders among them went out: "Walk away, and remain patient to your gods. This thing can be turned back."

38:7   "We never heard of this from the people before us. This is but an innovation."

38:8   "Has the remembrance been sent down to him, from between all of us!" Indeed, they are doubtful of My reminder. They have not yet tasted My retribution.*

38:29   A book that We have sent down to you, that is blessed, so that they may reflect upon its signs, and so that those with intelligence will take heed.

41:53   We will show them Our signs in the horizons, and within themselves, until it becomes clear to them that this is the truth. Is it not enough that your Lord is witness over all things?*

46:10   Say, "Do you see that if it were from God, and you rejected it, and a witness from the Children of Israel testified to its similarity, and he has acknowledged, while you have turned arrogant? Surely, God does not guide the wicked people."*

54:1   The moment drew near, and the moon was split.*

54:2   If they see a sign, they turn away and say, "Continuous magic!"

74:25   "This is nothing but the words of a human."

74:26   I will cast him in the Saqar.

74:27   Do you know what Saqar is?

74:28   It does not spare nor leave anything.*

74:29   Manifest to all the people.*

74:30   On it is nineteen.*

74:31   We have made the guardians of the fire to be angels/controllers; and We did not make their number except as a test for those who have rejected, to convince those who were given the book, to strengthen the acknowledgment of those who have acknowledged, so that those who have been given the book and those who acknowledge do not have doubt, and so that those who have a sickness in their hearts and the ingrates would say, "What did God mean by this example?" Thus God misguides whoever/whomever He wishes, and He guides whoever/whomever He wishes. None knows your Lord's soldiers except Him. It is but a reminder for people.

74:32 No, by the moon.*

74:33 By the night when it passes.

74:34 By the morning when it shines.

74:35 It is one of the great ones.

74:36 A warning to people.

74:37 For any among you who wishes to progress or regress.

74:38 Every person is held by what it earned;

74:39 Except for the people of the right.

74:40 In paradises, they will be asking

74:41 About the criminals.

74:42 "What has caused you to be in Saqar?"

74:43 They said, "We were not of those who offered support (or observed contact prayer)."*

74:44 "We did not feed the poor."

74:45 "We used to participate with those who spoke falsehood."

74:46 "We used to deny the day of Judgment."*

74:47 "Until the certainty came to us."

74:48 Thus, no intercession of intercessors could help them.

74:49 Why did they turn away from this reminder?*

74:50 Like fleeing zebras,

74:51 Running from the lion?

74:52 Alas, every one of them wants to be given separate manuscripts.

74:53 No, they do not fear the Hereafter.

74:54 No, it is a reminder.

74:55 Whosoever wishes will take heed.

74:56 None will take heed except if God wills. He is the source of righteousness and the source of forgiveness.

78:27 They did not expect a reckoning/computation.*

78:28 They denied Our signs greatly.

78:29 Everything We have counted in a record.

78:30 So taste it, for no increase will come to you from Us except in retribution.

83:18 No, the record of the pious is in Elliyeen.

83:19 Do you know what Elliyeen is?

83:20 A numerical book.

83:21 To be witnessed by those brought near.

Don't you ever wonder why is it always that monotheists who appreciate the divine signs and miracles, and it is always the hypocrites and the ignorant and arrogant people who mock or unable to see the miracles?

Finally, here is the summary of this debate:

You believe in a god who cannot provide any good argument against the unappreciative person who claimed that the Quran was man-made. Your god's only answer is "I will burn you in the hell." That is the value and capacity of your god: unable to refute the challenge of the skeptics with evidence. No wonder the Quran states, "They do not value God as He should be valued."

In your response, you highlighted the word FITNA and ONLY in 74:31. That tells me alot. It shows how you became blind to the rest of the verse, especially the part about the GOOD NEWS, and the following verses referring to it as ZIKRA (which has positive Quranic connotation). You see HELL and FITNA alone in those verses.

As for me: besides the Hell and Fitna, I ALSO see a GREAT NUMERICAL EVIDENCE, AN INTELLECTUAL REFUTATION OF THE CLAIMS OF THE DISBELIVER MENTIONED IN THE BEGINNING, A FULFILLED PROPHECY and GOD'S WISDOM. That is the difference. You see only HELL FIRE and FITNA. You believe a god who cannot provide evidence for the person who question the authenticity of His words; but he ONLY THREATENS TO ROAST them IN FIRE. This is your choice, based on your inner intentions. Intentions that might be residing in your subconsious level.

Besides, your article proves that you did not even read my response to Ayman, since I refuted to limit the reference of ONLY with Fitna. I gave an English example to make my point, and it seems that you are not aware of my counter argument.

So: Why should I repeat myself to you too? Why should I waste my time with you or other bigots who do not even read what they intend to criticize.

So Obvious and so Hidden

I have no personal qualms with you. ... I am not angry. I just cannot afford wasting my time for people who do not understand. My aversion is to your lack of understanding and appreciation of one of the greatest signs.

It seems that we are entirely in different universes. Though I see your point of view very clearly, but you have proven that you are entirely blind to my view. Subhanallah.

The blindness demonstrated by you and your comrades to one of the clearest signs in the world, is one the greatest miracles by itself. What a great test! What

386

a fitting test to the Wisdom of God! A miracle that shines in the horizon brighter than the Sun for those who appreciate, yet it is hidden from those who have certain disease in their intentions.

Whatever I do, nothing will change my friend. It is up to you to open your eyes. I hope that you will not be swayed by those whose hearts are closed. I hope that you willl beware of being with those who have regressed (74:37).

I will no more engage in debates on this issue for quite some time. Perhaps, I exceeded the limits of reasonableness in trying so hard to show the OBVIOUS to those who are so eager to cover their eyes, their ears, minds, and their hearts.

14
Intelligent People's Guide to Code 19 (5th Round)

IDMKHIZAR: Salam. Ayman insists on rendering 74:31 as follows:

> 74:31 We appointed angels to be guardians of Hell, and we assigned their number ONLY as a fitnah for the disbelievers (FULL STOP); to convince theppl of the book, to strengthen the faith of the faithful,...

But I do not agree with this rendering for I don't know how anyone could believe that a fitnah should increase his/her faith or that it should convince anyone not 2 mention how a fitnah would remove the doubts of the Jews n Christians. It doesn't make any sense whatsoever. Ironically the same people who reject code-19 like Zlatan and idolfree were initially convinced by code-19 thus they deny the rendering of the prophecy in 74:31 which fulfilled itself through them. How absurd? The following rendering is perfectly grammatically correct:

> 74:31 We appointed angels to be guardians of Hell, and we assigned their number (19) ONLY (1) as a disturbance for the disbelievers, (2) to convince the Christians and Jews (that this is a divine scripture), (3) to strengthen the faith of the faithful, (4) to remove all traces of doubt from the hearts of Christians, Jews, as well as the believers, and (5) to expose those who harbor doubt in their hearts, and the disbelievers; they will say, "What did GOD mean by this allegory?" GOD thus sends astray whomever He wills, and guides whomever He wills. None knows the soldiers of your Lord except He. This is a reminder for the people.

I can also prove from other verses in the Quran that the word 'illa' can well apply to more than just the first reason:

> 2:85 ...What should be the retribution for those among you who do this, EXCEPT (illa) (1) humiliation in this life, and (2) a far worse retribution on the Day of Resurrection? GOD is never unaware of anything you do.

> 2:111 Some have said, "No one will enter Paradise EXCEPT (illa) (1) Jews, or (2) Christians!" Such is their wishful thinking. Say, "Show us your proof, if you are right."

> 3:7 ...None knows the true meaning thereof EXCEPT (illa) (1) GOD and (2) those well founded in knowledge. They say, "We believe in this - all of it comes from our Lord." Only those who possess intelligence will take heed.

Note that the classic commentary of al-Baydawi also prefers this understanding. Also note that Yusuf Ali too acknowledges this fact in the footnote of 3:7:

"One reading, rejected by the majority of Commentators, but accepted by Mujahid and others, would not make a break at the point marked *Waqfa Lazim*, but would run the two sentences together. In that case the construction would run: 'No one knows its hidden meanings except God and those who are firm in knowledge. They say', etc."

So you see that this more logical interpretation was even accepted by hadithists aforetime n it is clear that the Arabic allows 4 both interpretations.

> 6:48 We do not send the messengers EXCEPT (illa) (1) as deliverers of good news, (2) as well as warners. Those who believe and reform have nothing to fear, nor will they grieve.

> 6:145 Say, "I do not find in the revelations given to me any food that is prohibited for any eater EXCEPT (illa): (1) carrion, (2) running blood, (3) the meat of pigs, for it is contaminated, and (4) the meat of animals blasphemously dedicated to other than GOD." If one is forced (to eat these), without being deliberate or malicious, then your Lord is Forgiver, Most Merciful.

> 6:146 For those who are Jewish we prohibited animals with undivided hoofs; and of the cattle and sheep we prohibited the fat, EXCEPT

(illa) (1) that which is carried on their backs, or (2) in the viscera, or (3) mixed with bones. That was a retribution for their transgressions, and we are truthful.

7:20   The devil whispered to them, in order to reveal their bodies, which were invisible to them. He said, "Your Lord did not forbid you from this tree, EXCEPT (illa) (1) to prevent you from becoming angels, and (2) from attaining eternal existence."

14:4   We did not send any messenger EXCEPT (illa) (1) (to preach) in the tongue of his people, (2) in order TO (li) clarify things for them. GOD then sends astray whomever He wills, and guides whomever He wills. He is the Almighty, the Most Wise.

Comment: Do you now suggest that clarifying the things with Quran was NOT a duty of the messenger? So these are TWO points although the 'to' came after the first point. Therefore 'illa' does not necessarily refer only to the first part/thing as u claim.

17:105 Truthfully, we sent it down, and with the truth it came down. We did not send you EXCEPT (1) as a bearer of good news, (2) as well as a warner.

20:109 On that day, intercession will be useless, EXCEPT (illa) (1) for those permitted by the Most Gracious, and (2) whose utterances conform to His will.

30:8   Why do they not reflect on themselves? GOD did not create the heavens and the earth, and everything between them, EXCEPT (illa) (1) for a specific purpose, (2) and for a specific life span. However, most people, with regard to meeting their Lord, are disbelievers.

103:3 EXCEPT (illa) (1) those who believe, (2) lead a righteous life, (3) exhort one another to uphold the truth, and (4) exhort one another to be steadfast.

So now this should be absolutely clear and I hope not to hear the 'illa ONLY applies to counts 19 being a fitnah' argument again. Moreover what do you make of when GOD says in the Quran that He alone revealed the reminder and He will preserve it? As we all might know a law cannot be proven to exist unless someone tries to break it. For example, when I make a law that you cannot leave this room and you stayed in it forever then nobody will know whether my law was true or not. You have to try and break it in then when you are thrown back into the room while trying to escape then people would know that the law exists. Same is with the Quran. People tried to tamper with it and the GOD proved that nothing can be done to the Quran since it is guarded by a Divine Mathematical Code. Also don't you see a clear pattern when you find all words in the Bismillah occurr a multiple of 19 times in the Quran without 9:128&129? Isn't that significant? Isn't the number of chapters, number of verses, and number of Bismillahs despite its absence on sura 9 being multiples of nineteen significant? I believe they are.

You May Cut OFF Their ONLY Rope to Hell, But…

EDIP: Dear brother Khizar: The unappreciative and arrogant ones who ONLY see CAHEEM (Hell), NAR (fire) and FITNA (confusion and discord) in chapter 74 have tried hard to blind other people by emphasizing the word ONLY in 74:31 while cutting off its full reference to the five functions of the number 19. I tried to show them that the word ONLY could refer to an enumerated list and I provided them an English sentence; but they did not get it.

Now, you came up with numerous verses from the Quran (2:85; 2:111; 3:7; 6:48; 6.145; 6:146; 7:20; 14:4; 17:105; 20:109; 30:8; and 103:3), conclusively refuting their ONLY novel argument against 19. Mashallah. Now you cut off their ONLY rope to their HELL and FITNA, but you will be surprised. They will immediately fabricate another rope or even many ropes in their imaginations. Hell is in their minds, and it already has confounded them.

Whatever you do, they will insist to see ONLY (1) magic, (2) hell-fire and (3) fitna (Interesting, their ONLY may only refer to numerous <u>negative</u> terms). And they will lie to others and perhaps themselves that their faith has increased because they see Hell and Fitna. They lie and deceive themselves even more, by claiming that Christians and Jews are accepting the Quran because they witness Hell and Fitna!

The number 19 is described in Chapter 74 alone with the following words:

- Hidden (*muddathir*, 74:1)
- Recorded in grooves (*nuqira fil naqur* 74:8)
- Signs (*ayaat* 74:16)
- **Overwhelming uphill** (*sauud* 74:17)
- Intellectual challenge (saqar (74:26)
- Precise (la tubqi wa la tazar 74:27)

391

- Universal open/screen (*lawahah* lil bashar 74:28)
- Their number (*iddatahum* 74:31)
- **Test** (*fitna* 74:31)
- Convincing evidence (*yastaqinna* 74:31)
- Increasing conviction (*yazdad ... imanan* 74:31)
- Leaving no doubt (*la yartab* 74:31)
- **Means of diversion** (*yudillu* 74:31)
- Means of guidance (*yahdi* 74:31)
- Reminder (zikra 74:31)
- One of the Greatests (*ihdal kubar* 74:35)
- Warning (*nazyr* 74:36)
- Means to rise to eminence (*tazkirah* 74:49,54)
- **Lion** (*qaswarah* 74:51)

Their denial and phobia against 19 is part of the fulfillment of the prophecy. What they see is only the ones written in Bold and they call it SIHR (74:24), and they are running from it like KHUMUR (74:50)! Let me correct it: they do not see the QASWARA clearly, since they are in panic like KHUMUR, they just see its shadow or tail and they run screaming! Ironically, while they are running from it they make very loud noise in attempt to scare others who marvel the QASWARA:

> "Run for your life!"
> "It is ONLY a magic!"
> "It is ONLY a trick!"
> "It is only Diversion!"
> "It is ONLY Hell!"
> "It is ONLY a fitnah!"
> "It is ONLY a ghost!"

If there were no people denying the obvious evidence of the Quran, it would not be an AYAT or one of the AYAAT, since AYAAT are witnessed only by the open-minded progressive people who appreciate God's signs. Read the last verse of chapter 74.

SYTALLS: Here is where we must part ways, because I also rejected the hadith without studying them. I only had to hear a few of them to know they went against what I'd read in the Quran. For Code 19, I don't need much time to study what's there and haven't read much about it, because it isn't particularly interesting, for me, mathematically or spiritually. I suppose my main question

392

would be this: Do you need this code to bring you closer to God or to help you understand the Quran?

Because I think there are enough powerful and beautiful things in the Quran that any one of them could be isolated and studied in depth from now to Judgment Day. But, while I think they are worthy of study and discussion, I would be very cautious about saying that one particular part would be critically important to my spiritual life. To me, that tears apart the totality of the Message and is very much like saying that one prophet is more important than another. Even reading over this thread, which is more than I've ever read about Code 19 since running across Khalifa's website a while back, I still fail to see a succinct and spiritually relevant answer to one essential question.... What is the point?

EDIP: Your "point" is one of the points. "Maza Aradallahu bi haza masalan" (**What does God mean by this**?) (74:31). Yes, your lack of understanding the meaning of the number 19 and expressing it by asking "**what is the point**?" is part of the prophecy mentioned in 74:31. You will know which group you belong if you read that verse. There you will see that some people like you will exactly miss the point and they will ask exactly the question you asked.

NATU: So it is clear that it doesn't matter whether you use "Only" or "Except" The effect is the same. However, I cannot be certain and I will not argue this through because my knowledge of classical Arabic is (currently) very limited, so I will not debate the exact definition of "only/except" with you without the correct knowledge (I'm only 18 remember).

EDIP: Dear Natu: I have acknowledged your young age and I expressed the fact that you have inshallah a lot of time to reflect on this position. My reaction is not to your person, but since I am hundred percent sure about the existence the great miracle and its prophecy and its role of sorting people out, I will not be shy of expressing those Quranic facts. It is not me but the Quran condemning the unappreciative who deny God's signs. If indeed 19 is one of the greatest signs, which I have no doubt, then why my expression of its consequence and the Quranic depiction of groups of people according to their reaction to it, yes why would that be considered a personal insult?

As you quoted yourself, you proved my point. In your previous article you highlighted and harped on that ONLY rope without knowing anything, since you followed Zlatan and Ayman. When it was cut off by clear Quranic evidences, instead of accepting your blunder you expressed your excuse of not knowing Arabic and thus not being able to make judgment on this simple issue. Simple, since you DO NOT NEED TO LEARN ARABICA to learn that word ILLA

(ONLY or EXCEPT) refers to ENUMERATED lists, not a single item as your big brothers claimed so. You could have easily looked at any translation of the Quran and reflect little on the dozens of verses. Acknowledging the fact had nothing to do with knowledge of Arabic.

However,

Instead of stopping your attack, which by your own admission, based on your ignorance, you have continued your attack to one of the greatest signs. Under the guise of "I do not know this" you cunningly switched from your ONLY rope to another rope.

But, it seems, you are unable to see your dance around the truth. Like all the unappreciative people of the past, you will have your own excuses and self-deception.

I do not know about your inner intention, your tomorrow, but I have no doubt that at this point you are one of the enemies of Truth, enemies of Truth who falsely assume that they are guided.

Then you quote one of my previous statement posted at free-minds.org, and accuse me of contradicting it with my accusatory and aggressive statements in my recent writings:

NATU: I would be grateful if the readers take some time to read the following which are a few thought I have collected recently. But before I present them I must make clear that I do not have any personal problems with the people on either side of the debate, and that my respect for anyone has not and will not decline because I have greatly benefited from everyone registered here at free-minds, and it would be highly unjust of me to take sides on this topic. Also, I do not wish to create any further unnecessary friction. I intend not to mock anyone's beliefs, nor do I intend to guide anyone and I sincerely hope nobody is offended by this. Peace is the key to salvation, and keeping that in mind I will refer to those who accept 19 and those who reject it as +19 and -19 respectively.

EDIP: You seem have amnesia. I did not start attacking the unappreciative gang posting at the free-minds forum until AYMAN posted his article for IDIOTS condemning us to be HYPOCRITES, TRICKSTERS, MUFTIN (those who have fallen into fitna), and MISGUIDED, and many other negative names.

394

The "Intelligent People's Guide..." article was written against this person. He started the attack and he got what he deserved. When you side with the aggressor, with the real deceiver, with the real twister of facts, then you get your share of criticism too.

No my dear, you cannot hide the chain of events. You are young and perhaps is getting a kick out of this debate; that is fine. As long as you are helping your big brothers in their vicious and deceptive attack against the Quran, you will not have my sympathy. May God guide us to the straight path.

AYMAN: Peace Edip, all.

> "I am leaving the decision to our 'idiot' and 'intelligent' readers. The intelligent ones will find the books about the history of numbers rather than relying on my or your selective scanning. I am not going to post the pages of the book here, for the following reason: My targeted audiences are not idiots; they will find the book on their own and they will see the chapters I referred to. They will also check other books to find the truth of the matter. I provided them with some information and gave them the name of the most comprehensive and authoratative book on the subject. But, if one day I am going to publish our debate as a book, then I might add some excerpt from the book to show the falsity of your claims."

Edip, when you decide to resume the debate, you need to give us a simple answer: **Does Irfah's book (or any other book) contains or refers to an Arabic pre-quranic inscription or manuscript where dotted letters were used as numerals, YES or NO?**

> "As for your supporters, they will be glad to see what YOU CHOSE to show them, and they will be blind to what you are hiding from them. Ayman conveniently shows a diagram from the beginning of the book, and somehow ignores to look at the most relevant chapters: 17, 18, and the chapter 19. What he is sharing with is us is ONE THOUSAND year before the revelation of the Quran. He is obviously scared to come closer to the era of revelation since he will face the ALPHABET LETTERS ALSO FUNCTIONING AS NUMERALS."

Edip, the diagram that I showed is not from the beginning of Ifrah's book. The table that I posted is from a different source. Although similar, the one in the beginning of Ifrah's book doesn't have the Nabataean numerals. If you claim that

395

the Nabataean script stopped being used ONE THOUSAND years before the revelation of the great reading, then you can add this to your long list of extraordinary claims.

The Nabataean kingdom flourished until the 4th century CE. This is about 200 years before the alleged date of birth of the prophet NOT ONE THOUSAND years as you claim. Moreover, although the Nabataeans ceased to be a political force in the 5th and 6th centuries and the Christian Ghassanids became the more dominant political force, the Nabataeans as a people didn't simply disappear into thin air. The Nabataean people and culture continued to exist.

We know that the Nabataean culture, although in decline, was alive and well at the time of the revelation of the great reading not only based on archeological evidence but also based on internal evidence from the great reading itself. For example, Allat, Aluzza, and Manat are all Nabataean idols that are mentioned in the great reading.

Although Arabic is a Southern Semitic language that descended from Northern Arabian languages such as Safaitic, the Arabic script descended from the Nabataean script. Hence, pre-quranic Arabic inscriptions such as the Namara inscription were written in the Nabataean script. As the picture that I posted clearly shows, the Nabataean script used tallies and NOT numerals for numbers.

> "He might wants us to excuse him for him not having time to go that further in the book, but even in the introduction, the author of the book informs the reader about this issue. Here is a paragraph from page xxi of the introduction: "
>
>> "Given their alphabets, the Greeks, the Jews, the Arabs and many other peoples thought of writing numbers by using letters. The system consists of attributing numerical values from 1 to 9, then in tens from 10 to 90, then in hundreds, etc., to the letters in their original Phoenician order (an order which has remained remarkably stable over the millennia)."

The order of Greek, Jews, and Arabs is actually the correct chronological order. However, "thought of writing numbers by using letters" is a little misleading, because they actually adopted the practice from one another. So the Greeks thought of it first around 3rd-4th century BCE (the earliest examples are coins (266 BCE) and papyri (311 BCE)) and then the Hebrews copied it from the Greek around the first century BC (the earliest example is a coin dated to 103

BCE) long after the revelation of the Torah. Similarly, the evidence shows that Arabs copied the system from Hebrews long after the revelation of the great reading (e.g., the gauge of Ahmad b. Muhammad al-Isfahani in one of your links is dated to the 13th century CE, or 600 years after the revelation of the great reading!). Here is what Ifrah says:

"The oldest examples of the Hebrew system go back only to the beginning of the first century B.C. or, at most, to the last few years of the second century."

So Ifrah actually inadvertently proves beyond doubt that there CANNOT be a Code 19 or any other code in the Torah based on the Hebrew Abjad system.

Isn't it a fact that the Torah was revealed in an earlier millennium, long before the invention of the Hebrew Abjad system, YES or NO?

> "As I told, I do not have time to scan chapters of an excellent book that is available at libraries. You may declare your victory if you wish. I should not care for someone who is so stubbornly insisting to turn a miracle to hell-fire! What you choos is what you will deserve. If I find time in the future, I will, for the las time, refute your charges, expose your fallacious arguments and red herring, God willing. Who knows, I might find more time to scan the pages from the chapters I mentioned. Anyone who studies Aramaic, Hebrew and Arabic will find out the similarities of the language and their alphabet. However, by invoking the tribal emotions of his audience against Jews, Ayman wants to disconnect historical relationship between the two languages, their alphabet, and their numbering system."

As I said many times, if you have any historic evidence, such as a pre-quranic Arabic inscription, that shows Arabic letters being used as numbers then please bring it and close this issue.

> "It is an interesting coincidence that Ifrah dedicated Chapter 19 of his book to the ABJAD numbering system. Ayman is rejecting the undisputed fact that Arabs adopted Hindu numerals long after revelation of the Quran."

I didn't reject the fact that Arabs adopted the Hindu numerals long after the revelation of the great reading. Show me where I brought up Hindu numerals. You are either building a straw man or demonstrating your inability to properly read.

397

> "He is however, unable to suggest an alternative: What were Arabs then using as numerals during the time of revelation of the Quran, not a thousand year earlier?"

I already pointed you to the fact that Arabic script is based on the Nabataean script, which used tallies for numbers and is dated very close to the revelation of the great reading. That is what they were using as numerals. I already provided you with a picture of the numerals of the Nabataean script.

> "Perhaps, he wishes us to believe that they were drawing a line for each number! For forty, forty lines, or a mixture of lines and symbols!"

I noticed that you used Roman numerals to denote the page number from Ifrah's book. I hope that unlike 19ers falsely claim on the Submission.org web site you realize that Roman numerals rely on tallies and NOT letters. Nabataean tallies are somewhat like Roman tallies. They used straight line tallies for numbers 1-4. While Roman numerals were connected at the top and the bottom, those lines were sometimes connected at the bottom (for example, to form a U shape in the case of a 2 or a W shape in the case of a 3). Now when you come to the number 5 in Nabataean, you notice that they used a zigzag kind shape. The number 6 is a zigzag and a vertical line for the number 1. The number 7 is a zigzag and a U (which as we saw is two vertical line tally connected at the bottom). The number 10 is a curve. The number 20 is two curves connected together. The number 21 is two curves connected together and a vertical line (the numeral for one). This is similar in principle to the Roman XXI tally that Edip himself used above but is a totally different system to letters used as numerals.

> "Or, he would hope that we would just forget about numerals, and think that they did not have any numerals. That they were using the names of the numbers. Nothing else! According to Ayman, during the era of revelation, Arabs who were trading on three continents had no clue about numeral systems, which were used by many nations around them!"

I don't know what Edip has against tallies. They are much more efficient and logical than the alphabet-based systems. In fact, if we look at how the shapes of the numerals that we use today came about, we will see that 1 has its origin in a vertical straight line tally and 2 was a two horizontal lines tally connected by a diagonal, and 3 was a three horizontal lines tally connected by a vertical line on one side. Edip used the Roman tallies himself in giving the page number as XXI. You tallied two tens (XX) and a one (I). The advantage of tallies is that they are more intuitive and they make addition and subtraction much easier. Thus, they are much more universal and suitable for trading then some foreign alphabet-based systems. Given that the Romans were a superpower and probably the largest trading partner of pre-quranic Arabs, it is natural that the Nabataean tally system is very similar in principle to the Roman tally system as we saw.

"For those who wish to learn that Arabs were using their original alphabet order ABJAD before they adopted the Hindu numerals in 760 AC, please check the following links before getting hold of Ifrah's book:"

http://en.wikipedia.org/wiki/Arabic_alphabet
http://en.wikipedia.org/wiki/Abjad
http://www-gap.dcs.st-
and.ac.uk/%7Ehistory/HistTopics/Arabic_numerals.html
http://alargam.com/english/ragm364.htm
http://www.counton.org/museum/gallery2/gal2p3.html

All the links that you give do not provide any evidence nor give any information on when Arabs copied the Hebrew practice of using letters as numerals. As I said, there is no pre-quranic or early post-quranic evidence of that. In fact, as I demonstrated many times the evidence points otherwise because we have pre-quranic and early post-quranic inscriptions and manuscripts that are dotted and that do not use dotted letters as numbers.

The only thing that remotely may resemble evidence that you provide is the gauge by Ahmad b. Muhammad al-Isfahani in the last link (Inventory no. 44991 at the Oxford Museum). However, as I said earlier much like the Hebrew Abjad system which was invented in a later millennia long after the revelation of the Torah, this gauge is dated to the 13th century, a full 600 years after the revelation of the great reading.

A lot of information was exchanged during this debate and I know that it takes time to digest all of this. I hope that you will resume the discussion after your frustration had subsided and you had a chance to think more clearly. Peace on who followed the guidance.

EDIP: As I said Ayman, I am working hard to beat a deadline for a book. I am recording your postings and inshallah when I find time I will give them proper response. Since it was you who started this debate by your "Idiot's Guide" you might have ample time and motivation to continue more. But, I have some projects to finish... So, my delay in responding you should not be interpreted as disrespect to you or my acquiescence to your criticism.

Meanwhile, I made some changes in the BOOK section. I posted your first, second, and third articles as independent articles, so that readers can find your articles intact without been swallowed by responses.

I will inshallah present this debate to my best ability. If you find any unfair presentation, please let me know so that we can make the necessary corrections. Peace.

AYMAN: Peace Edip, all. As people may know, Edip has decided to stop (or I hope pause) the debate. I have no issue with this because a lot of information was exchanged during this debate and I know that it takes time to calmly digest all of this. I hope that he will resume the discussion after calmly reviewing the information and having had a chance to think more clearly.

Before I give Edip some questions to consider, I would like to share a story. A couple of nights ago when I came home, I found some literature that a Baptist missionary left at my door. Although one may be tempted to throw away such material immediately in the trash, I encourage people to browse it, at least to see the marketing strategies used. The devil is not very smart or creative so he uses the same strategies and only changes the names of things.

Briefly, the Baptist pamphlet goes like this: "GOOD WORKS cannot save you. RELIGION cannot save you. SINCERITY cannot save you. You need to receive CHRIST to be saved."

Where else have we heard such arguments? People who do good works, who obey and submit to The God, and who are sincerely trying to get closer to the truth are constantly being told that all of the above doesn't matter and that they need to receive CODE 19 to be saved.

When you ask Baptists, what did people do to be saved before Christ was born? You get the same blank stare that you get when asking 19ers about what people did before 1974.

One can learn a great deal from seeing how the count of 19 is nothing but a "fitna"/trial/affliction for the unappreciative. Indeed, The God guides whom He wills and misguides whom He wills.

Edip, let me conclude by giving you some simple questions to think about so that you are better prepared next time:

1. Did Abraham need a Code 19 to submit to The God, YES or NO? Did he even need this honorable reading, YES or NO?

2. Does 74:31 clearly say that the count of 19 is ONLY a "fitna"/trial/affliction for the unappreciative, YES or NO?

3. Isn't it a fact that Gematria's origin is the assignment of meaning to numerical measurements of pagan idols, YES or NO?

4. Does Irfah's book (or any other book) contain or refer to an Arabic pre-quranic inscription or manuscript where dotted letters were used as numerals, YES or NO?

5. Isn't it a fact that the Torah was revealed long before the invention of the Hebrew Abjad system in another millennium, YES or NO?

6. Would you appreciate The God and His book without a Code 19, YES or NO?

7. 19ers claim that Code 19 preserves the great reading. Can you reconstruct the great reading starting from Code 19, YES or NO?

8. Does Code 19 provide guidance, YES or NO?

9. Is Code 19 inconsistent, YES or NO?

10. My last question to you Edip is about what will happen if someday you stop trusting in Code 19. Would you stop trusting in the book of The God, YES or NO? Would you stop trusting in The God, YES or NO?

Think about those questions until we meet again. Meanwhile, I would appreciate it if you provide in your book a link to my "Idiot's Guide to Code 19" article on Free-minds.org and to the 19.org and Free-minds.org threads where the debate took place. Peace on who followed the guidance.

EDIP: I do not want to make this already lengthy debate longer. Thus, I will let the readers and researchers judge about the accuracy of our claims regarding ABJAD system. Here, I will briefly respond to your 10 questions:

Q-1: Did Abraham need a Code 19 to submit to The God, YES or NO? Did he even need this honorable reading, YES or NO?

A-1: No, but you are not Abraham and you did not live thousands years ago. Abraham received divine revelation. Besides, he asked for empirical evidence from God and God provided him with a demonstration how He revives the dead. Ironically, you picked the perfect example to incriminate yourself:

> 2:260   Abraham said, "My Lord, show me how you resurrect the dead." He
> said, "Do you not already acknowledge?" He said, "I do, but to assure
> my heart." He said, "Choose four birds, then cut them, then place

parts of the birds on each mountain, then call them to you; they will come racing towards you. Know that God is Noble, Wise."

You refuse to witness the mathematical miracle of the Quran by saying "I already believe in the Quran; I do not need miracles." Do you have stronger faith than Abraham does?

If you had a little knowledge of the Quran or had respect to the Quran you would not ask such a question. Ironically, I had reminded you those verses and you acted deaf and blind against them. Interestingly, you are repeating exactly the same excuse used by Pharoah to blind himself to witness the signs shown by Moses and Aaron:

20:47	So come to him and say, "We are messengers from your Lord, so send with us the Children of Israel, and do not punish them. We have come to you with a sign from your Lord, and peace be upon those who follow the guidance."
20:48	"It has been inspired to us that the retribution will be upon he who denies and turns away."
20:49	He said, "So who is the lord of you both O Moses?"
20:50	He said, "Our Lord is the One who gave everything its creation, then guided."
20:51	He said, "What then has happened to the previous generations?

Modern polytheists and unappreciative people, when they do not have the intellectual courage to accept truth, in order to avoid acknowledging the truth, ask exactly the same question Pharaoh asked millennia ago. While delivering the monotheistic message of the Quran, while sharing the signs and miracles of the Quran, if you encounter similar excuses, you should use exactly the same answer given by Moses in verse.

2:106	We do not duplicate a sign, or make it forgotten, unless We bring one which is like it or even greater. Did you not know that God is capable of all things?
2:107	Did you not know that to God belongs the kingship of the heavens and earth, and that you do not have besides God any guardian or supporter?
2:108	Or do you want to ask your messenger as Moses was asked before? Whoever replaces acknowledgement with rejection has indeed strayed from the right path.

Here are some other verses diagnosing your problem:

6:4	Whenever a sign came to them from their Lord, they turned away from it.

402

| 6:5 | They have denied the truth when it came to them. The news will ultimately come to them of what they were mocking. |

26:4	If We wish, We could send down for them from the heavens a sign, to which they would bend their necks in humility.
26:5	Not a new reminder comes to them from the Gracious, except that they turn away from it.
26:6	They have denied, thus the news will come to them of what they used to ridicule.

27: 93	"Say: 'Praise be to God, He will show you His signs and you will know them. And your Lord is not unaware of what you do.'"
38:7	"We never heard of this from the people before us. This is but an innovation."
38:8	"Has the remembrance been sent down to him, from between all of us!" Indeed, they are doubtful of My reminder. They have not yet tasted My retribution.

29:1	A1L30M40*
29:2	Did the people think that they will be left to say, "We acknowledge" without being put to the test?
29:3	While We had tested those before them, so that God would know those who are truthful and so that He would know the liars.
29:4	Or did those who sinned think that they would be ahead of Us? Miserable indeed is their judg-ment!

78:27	They did not expect a reckoning/computation.
78:28	They denied Our signs greatly.
78:29	Everything We have counted in a record.

| 74:56 | None will take heed except if God wills. He is the source of righteousness and the source of forgiveness. |

Q-2: Does 74:31 clearly say that the count of 19 is ONLY a "fitna"/trial/affliction for the unappreciative, YES or NO?

A-2: This is a faulty question since you are cutting the statement and hiding the rest. We have discussed this in the past and you were refuted my many examples from the Quran. You are free to blind yourself with the smoke of hell in your mind.

403

Q-3: Isn't it a fact that Gematria's origin is the assignment of meaning to numerical measurements of pagan idols, YES or NO?

Q-3: None knows this for fact; and it is immaterial. The same could be said for many human languages that were developed by pagans and that does not make their usage a sin. God sent messengers to pagans to deliver His message in their own language! Many of the good things we are using today could have been invented by pagans, atheists or Trinitarian Christians! Do you refrain from using them?

Q-4: Does Irfah's book (or any other book) contain or refer to an Arabic pre-quranic inscription or manuscript where dotted letters were used as numerals, YES or NO?

A-4: First, after witnessing the examples of ABJAD in the Quran (in divine attributes and the prophetic timing of the code and the numerical values of the 19 letters in the first statement of chapter 74) are sufficient to prove this issue beyond doubt. It is now clear that you do not have Ifrah's book. Here is some more information:

> "The order of the twenty-eight letters of the Arabic alphabet, in its Eastern usage, is quite different form the order of the letters in the Phoenician, Aramaic or Hebrew alphabet."

> "A glance at the names of the first eight Arabic letters compared with the first eight Hebrew letters shows this straight away:...."

> "We would expect to find the twenty-two western Semitic letters in the Arabic alphabet, and in the same order, since Arabic script derives from archaic Aramaic script. So how did the traditional orde of the Semitic letters get changed in Arabic? The answer lies in the history of their system for writing numbers."

> "The Arabs have frequently used a system of numerical notation in which each letter of their own alphabet has a specific numeric value (Fig. 19.3); according to F. Woepke, they 'seem to have considered [this system] as uniquely and by preference their own." (Ifrah, Chapter 19, p. 241-242)

Q-5: Isn't it a fact that the Torah was revealed long before the invention of the Hebrew Abjad system in another millennium, YES or NO?

A-5: You do not have conclusive evidence to label your assertion as "a fact". Lack of archelological evidence cannot be evidence itself.

Q-6: Would you appreciate The God and His book without a Code 19, YES or NO?

A-6: Of course, I do appreciate it and I have written books on this subject. But it is you who do not appreciate it as a "kitabun marqum" numerically-coded book. It is you who is escaping from the number 19 like a zebra.

Q-7: 19ers claim that Code 19 preserves the great reading. Can you reconstruct the great reading starting from Code 19, YES or NO?

A-7: If I am a 19er then you are 19-less. You can direct the very same question to anyone who asserts that the Quran is preserved by God. You better educate yourself about the "error-correction" nature of codes. As it seems, you missed the entire point regarding the preservation of the Quran via code. You either have comprehension problem or you have not read my previous responses to you.

Q-8: Does Code 19 provide guidance, YES or NO?

A-8: Yes. Read 74:31.

Q-9: Is Code 19 inconsistent, YES or NO?

A-9: Yes. But not according to idiots, bigots, ingrates, and arrogant people.

Q-10: My last question to you Edip is about what will happen if someday you stop trusting in Code 19. Would you stop trusting in the book of The God, YES or NO? Would you stop trusting in The God, YES or NO?

Q-10: First, I have other scientific and philosophical evidences and arguments for both God's existence and the authenticity of the Quran, but the number 19 is the most powerful one. Perhaps, I should not expect a better question from a person who calls his refutation of code 19 as "The Idiot's Guide". Let me use a similar hypothetical to expose the problem with your question:

My last question to you Ayman is about what will happen if someday you stop trusting the meaning of all Arabic words. Would you stop trusting in the book of The God, YES or NO? Would you stop trusting in The God, YES or NO?

405

Though, losing trust in the meaning of Arabic words is easier than losing the trust in a proven mathematical system and related prophecies, your answer to my question would be my answer too.

Eggs might appreciate under the sand, but...

Ayman is hardening in his disbelief. He has no clue that any Christian or any onion-worshiper can also increase their faith by witnessing a party among them falling into their own version of a "fitna!"

Seeing someone being misguided does not necessarily mean that you are guided. Like Ayman, all the blind followers of Sunni and Shiite sects and the members of numerous of cults too claim that their faith has increased by seeing us falling into fitna of 19. How interesting! What a wonderful way of increasing one's faith!

Here are just two facts for Ayman:

- The number of letters in the first verse of the Quran: 19.
- The number of chapters of the Quran: 114 (19x6).

Ayman might claim it to be a MAGIC, a FITNA, or whatever. But he cannot deny these two simple but IMPORTANT Quranic facts. Ayman wants us to believe that God designed the Quran on the number 19 so that ignorant followers of various Sunni and Shiite sects, various sects of Christians and Jews, atheists, and arrogant people like Ayman increase their faith by not witnessing it! This is the zenith of absurdity.

These two examples out of many should be enough for him "to increase his faith" in his own FITNA of NOT BEING ABLE TO REFLECT, THINK and connect the dots that God wants us to connect by OUR REASON. Ayman thinks that appreciating God's signs is putting one's head in the sand! It would be an appreciation if the one hiding under the sand were an ostrich egg, since it could hatch a baby within several weeks! As for a human head under the sand, well my dear friend, it does not appreciate there!

16

"Attention All Code 19 Discoverers:

Please share your failures"

19.org/forum

9 June 2008 – 27 July 2008

(The following is an excerpt from the long discussion thread at the forum of www.19.org that had numerous participants.)

AYMAN: This is an open message to all honest Code 19 discoverers. We always hear about your amazing discoveries of Code 19 "precise mathematical miracle" usually followed by a self congratulory "praise the lord" or "allahu akbar".

How come we never hear about your failures (followed by "darn it" or whatever)?

I have a theory and I hope that you would honestly help me see if it is true or not. My theory is that for every success celebrated, there are 18 failures that are never mentioned.

Now, if you are using a computer program then you have to be careful because a program that presents you with the successful results is in fact masking the failures. For example, if your program goes through iterations or loops, then please keep track of how many loops it went through until it achieved the result. This can be easily done by adding a counter to the loop and displaying the count with the result. So for example, if the program went through 19 loops to get the

1 successful result then you know that you had 18 failures and you can post them here on this thread.

This thread is not for debating Code 19. The thread should speak for itself. If honest Code 19 discoverers share their failures then we would be able to see the success to failure ratio without need for a debate. If no one shares their failures then the question of "how come we never hear about your failures?" still stands. I hope that everyone can see that my request is reasonable.

MODERATORS: Please make this a sticky subject so that we can see how many honest Code 19 discoverers will respond so that we can put this issue to rest once and for all.

EDIP: Dear Ayman: I counted the frequency of the following words:

> SHAHR (Month): 12 times.
> YAWM (Day): 365 times.
> YAWMAYN (Two days): 3 times; and AYYAM (more than three days): 27 times.
> All the forms of the word Day: 475 times.

Only one is multiple of 19!

Your invitation is both reasonable and silly, depending how you apply it.

AYMAN: I am not applying it in anything and this is not a thread for debating Code 19. The answers should speak for themselves. One of two things are going to happen:

1. Honest code 19 discoverers are going to come out and list their trials that failed to get a 19 divisible. We can then objectively assess the ratio of failure to success.

2. There are no honest code 19 dicoverers and hence all we will see are touting of successful trials that were already well publicized elsewhere, continued complete cultish silence on failed trials and the usual accusation of being blind for asking a simple question.

TARIKH: Peace, well if u read Ahmed Deedat's early book on code 19 (link found mo some1's post in this forum), you will understand that the code is more overwhelming and evident when left to simplicity. Can you deny the fact that the 1st verses revealed contain code19? That bismillah contains code 19? Should we count the number of letters in bismillah for you? There are no arithmetics needed to count bismillah. What more can u ask? That 114 suras contain 19? Etc

AYMAN: So what? There are many verses where we can count 19 letters such as 105:2. Why is it not significant in this case? If we go by letter count then the largest occurrence of count of letters in verses is not 19 letters but some other count.

Again, this is all random. The god clearly tells us that had the great reading been from someone other than the god then there would have been much inconsistency in it. Code 19 is full of inconsistency so it cannot be from the god but is entirely manmade numerology. This is how numerology works. Read up on it.

Again, this is not a debate about Code 19 or any numerology. People can do their own research on numerology and easily figure out what it is all about.

The objective of this post is to see if some honest Code 19 discoverers (if such people exist) would come out and share their failures (which everyone here knows for sure exist). Not a single 19.org member coming out in over 10 days since I posted this and honestly sharing failures is very telling in itself.

EDIP: Dear Tarikh: Thank you for expressing this simple but crucial point. I had in the past discussed this issue with Lomax and Ayman extensively. I pulled their attention to the fact that ALL EXAMPLES ARE NOT NECESSARILY EQUAL IN DEGREE/IMPORTANCE/RELEVANCE. Furthermore, categorization of examples is also important.

Ayman might wish to turn himself deaf to the architect who built a mansion to promote the flag of the European Union with the number 12 screaming everywhere.

On the intersection of 12th street and 24th Avenue with house number of 1212 with:

      12 Windows
      24 Doors
      24 Rooms of 12*12 or 12*24 feet square space
      48 Pillars
      48 Trees

Now, Ayman wishes to cover those and more patterns of 12 by asking people to list their counts of the number of bushes (since he knows the ambiguity will lurk), or the number of bird nests on the trees, or the number of bricks etc.

Sometimes, the pattern is so obvious for those with DISCERNING EYES, if people like Ayman lists even thousands of things that do not follow that pattern, still the pattern will not lose any of its initial significance.

Ayman is blind to one of the greatest miracles, and wants to blind others by distracting them from what they have seen or from what they could have seen. He is doing his job:

> 7:146   I will divert from My signs those who are arrogant on earth unjustly, and if they see every sign they do not acknowledge it, and if they see the path of guidance they do not take it as a path; and if they see the path of straying, they take it as a path. That is because they have denied Our signs and were heedless of them.

AYMAN: You know very well that I am asking for examples of trials on the great reading. There is nothing else that Code 19 hunters hunt the code in. Isn't all the great reading equally important?

If you have some failed trials then please share instead of closed minded cultish brushing of failures under the rug. If you don't have any failed trials then good for you because I guess you never wasted your time in hunting for Code 19 in the great reading and hence never got the inevitable failed trials while you were at it.

As a leader of this forum, it is your responsibility to encourage honesty and sharing of information instead of discouraging it, unless of course this forum has turned into a cult.

EDIP: A cult most likely would not allow someone like you even post such a challenge in their forums, Ayman. I tried to expose the trick in your invitation.

None is bound by my opinion and I never assumed to be the leader of this or any other forum. I prefer blind opponents like you to blind followers. So, please feel free to continue looking for others to share your ingrate attitude towards one of the greatest signs. You will find many, since those who witness divine signs have always been a minority.

AYMAN: How is asking such simple and obvious question being "blind"?

If no Code 19 discoverers (including you) come forward with their inevitable failed trials, then what does this say?

Cults are not just about censorship. Cults are primarily about brushing failures and falsehood under the rug and give the false impression to the naive that so-called "miracles" are everywhere.

Cults feel threatened by such a simple and obvious question and they attempt to kill inquiry by accusing of blindness anyone who asks about things that they know for sure are there but are in denial and want to conveniently forget about.

Edip, you should have asked the same question that I am asking a long time ago. If anything you should be thanking me for finally bringing up this question for the sake of truth.

EDIP: Ayman wishes to portray me as if I am scared of his challenge. In the past I had lengthy arguments with those who are blinded themselves to the miracle, including Ayman.

For those who are interested in this topic, here are some of the early articles and debates:

[The links are omitted since we lost them in a technical error but they are all published in the previous chapters of this book]

AYMAN: Peace Edip, Why can't you answer the direct, simple and obvious request in my first post?

I don't think that you don't understand the question. I think that you are intelligent and you knew the answer from the first time you read the question. The reason that you don't answer is that you can't answer. You can't not because you don't know but because you know. You know that honestly sharing the Code 19 discoverers' failures will reveal to everyone the proportion of success to failure.

Bahman is more honest because although he didn't provide specifics at least he shared on the Free-Minds forum that he failed on 25% of the trials to find Code 19 in the great reading.

Is this not the 19.org forum? Where have all the honest Code 19 discoverers gone? Do they need to be reminded that Code 19 should not be taken as a partner with the god and thus they shouldn't think that sharing failed trials is somekind of blasphemy against their "perfect" Code 19? What are you all running from?

EDIP: Dear Ayman: I thought that my first and second brief answers made my response clear to you. Let me try another way.

If you see a series of numbers in an IQ test asking you to find out the next number, what do you do? You try various patterns... For instance, try this one, which I just created:

5, 11, 28, 53, 126, 175, ?

I could make it a bit tougher by skipping some and jumbling them, then including a non-conforming number and ask you to find out the non-conforming one.

411

So, what is the next number in the above series? If you fail in your first trial, you try another one, until you find the pattern. Don't you? Then when you find the pattern, if it is a bit complicated and/or made of a long series, you may not have any doubt about the presence of the series and about the intention of the IQ designer.

Now, after you find the pattern and get convinced about its deliberate design, if I challenge the presence of the pattern by pointing at your numerous failed trials, would you take me seriously? Would you waste your time listing those failed trials? Would you care about their numbers?

So, please reflect/think/ponder a little about your objections before you ride on your highly ingrate zebra that turns innumerate and blind when encounter such an extraordinary numerical pattern in the Quran.

HED: Ayman's argument is based on mathematical probability theories to prove the existence or inexistence of a pattern, and your argument is based upon an existing pattern/rule to find, so both are good arguments.

But Ayman must believe in the pattern in order to find the evidences, he must let his soul and heart speaking like he has seen the truth of God, maybe he will accept the evidence of 19.

Ayman, I suppose you believe in God and his his messengers, but you don't see - mathematically - any God or messenger. So don't you agree that we must leave mathematics for a moment?

Noor had the good idea to speak about "observation" and not mathematics when dealing with 19 findings, like the fact that there are 19x6 suras.

EDIP: Dear HED, of course I am aware of the issues related to probability laws, and I have been critical of those who juggle and manipulate numbers by arbitrarily putting some numbers next to each other in various ways until they find a divisible number. They think they have discovered a diamond, while in fact they have just found a piece of glass. They do not wonder the probability of those calculations. I have written numerous articles on those and a few of them can be found on the net.

But, if you pay attention to my latest post, which gives an example of a series, you will learn that there are differences. When the existence of a pattern becomes clear beyond probability, then all the trials and errors leading to the discovered pattern becomes mute, since you will learn that those trials and errors were only the products of your ignorance, your inability to distinguish the relevant events from the irrelevant ones.

412

So, Ayman is closing one of his eyes or standing miles away from a 3-D random dots stereogram and then complaining of not witnessing any picture in the frame. He will never be able to see it unless he gives the possibility of such a pattern and purifies his mind from the arrogance and prejudices.

Below is my response to a recent posting on this forum:

HATI: If you write all verse numbers from 1:1 to 27:30 side by side including in each sum of sura numbers & sum of verse numbers:

1 1 1 2 1 3 1 4 1 5 1 6 1 7 728 2 1 2 2 2 3 2 4 2 5 2 6 2 7 2 8 2 9 2 10 2 11 2 12 2 13 2 14 2 15 2 16 2 17 2 18 2 19 2 20 2 21 2 22 2 23 2 24 2 25 2 26 2 27 2 28 2 29 2 30 2 31 2 32 2 33 2 34 2 35 2 36 ... 26 227 590225878 27 1 27 2 27 3 27 4 27 5 27 6 277 7 27 8 27 9 27 10 27 11 27 12 27 13 27 14 27 15 27 16 27 17 27 18 27 19 27 20 27 21 27 22 27 23 27 24 27 25 27 26 27 27 27 28 27 29 27 30

The above number is a multiple of 19!

EDIP: Dear Hati: This is an arbitrary calculation, since there are too many alternative ways and too many other numbers you could involve.

Since, your way of ordering a select group (chapter number followed by verse number and then followed by their sum) are not supported enough by similar way of ordering those numbers, in other words, since it does not demonstrate a clear doubt-free pattern, your observation may be just a mere coincidence.

AYMAN: Peace Edip, [you asked] "So, what is the next number in the above series? If you fail in your first trial, you try another one, until you find the pattern. Don't you?"

No, this is not how scientific discovery works. In your example, if there is a pattern then it can be found analytically and only because the answer is in the question. In other words the IQ test has already TOLD you that this is a series and asked you to find the next number. So it had already given you that preconception (not a very intelligent IQ test).

The question could just as easily be about points along an oscillating curve where the pattern could be for example:

5, 11, 28, 53, 126, 175, 195, 175, 126, 53, 28, 11, 5

Or there could be no pattern and the next number is larger than the one before it by a completely random amount.

Your not so intelligent IQ test is not about the answer it is about the question. The question predetermines the answer. This is not how real scientific inquiry works.

413

Ironically, you are right that your IQ test is similar to Code 19 in that it is also about the question predetermining the answer. All code 19 hunters ask themselves the same question: "Can you please find Code 19 in the great reading?" They then go and do exactly what the question asks by following the trial and error procedure that you outlined. They keep trying different methods and data sets combinations and they present the successes because that is what the question is asking. They don't present the inevitable 18 failures for every 1 success because this is not what the question is asking. The question has already predetermined the answer.

> "Then when you find the pattern, if it is a bit complicated and/or made of a long series, you may not have any doubt about the presence of the series and about the intention of the IQ designer."

As you saw, the intention of the IQ test designer is already apparent in the question. The answer is completely useless on giving me anymore info on the intention of the designer.

> "Now, after you find the pattern and get convinced about its deliberate design, if I challenge the presence of the pattern by pointing at your numerous failed trials, would you take me seriously? Would you waste your time listing those failed trials? Would you care about their numbers?"

It depends on the question. If your question is "find the Code 19 pattern" then you have already assumed a priori the presence of a Code 19 and will only show the successful trials. But the real question is "find if there is a Code 19 in the great reading or not" and therefore the failed trials are just as important as the successful ones. No Code 19 hunter has ever tried to find the honest answer by honestly sharing the MANY MANY failed trials (which you admit to the fact that they are present and are far more than the successful ones).

All this is naturally expected because as you admit Code 19 methodology is trial and error and all trial and error is solution-oriented.

> "So, please reflect/think/ponder a little about your objections before you ride on your highly ingrate zebra that turns innumerate and blind when encounter such an extraordinary numerical pattern in the Quran."

It is ironic because another term used to describe your Code 19 hunters trial and error approach is "blind variation and selective retention". It doesn't care about why the solution works and it requires no knowledge. At the end of the day you don't know if the result you got was because of a pattern or was completely random. This is also why trial and error solutions cannot be generalized. This is

414

why everyone here knows that you will never find a consistent method across Code 19 hunters or even within the results of each hunter.

> "Of course I am aware of the issues related to probability laws, and I have been critical of those who juggle and manipulate numbers by arbitrarily putting some numbers next to each other in various ways until they find a divisible number. They think they have discovered a diamond, while in fact they have just found a piece of glass. They do not wonder the probability of those calculations. I have written numerous articles on those and a few of them can be found on the net."

Dr. Khalifa is equally one of those who juggle and manipulate numbers by arbitrarily using different methods and tampering with the data set until they find a divisible number.

You and no one has ever contested the fact that Dr. Khalifa has used at least four different methods to get at his preconceived divisible number of the so-called initials as I outlined in this article:

http://www.free-minds.org/articles/science/ayman19.htm

Not only this, but in addition Dr. Khalifa has clearly tampered with the data set until he found his preconceived divisible number.

So you can count Dr. Khalifa as one of those who think that they have discovered a diamond, while in fact they have just found a piece of glass.

> "But, if you pay attention to my latest post, which gives an example of a series, you will learn that there are differences. When the existence of a pattern becomes clear beyond probability, then all the trials and errors leading to the discovered pattern becomes mute, since you will learn that those trials and errors were only the products of your ignorance, your inability to distinguish the relevant events from the irrelevant ones."

But the trial and error method itself is an indication that the discoverer is ignorant of the relevant from the irrelevant. Trial and error by definition gives him a solution but doesn't give any additional insight as to why he got the solution. Thus, the Code 19 discoverer remains just as ignorant about what is relevant and what is not relevant even after finding his preconceived solution.

> "This is an arbitrary calculation, since there are too many alternative ways and too many other numbers you could involve. Since, your way of ordering a select group (chapter number followed by verse number and then followed by their sum) are not supported enough by similar

way of ordering those numbers, in other words, since it does not demonstrate a clear doubt-free pattern, your observation may be just a mere coincidence."

I bet that Hati tried chapter numbers followed by verse numbers by itself and several other such failed trials. Had he honestly presented those other failed trials then everyone would have seen if his successful trial is well within the normal probability of the success of finding a 19 divisible among a random set of numbers. Do you now see why I am asking this simple question and see why it is reasonable?

HED: Dear Edip, My point of view it that if anyone uses mathematics in order to prove anything concerning code 19, he will loose the battle, and the legitimacy of code 19, like if he tries to prove the existence of God by mathematics.

Using the word "observation" -as you did- is a way to maintain code 19. The observation leads to belief. And the belief leads to more observation of things!

EDIP: Dear HED: I disagree with you. The mathematics is clear and powerful. However, similarly the blindness and dogmatism of those who reject to witness the miracle too is clear and powerful.

I have given lectures on this to groups of math professors, in American Universities and on Turkish TV programs.

If you read my previous length discussion with Ayman, you will find out that his arguments fall apart when they are analyzed. But, I think a better way of exposing his problems is engaging in oral debate, since oral debate has a quality that these forums do not have: impeachment. By using Socratic Method, it is much easier to impeach ignorant yet arrogant opponents of the message.

Here is my invitation to Ayman: If you are in the State, let's meet somewhere and record our debate on this issue.

JONNY: Peace again Ayman,

Edip wrote:

But, if you pay attention to my latest post, which gives an example of a series, you will learn that there are differences. **When the existence of a pattern becomes clear beyond probability, <u>then all the trials and errors leading to the discovered pattern becomes mute,</u> since you will learn that those trials and errors were only the products of your ignorance, your inability to distinguish the relevant events from the irrelevant ones.**

416

Then you(Ayman) responded:

> But the trial and error method itself is an indication that the discoverer is ignorant of the relevant from the irrelevant. Trial and error by definition gives him a solution but doesn't give any additional insight as to why he got the solution. Thus, the Code 19 discoverer remains just as ignorant about what is relevant and what is not relevant even after finding his preconceived solution.

Ok then tell me how did we verify the theory of evolution, big bang, age of the earth, age of the universe etc? Do you think trial and error was not evolved? Do you think there weren't a huge amount of failures? But after a pattern emerged it showed that these theories were highly probable and there's no need to elucidate the failures. They are there of course as are the code-19 ones so many proposed by people all over the forum which both in my view in brother Edip's have zero value i.e. they are pieces of glass. So do you say that since there's already a proposed "theory of evolution" now from all the trial and errors only the good ones are selected in favor of the theory?

When RK started his work he did not even expect a pattern based on code-19 neither did he hypothesize such. It was only after Hed already found statistical significant FROM THE INITIALS that he even started hypothesizing this.

Also the pt to note is a good theory makes predictions which later turn out to be true. This is what showed evolution to be true and this is what showed code-19 to be there in the Quran. Edip has made two major predictions one involving the word "Bastatan" being spelled with sin instead of "sad" and another that after finding that the Gematrical Values of only four ATTRIBUTES match the no of counts of the words in the Bismillah, as counted by RK n Abdul Baqi before him, there must be another attribute whose count in the Quran must equal 19, SINCE the word "ISM" is NOT an attribute, in order to complete the sequence. And YES it happened. The attribute was "Shahyd" occurring 19 times counted by the same rules. So you see not only did this verify the code-19 pattern in the Quran BUT it also verified the counting method which RK and Abdul Baqi had used with the words in the Bismillah. GOD Bless!

EDIP: Jonny: By highlighting the gist of the entire argument from my response, and by referring a scientific theory, which contains numerous failed trials and errors by scientists (and interestingly includes some fraudulent claims like Piltdown), you demonstrated without doubt the absurdity of Ayman's original claim.

Ayman is a fanatic ingrate, and he will never see what we see. He will never be objective about simple facts. This is an article of the great prophecy in 74:31.

417

Many smart and educated people with problems in their hearts will betray their intelligence and education just to deny the existence of one of the greatest divine signs. Ayman, as it appears, has doomed himself to witness hellfire and smoke, rather than the miracle we are witnessing. What is worse that he is proud to witness the hellfire and smoke! "The hell has already surrounded the ingrates." Subhanallah! Mashallah.

AYMAN: I see that you wrote this before our debate where you acknowledged the inaccuracies in Dr. Khalifa's translation of 74:31 and provided a better translation that shows the clear cause and effect relationship in the passage. You should go back to our debate and read your improved translation and see that "the count of 19 (or if you prefer "the number") is ONLY a "fitna"/trial for the disbelievers/unappreciative". Everything that comes afterwards in 74:31 is a result of this whole underlined statement. Of course, "fitna" is something that is attractive. Ask yourself, who thinks that 19 or counting 19 is attractive?

EDIP: Well, I recommend the readers of this debate to read our first debate, which beame a book, titled "The Intelligent People's Guide to Code 19" at 19.org. I will quote from the chapter 28, which deals with this argument. There you will see how delusional Ayman is. Here is an excerpt from that chapter:

The word ILLA (only) includes all the reactions described by the verse. Let's reflect on the following English statement:

"Putting the hat on her head was ONLY to show her my love, to declare our friendship, and to adorn her head, to protect her from sunshine, and to increase Ayman's confusion; he will say 'what did Edip mean by this'?"

Ayman will argue that, "Edip put the hat on his daughter's head, only to show her his love. Because Edip loved her, her head was adorned by the hat. Similarly, because the sun and her head witnessed that Edip loved her they cooperated to protect her from the sunshine, and witnessing that Edip was capable of loving, Ayman will get confused." This is the translation of Ayman logic on 74:31!

...

I discussed this issue enough and I will rest my case here. As for Marie, she was misled by you since she thought that we were considering the first word Chapter 78. You did not correct her by saying that it was you who fabricated such a claim in our name. Now, you have the audacity to criticize me for correcting her against your false accusation.

> 74:49   Why did they turn away from this reminder (ZiKRaY)?
> 74:50   Like fleeing zebras,

74:51 Running from the lion?

74:52 Alas, every one of them wants to be given separate manuscripts.

74:53 No, they do not fear the Hereafter.

74:54 No, it is a reminder (taZKiRah).

74:55 Whosoever wishes will take heed.

74:56 None will take heed except if God wills. He is the source of righteousness and the source of forgiveness.

AYMAN: You can't escape having to put a "to" to make it work and there is no "to" in 74:30-31. Good night Edip and may the god have mercy on us and save us from this "fitna".

(The following is Ayman's response to Hamza, another participant of the debate)

HAMZA MUTLU: Considering 6 pairs of attributes of Allah with the same GV, the probability that GV's of only 4 among the remaining 117 (=123-6) attributes to be 19 divisibles is: $(1/19^4 \times 18^{113}/19^{113}) \times 117!/4!x113!$ Which is around 1/8 and as 2698/19=142, the probability that GV's of these 4 attributes to be 19,2698,57,114 is: 4! x 138! / 142! = 1/16234505. So the probability under consideration is: 1/8 x 1/16234505 which is roughly 1 / 130 millions. Meaning: either RK found these GV's after trying TOO MANY MANY words (actually tens of millions of words) or he was as lucky as a national lottary winner.

AYMAN: So according to your logic, winning a lottery is a "true 19 miracle". People win the lottery evey week so I guess we should make them saints and messengers since according to you they performed "miracles".

The same "probabilities" that you are talking about can be said about any numerology result. You don't seem to know about numerology so I would suggest reading up on it first.

I also don't think that you understand the problem of identifying a pattern in the great reading or pattern recognition in general. When using trial and error as 19ers do, the question in pattern recognition is not about "winning" (or finding a result), the question is about whether "winning" (or finding the result) in the great reading has a higher chance than in a random sample. So knowing the failed trials is just as important as knowing the successful ones.

HAMZA: As for why GVs of 4 words of Basmalah not 19, 2698, 57, 114, if they were so then the table of attributes of Allah would not be

419

possible, since this would necessitate that GVs of "Shaheed" and "Bism" to be the same.

AYMAN: I thought that the whole idea of "miracle" was not about what is "possible". Certainly, a much more impressive code would have used words in the Basmalah whose count directly corresponds to their GVs and are all divisble by 19, not the dubuously connected counts and GVs that you have now.

> **HAMZA**: On the other hand I'm always cautious about this "code 19". We all know that quran aloners believe due to their understanding from several verses that Quran is fully coded and protected in this way, and with this preconception they accept every individual interesting match they found in Quran by trial and error as part of this "code" regardless of its statistical importance, so I'm not sure if such a code really exists or it is just an invention of these people as their "subjective miracle"

AYMAN: Only people who are attracted to the temptation/"fitna" of counting 19 believe in a Code 19. No one else does. Unfortunately your point may only disprove "code 19", and not the true 19 miracle. There is nothing true about any numerology. The count of 19 is only an attractive temptation for the disbelievers who don't believe in the great reading without somekind of false miracle. Peace, Ayman

EDIP: Ayman's response above puts period to the debate. In sum, he argues that:

> "19 is mentioned in the Quran only as a numerical fitna for those who count it, and God provides them with a winning lottery ticket by accident!"

This is the zenith of absurdity and is a fitting place for this arrogant disbeliever who has been fighting against God's great sign without knowledge and without guiding light.

I would like to thank Hamza Mutlu for posting such a calculation that conclusively exposed this twisted mind who did not show courage to accept my invitation for an oral debate.

17
Even More Running...

74:32   No, by the moon.
74:33   By the night when it passes.
74:34   By the morning when it shines.
74:35   It is one of the great ones.
74:36   A warning to people.
74:37   For any among you who wishes to progress or regress.

ZLATAN: 19 Questions for the believers of code 19.

EDIP: 19 Answers for the disbelievers of code 19!

ZLATAN: Where Quran says that there is code 19 embedded in it?

EDIP: 74:30-35 verses in Chapter named The Hidden One IMPLIES and the Quran, with its 114 (6x19) chapters, with its 19 lettered Basmalah and many other units, confirms the implication and fulfills the prophecy of 74:30-35.

According to 74:31, when the secret is unveiled, hypocrites and disbelievers will not understand the implication of 19. According to these prophetic verses, this number will disturb them. Number 19, indeed is an intellectual punishment for hypocrites and disbelievers.

ZLATAN: Where Quran says that 19 is a Sign?

EDIP: 74:31-35 and 7:146 and 46:10.

ZLATAN: Where Quran says that J day will happen in the year 2280?

EDIP: Nowhere in the Quran you will find an explicit statement regarding this date! It is a personal speculation based on some hints and indication of certain verses. I had come to the same conclusion independent of Rashad, albeit years after his original "discovery." Nevertheless, I admit that this information is not explicit in the Quran and it requires some kind of deeper study to understand it.

ZLATAN: Where, in Quran, God ordered that words of Quran and its letters should be counted?

EDIP: Nowhere you will find such an order as nowhere you will find a prohibition for such a task. The Quran tells us to study the Quran and informs us that it is revealed by God's knowledge. The Quran is also described as an AYAT, sign or miracle. One of God's attributes is HASEB and SAREE'UL HESAB. He also has "counted everything in numbers" (72:28).

Discovering or appreciating mathematical patterns in the Quran is a matter of intellectual curiosity and accepting it is a matter of sincerity you 9, 14, 14, 21, 13, 5, 18, 1, 20, 5; 4, 9, 19, 2, 5, 12, 9, 5, 22, 6, 18. Over it is twenty-six! (If you understand this, you will also catch an error!)

ZLATAN: Where, in Quran, God said that anything and everything in Quran should be counted and divided by 19?

EDIP: Where, in the Quran, God said that anything and everything in the Quran should be considered as just literature? Where, in the Quran, God said that God communicates with us only in letters?

- Where, in the Quran, God said that "you have rejected My revelations/signs/miracles (ayaat), before acquiring knowledge about them. Is this not what you did?"?

- Where, in the Quran, God said that "Praise be to God; He will show you His signs, until you recognize them. Your Lord is never unaware of anything you do?"?

- Where, in the Quran, God said that "He is fully aware of what they have. He has counted the numbers of all things"?

- Where, in the Quran, God said that "Do you know what 'Elleyyeen is? A numerically structured book (kitabun marqum). To be witnessed by those close to Me"?

- Where, in the Quran, God said that "This is one of the greatest. A warning to the human race. For those among you who wish to advance, or regress"?

- Where, in the Quran, God said that "I will divert from my miracles/revelations those who are arrogant on earth, without justification. When they see every kind of miracle they will not believe"?

ZLATAN: Where Quran says that Rashad Khalifa is a messenger of God?

EDIP: What? Wasn't Rashad assassinated by Sunni terrorists? Was Rashad reincarnated somewhere? How can a dead person be a messenger?

422

ZLATAN: Where Quran says that, after revelation of Quran, God will send a Rashad Khalifa to make clear and explain some of His verses?

EDIP: Nowhere, but, the Quran advices believers to learn from each other. The verses and words of the Quran explain themselves. A student of the Quran might invest his time and intellectual energy in studying some Quranic issues and might be in better position to inform others about the meaning of a particular word or verse.

ZLATAN: Where Quran says that we will find some divinely inspired explanations of its verses in the book called appendices of Rashad Khalifa`s (Authorised!?) translation of Quran?

EDIP: Nowhere. The Quran is detailed, complete and is the only source of God's religion.

ZLATAN: Can you answer to any of these questions above with a clear, straightforward and unambiguous verse of Quran?

EDIP: Perhaps you should have put this request in the end. I tried; but I would like to remind you that the Quran contains some multiple meaning verses to expose hypocrites and disturb disbelievers. Over it is Nineteen!

ZLATAN: Isn't Quran fully detailed book?

EDIP: Yes, therefore A.L.M.; A.L.M.S.; K.H.Y.'A.SS.; Y.S.; H.M.; 'A.S.Q.; Q., etc. Yes, therefore 2:23

ZLATAN: Why is it so hard to find a SINGLE clear verse in Quran to support theory of code 19?

EDIP: After 1974, the single clear verse is "over it 19" in Chapter 74. However, the entire Quran is full of explanation and implementation of this prophecy.

> 74:30 Over it is 19.
>
> 74:31 We appointed angels to be guardians of Hell, and we assigned their number (1) to disturb the disbelievers, (2) to convince the Christians and Jews, (3) to strengthen the faith of the faithful, (4) to remove all traces of doubt from the hearts of Christians, Jews, as well as the believers, and (5) to expose those who harbor doubt in their hearts, and the disbelievers; they will say, 'What did God mean by this allegory?' God thus sends astray whomever He wills (or, whoever wills), and guides whomever He wills (or, whoever wills). None knows the soldiers of your Lord except He. It is a reminder for the people."

Traditional commentators of the Quran had justifiably grappled with the understanding this verse. They thought that disbelievers would be punished by 19 guardians of hell. That was fine. But they could not explain how the number of guardians of hell would increase the faith of believers and convince the skeptical Christians and Jews regarding the divine nature of the Quran. Finding no answer to this question, they tried some explanations: the Christians and Jews would believe in the Quran since they will see that the number of guardians of hell is also nineteen in their scripture. Witnessing the conversion of Christians and Jews, the faith of Muslims would increase.

This orthodox commentary has three major problems. First, neither the Old, nor the New Testament mentions number nineteen as the guardians of hell. Second, even if there were such a similar statement, this would not remove their doubts but to the contrary, increase their doubts since they would consider it one of the many evidences supporting their claim that the Quran had plagiarized many stories from the Bible. Indeed, there are many Biblical events are told by the Quran, though occasionally with some differences. Third, none so far converted to Islam because of guardians of hell.

Scholars who noticed this flaw in traditional commentaries, such as Fahraddin el-Razi, in his classic commentary offered many speculations, including that the number nineteen indicates the nineteen intellectual faculty of human being. Tough it is a clever interpretation, but it fails to explain the emphasis on the number nineteen itself and it also fails to substantiate the speculation.

The following verses emphasize the crucial function of number nineteen:

74:32   No, by the moon.
74:33   By the night when it passes.
74:34   By the morning when it shines.
74:35   It is one of the great ones.
74:36   A warning to people.
74:37   For any among you who wishes to progress or regress.

ZLATAN: Why do you have to rely upon ambiguous verses while defending code 19 and messengership of Rashad Khalifa?(3/7)

EDIP: The "ambiguous" or multiple meaning verses can be understood by "those who are deep in knowledge" (3:7). Hypocrites will not understand their meaning (3:7; 74:31). (See my article titled: Glorifying Unlimited Ignorance via Distorting the Meaning of "Mutashabih" of 3:7.)

ZLATAN: Have you checked all the counts and calculations, that you consider miraculous and that you believe in, by yourself, or you rely upon hear-say?

EDIP: Yes, I did check every count that I deem to be the part of the miracle. I have started this since 1980 when I was in Turkish prison as a political prisoner.

ZLATAN: Did you count 40.000 letters ("initials" in initialed chapters), and how many times?

EDIP: I counted and recounted the frequency of initials except the letters A.L.M. I have verified and examined my count by comparing it to the count of others.

ZLATAN: Have you ever heard about messenger of God who announced his messengership RETROACTIVELY to people as Rashad Khalifa did?

EDIP: I fail to understand the importance of this issue. If you were contemporary of Muhammad then your question would be: "Have you ever heard about messenger of God who claimed to have BOOK as MIRACLE as Muhammad did?" As if there is a universal law requiring the messengers to be the clones of each other!

ZLATAN: Did you know that appendices of RK`s translation of Quran are full of manipulations with counting criteria, errors and mathematical properties that is, insignificant calculations and counts?

EDIP: Yes, I do know. I also do know that among those arbitrary calculations there are very significant and incredible mathematical diamonds. Neither I swallow every calculation as part of the mathematical code because Rashad suggested it, nor I reject the diamonds.

ZLATAN: Did you ever read a serious critic and reevaluation of Rashad Khalifa`s work (calculations, counts etc.)?

EDIP: Yes, I did read more than a dozen books and hundreds of articles that are published in English, Turkish, Arabic and Persian and I have answered most of their criticism in my Turkish book, "Over it 19," which I am now writing its English version. Besides, I have a lengthy argument with Daniel Lomax who came up with a much better opposition and skeptical evasion than anyone has done so far. (See: Running Like Zebras). If you want to increase your knowledge to sophisticate your unappreciative attitude against the mathematical miracle of the Quran, here is the list of some of those books:

TURKISH:
1.   Kuran-i Kerim ve Ondokuz Efsanesi, Mahmut Toptas- Hikmet Zeyveli-Dr. Orhan Kuntman-Sadrettin Yuksel, Inkilab Yayinevi,1988, Istanbul.

425

2. Yuce Kitabimiz Kuran, Dr. Tayyar Altikulac, Turkiye Diyanet Vakfi, 1988, Ankara.
3. Kuran'in Anlasilmasina Dogru, Doc. Dr. Salih Akdemir, Mim Yayinlar , 1983, Ankara.
4. 19 Muslumanlarina 19 Cevap: Insanlar da Kayar, Emine Ozkan Senlikoglu, 1992, Istanbul.

ENGLISH
5. The Qur'an's Numerical Miracle: Hoax and Heresy, Abu Ameenah Bilal Philips, Al Furqan Publications, 1987, Riyad.
6. On Dr. Khalifa's Theory of the Nineteen in the Quran, Dennis G. Lomax, Marjan Publications, 1994, San Quentin, California.
7. Al- ttihad: A Quarterly Journal of slamic Studies, Anis Ahmad (The Miracle Called Quran At the Mercy of Charlatans), Muslim Student Association of United States and Canada, April 1978, Brentwood, Maryland.

ARABIC:
8. Tisata Ashara Meleken (Ondokuz Melek), Husayn Naci, Muhammed Muhyeddin, Az-Zahra Lil'Alemil 'Arabi, 1895, Kahire.

ZLATAN: Why do you not believe in Bible code(s)?

EDIP: Well, I read the Bible Code and some articles defending it and some articles critical of it. By applying his method equa-distance of finding words in a text to any volumous book one can generate some meaningful words and even some related ones. An excellent article, "Hidden Messages and The Bible Code" written by David E. Thomas, published in Skeptical Inquirer, Volume 21, No. 6, November/December 1997, clearly demonstrated that the claims are not warranted. Below is one of my correspondences with the author of the article.

Dear Mr. Thomas:

As a reader of Skeptical Inquirer and Skeptic magazines, I followed your articles on the so-called Bible-Code, and I enjoyed them. Though my mind is much closer to skeptics and agnostics rather than to the majority of "believers" of religions, I am a monotheist and have more than enough knowledge and reasons to believe that the Quran is the word of God.

I am currently writing a book on the Code 19 of the Quran and I would like your permission to include your article(s) on the Bible Code in the appendices section. Sure, you should have all the reasons to suspect of the abuse of your article for "another code" or numerological manipulation. However, I will offer you this:

If you are interested, you may write an article evaluating the Code 19 and I will put it in the appendices section together with your article on the Bible Code.

I had in the past a two-round arguments on this topic with Carl Sagan. I compiled our arguments in a booklet under "The Prime Argument." As for Martin Gardner, though he is one of my favorite authors, his treatment of the Code 19 in Scientific Americana, September 1980 was amusing or interesting at best, and his recent articles in the Skeptic were written with a superficial sarcasm.

If you want to learn about my work and me, I invite you to visit my website: http://www.yuksel.org

At 9:45 PM -0000 8/16/99, Dave Thomas wrote:

Dear Edip Yuksel,

Thank you for the e-mail note. I also received a phone call, where you left your phone number in Arizona.

I'm interested in your proposal, but I need to find out some more details.

For one, I probably will have to get permission from CSICOP for re-printing the Skeptical Inquirer articles. Who is the book's publisher? When is it planned for release?

Regarding writing an analysis of "Code 19", how long would I have to research such an article? (It took me about 4 months to delve into the Bible Code to my satisfaction; much less than the 5 years that Drosnin studied it, but still, some time.)

I have browsed your website, and will peruse the section on the 19 Code.

Regards,
Dave Thomas

ZLATAN: Why do you not believe in hadiths of Muhammad of which many are scientifically confirmed as true?

EDIP: Which one? When I was a proponent of hadith I dug all the six authentic books and many other secondary hadith sources to find scientifically interesting hadiths, what I found was hundreds of nonsense, false and ridiculous information and a few interesting hadith. The most scientific one was about dipping the fly completely into your soup if it has fallen into it by accident. The other one was about his advice to drink urine for cure. If you know better ones please let us know. We do not want to be deprived of scientific information!

ZLATAN: Why do you think that the sunnah and hadith which Rashad Khalifa ascribed to God, is any better from the sunnah and hadith which Bukhari & co.`s ascribed to God?

EDIP: Anyone who thinks that way cannot be the follower of the Quran alone and ironically is in contradiction with the basic teaching of Rashad Khalifa.

What is your proof that Chapter 98 is God's Word?

June 2004

EDIP: God tells disbelievers who wants to see SIMILAR ayat (sign/miracle), tells them that the only ayat given to Muhammad is the Quran. I am flabbergasted that people here pretending to be reasonable in rejecting narrative of miracles can also profess belief in the Quran. WHY do you believe in the Quran, the whole Quran?

Yes YOU, those who reject the mathematical miracle of the Quran with ignorant excuses and outmost arrogance! Why do you believe in the Quran? What is your PROOF for believing to such an EXTRAORDINARY CLAIM? Yes, claiming that "this book is authored by God" is as great, if not greater, MIRACLE as parting the Red Sea. These pretentious people demand proof for one EXTRAORDINARY CLAIM, but never show the same scruple regarding the other EXTARORDINARY CLAIM. What is wrong with your logic?

What is YOUR PROOF that any chapter in the Quran, say, the CHAPTER 98 is God's WORD? You do not have any proof, and in fact you do not believe in the Quran. All you do is lip service.

The root of the problem is lack of trust in the Quran. These people treat the Quran's statement about a miracle like they treat any other person's statement about a miracle. They ask from BOTH parties to produce ADDITONAL proof and thus they show that the level of their trust in the Quran is as much as their trust in any other person. When they hear God's word telling them that he gave this or that miracle to this or that prophet, they object God like a lawyer, "give me a proof to believe in your claims!" Ironically, they misuse the verse 2:111 for their disbelief and denial of God's assertions!

So, the real problem with these folks is their lack of belief in the Quran. They are not psychologically or socially ready to confront the hidden disbelief in their heart, thus, they are trying to find a WAY IN BETWEEN.

Please reflect on 74:31 and see which groups do not appreciate the message of the number 19. These people belong to one of those groups. God INFORMED you about them and you should appreciate such a knowledge.

IDRIS: peace, lol so all the people that contained the God's reading before 1974 had no reason to believe in it, they might as well had just thrown it behind their backs until someone came along and discovered the alleged miracle, what about the scientific miracles, what about the prophetic miracles, what about the reading's logical supremacy? Peace

EDIP: Dear brother Idris, I think, instead of LOLing your head you should cry (See 53:60). You resorted to the EXCUSE NUMBER 9 listed in my article, titled, *Which one do you see: HELL or MIRACLE?* Forget about hell or miracle, have you seen the article? ☺ Perhaps your eyes do not see certain titles!

Here is my brief answer to your excuse: We do not know the condition of ALL THE PEOPLE CAME BEFORE us and we cannot judge them. They are responsible for what God has given to them and we are responsible for what God has given to us. We cannot put ourselves in their shoes, and they should have had not either. Each group is responsible for their own with their own circumstances.

Using the previous generations as an excuse to reject a sign, to deny God's blessing, to witness a great miracle is a sign of dogmatic disbelief as it is exactly mentioned in the Quran. Please read verse 20:51. Your excuse to BLIND yourself to and DEPRIVE yourself from one of God's great signs (74:30-37) is EXACTLY the same as the person who had rejected to appreciate God's sign (20:47-51). My answer to you is the answer given to your role model in verse 20:52.

As for my question... Why are you mute? If you CLAIM to believe in the Quran, tell me why? If you claim to believe in the WHOLE QURAN then please share with us your reason why you think that Chapter 98 is God's word?

429

If you do not have a good reason for that you will be held responsible for being unappreciative of God's mercy who provide us and you with a great reason. You should not fight against one of the greatest miracles, you should not brother Idris. Peace, Edip

In my article you saw what you wished to see

June 6 2004

Dear Yankeedoodle:

Please remember the computer made 3-D stereographic or holographic images. To see the message or the picture hidden among the random dot-pattern images, you must first believe the possibility of that there might be some hidden picture and then you must follow some instructions, stay in a particular distance, focus accordingly, and spend some time. Some people who have witnessed to such a hidden message and picture would be sufficient to create curiosity in you. If you have two eyes and follow the instructions correctly you will join the ranks of those who have witnessed the picture or the message.

Let's assume that you and some of your friends have already SEEN the message/picture hidden among the random dots, and assume some other friends make fun of you and do not even consider to spend little TIME and ATTENTION to focus on the stereogram, do you think that these friends of yours will have any chance to witness what you have witnessed? What if the stereogram contains the code for a great reward for only those who could SEE the picture? Would your friend have any right to complain for not receiving the reward? Assume that some of your friends finally decided to give a chance and spent a few minutes staring at the stereogram but this time disregarding an important instruction. Instead of observing the stereogram with TWO EYES they decided to observe it with ONE EYE. Do you think they would have any chance to see what you saw? Do you think they would have right to complain when the rewards and penalties are announced?

You failed to see the positive part, the good news in my article. You blinded yourself just from the beginning and you wrote a lengthy excuse for why YOU and OTHERS should not understand what the Chapter 74 really mean. You have all the freedom to pick SAQAR as you have the same freedom to pick one of the greatest MIRACLES. Unfortunately, it seems that at THIS POINT you are not ready to OPEN YOUR mind to see and hear the MESSAGE (31:7; 45:8; 27:81).

I hope the noise you are making will remain at this level. However, if my past experience at this forum is a guide, and if the depiction of unappreciative people in the Quran is correct, soon the NOISE MAKERS will fill this section with IRRELEVANT responses hoping that it will divert people from focusing on the MAIN message.

Well, hopefully, the free-minders will hear the message regardless of the noise level.

Truth-seeking people do not care what the majority of people do. If someone claim to believe in the Quran and IGNORES a Quranic challenge and test with a DOGMATIC ARROGANCE, that person should ask himself or herself this question: "Do I really believe in the Quran?" As for original new arguments regarding the 19-based miracle, I have many, but I am sure that you will reject and ignore them without even looking at them.

Please give yourself a chance. Give the possibility that you might be self-righteously rejecting a divine sign. Think that many in the past rejected and were self-righteous regarding their rejection. As for me, I cannot have doubt of even as small as the weight of an atom from what I have WITNESSED.

On 15:9

June 6, 2004

As for your observation on 15:9, you are correct. In fact, this very short verse contains 5 emphases on "WE". As far as I recall, it is the number ONE verse in the Quran regarding the density/power of emphasis.

inNA:	Surely, **WE**
NAHNU:	**WE**
nazzalNA:	**WE** revealed;
wa inNA:	and surely, **WE**
lahafizUN:	**WE** are preserving...

So, what does it mean? It means this: It is NOT YOU, but WE who will preserve the Quran. We learn that indeed the mathematical code based on the number 19 has preserved the Quran, which is in "the hearts of the knowledgeable people."

431

I have written a lengthy chapter on this issue in my Turkish book ON IT 19, and inshallah I am in the process of translating it. I might have some English articles on this issue somewhere on the net, though.[29]

Virus Detection Program

June 7, 2004

Dear Zenje:

Whether we like it or not; the Creator of the universe decided to create some creatures that could make independent decisions from their original program. For some reason that we do not know the Omniscient and Omnipotent Creator is testing some of these creatures with their given ability to choose. He will later discard those programs that are infected with viruses and will pick those who made good and proper decisions.

Put it in computer terminology, Code 19 is one of those virus detectors. Its release year is 1974. It diagnoses those with virus and those who are healthy.

However, many people do not want to face the fact that the HELL they see all over the Chapter 74 is in fact the product of their own infected mind?

Miracle?

June 7, 2004

VINCENTIA: Notice how Edip quotes 74:35 as ONLY "It is one of the Greatest" BUT on the 19's web site it is wrongly quoted as "It is one of the greatest Miracles."

He has made an attempt to correct the wrong translation on 19's web site by leaving of the word "Miracle" BUT at the same time he has implied that the verse STILL MEANS Miracle as you can tell by his opening statement. Why did you leave out the word "Miracle" Edip, when you normally include it in your quotes? Is this an attempt to deceive or have you finally came to the

---

[29] This issue and many others reletead to the code 19 have been already discussed in *NINETEEN: God's Signature in Nature and Scripture*, Edip Yuksel, Brainbow Press, 2011.

understanding that the word "Miracle" does NOT appear in 74:35 and in which case your position on this verse in the past has been wrong.

EDIP: Dear Vincent: I do not remember that I ever translated the verse 74:37 with the word MIRACLE as part of it. My Turkish translation of the Quran MESAJ is available both online and in book format and there too you will not see the word miracle: "Bu büyüklerden birisidir." (See: www.quranix.com). BU means THIS, BÜYÜK means GREAT, and BİRİ means ONE. Miracle in Turkish means MUCİZE and you won't see it there. You can check these three Turkish words in my translation via an internet Turkish dictionary, if you wish.

However, occasionally in my articles I might have put the word MIRACLE in PARANTHESIS, indicating its implication. Obviously the THIS in the verse is referring to something. Some see it as HELL and I see it as MIRACLE. What you did was either an erroneous or unfair criticism:

For instance, I never used the word PORTENT myself and I am not even comfortable with that word, since it seems archaic for my taste of language. However, I was referring to Yusuf Ali and I introduced that "This number IS ALSO CALLED...." So, it is not my translation as you want to portray. Perhaps, I should have expressly mentioned the name of the translator, but sometimes when we are in rush in Internet environment we are not able to be thorough and meticulous.

Second, you quote from submission.org or from Rashad Khalifa. You easily equate me with them. Perhaps, you thought: "since Edip believes some of the same things, then He must be the cult member and so he must agree with all what Rashad had claimed." If you check my website you will see that I do not accept Rashad's translation infallible and I have written articles expressing my disagreements on many points.

Third, you quote from my book *Running Like Zebras*: While escaping from one of the greatest miracles (74:35). As you see I am not literally translating the verse since it is not in quotation and since I am also integrating another verse. In another word, I am referring to it by paraphrasing. They are two different things. In my articles when I make an assertion regarding Islam, I usually provide verse numbers that relates to that assertion. Since I am not suggesting nor should be assumed that my statement is the LITERAL TRANSLATION of the verse, none should accuse me of inserting a word or using a different word, etc.

433

Forth, though the word MIRACLE is not literally mentioned in the verse 74:37, considering the context of the verse, I AM HUNDRED PERCENT SURE that it refers to the miraculous function of the number 19 in the Quran.

By nitpicking on me, you avoided to read the article and chose to ignore one of the greatest miracles. It seems that your friendship with those who make mockery of one of the greatest miracles is more important than to pass the divine test.

INTERESTING, and even INCREDIBLE

June 8, 2004

In order not to keep too much space on this forum, I posted here the link to my recent article. The title of the article was: WHICH ONE DO YOU SEE: HELL or MIRACLE?

But many who are still accusing me of UNDERSTANDING the IMPLICATION of 74:37 as the MIRACLE have not even bothered to click and read my REFUTAION OF THEIR HELLISH misunderstanding.

I know ALL your arguments (and the ones that even you are not aware) on behalf of HELL, which is evident from my article. And I am ready to learn if any new argument you might come up with. However, many who chastise me here do not know ANY of my arguments and do not even care to learn them by just one click. Is this how you call yourself FAIR and open-minded people? Is this the way you discuss issues?

Translation, Parenthesis, Quotation Marks

June 8, 2004

VINCENT: You seem to be a very confused person as you spent most of your reply denying that you every said 74:35 translates as "the greatest miracle" and then you end your post by saying you are 100% sure it does mean miracle.

EDIP: Dear Vincent: Here again: The word MIRACLE is not EXPRESSED LITERALLY in verse 74:37; but it is IMPLIED CONTEXTUALLY and EVIDENTLY. Let me summarize it for you.

So, it is okay to TRANSLATE it as:

434

"It is one of the greatest (miracles)." (74:37)

But it is NOT okay to TRANSLATE it as:

"It is one of the greatest miracles." (74:37)

And it is okay to CLAIM that:

Nineteen is one of the greatest miracles (74:37)

And it is okay to CLAIM that:

Nineteen is one of the greatest misfortunes (74:37)

I cannot explain this issue simpler than that and this is my last correspondence on this issue. Please read and understand your opponents' answers before jumping on the keyboard. It does not require too much brain power to see the difference and reason behind them. Also, please get a dictionary and look at the use of PARANTHESIS in translation and the use of QUOTATION MARKS in general.

Proof

June 9, 2004

Dear Zenje: What is your EVIDENCE, PROOF, REASON to believe and assert that 9:128 and 9:129 are God's words. Without resorting to a number, in your case, the number of people who believe so, can you justify your belief in such a claim?

If you are going to make your argument based on the present manuscripts of the Quran, please think twice before jumping to respond.

Verse Numbers in Ancient Manuscripts?

June 10, 2004

> "The verse numbers are arbitrary. Allah did not send down verse numbers. He simply sent down a series of phrases or an entire surah at any one particular point. The later scholars put the numbers on the mushaf according to where most people stopped. I am almost sure that the verse numbers in both the Hafs and Warsh also post-date Imam al-

435

Hafs and Imam al-Warsh" source:
http://www.guidinghelper.com/qna/quran.html

The assertion above is absurd. Any MEANINGFUL combination of letters, words, sentences, and chapters immanently contains NUMBERS, whether the numbers are expressed or not.

For instance, someone may focus on the sentence above and refer to the word NUMBERS as the "2nd word in all-capital letters" or the "17th word". The very order of the words implies numbers and the number 17 is NOT ARBITRARY. The words in millions of books written in contemporary English books are almost entirely from the same pool of limited words found in a single-volume dictionary. However, it is the order of those words that makes all the difference. It is the order of words in the Quran that makes it divine, not the Arabic words themselves, which were also used by the opponents of the Quran. If there is ORDER there is NUMBER. That's simple!

Imagine that I wrote several pages containing about 238 different sentences, and an observant person noticed that every 17th word in my every other sentence starts with the letter "C". Can't that person easily refer to the 17 C's by stating that "Every 17th word in every even numbered sentences of this article starts with the letter C."? Do you see?

Besides, the Quran describes itself as *Kitabun Marqum*, which means NUMBERED BOOK, or NUMERICALLY CODED BOOK.

Functions of Code 19

June 10, 2004

SOMEONE: Peace be upon you, To Edip Yuksel, I like your rigorous way for finding the meanings of the words, the translation you provide fits better in context. However, I still see a problem with counting "Alif" in the concerned chapters. Can you tell how do you deal with this issue?

Also, in your article, you did not tell about the implications or the outcomes of your research. I'm not talking about "miracles", I'm talking about practical achievements in acknowledging the mathematical code. Thank you

EDIP: Dear Someone: I have similar problems with the count of Alifs, as I acknowledged decades ago in my books. For instance please read my arguments on this issue published in *Running Like Zebras*. There you will find FOCUSED and INFORMED arguments between me and Abdurrahman Lomax, whom I consider one of the few who well understands his misunderstanding or unappreciative of the miracle.

Counting Alifs require at least a three or four moth-long research by a team of scholars who would gather copies of oldest manuscripts (mostly partial versions) and critically analyze them and then reconstruct the entire manuscript with Alifs restored like its origin.

The VERY discrepancies in the spelling of Alifs, other letters and even words, is an indication that the Quran was not preserved as ink and paper/leather but as KNOWLEDGE, since the Quran is also defined as being in the heart of the those who have knowledge.

As for practical achievements of fulfillment of the code 19 prophecy, I will list some. I have limited time, so it will not be a thorough evaluation:

1. It proved that the Quran is not an ordinary book.

2. It increased the faith of believers and made it easy for them to question the established traditions and reject the authority of hadith and sunnah. Without the code 19, we would have difficulty in satisfying our mind and heart regarding the preservation of the Quran. Just think about the following question: "How can you trust in the Quran since it was also compiled, narrated and written by those people whom you have no trust. If they could fabricate so many hadiths in the name of the prophet, surely, they could fabricate some verses, even chapters attributing to God, like Jews and Christians. The very verses 15:9 and the like could be also the fabrication of that generation that burned the original Quran, added and subtracted from the Quran. Via these additions they wanted to convince people that the book they concocted was in fact protected by God...." Those who prefer HELL to the MIRACLE cannot answer this challenge. (Sure, they will try answer, but their answer would be filled with logical inconsistencies and flaws.)

3. It brought explanation to some mutashabih (multi-meaning) and prophetic verses, and in some cases enriched the meaning of many verses of the Quran. For instance, it explained the meaning and function of combination of alphabet letters initializing the chapters. Besides these initial letters, I have provided more than a dozen examples in my Turkish book Uzerinde 19 Var (ON IT 19.)

437

4. The MATHEMATICAL patterns, the deductions and inductions involving their comprehension trained and taught us to be RATIONAL. We have to use our reasoning in religious matters like we use in math and statistics..

5. It supported the Quranic claim that the majority of people in fact do not seek nor follow the truth and exposed that billions of believers are in fact not believers in Truth, but believers of their culture or followers various bandwagons of a particular group. (27:82)

6. It showed us that there is indeed SATAN who has hypnotic power over his constituencies. Blinded by their masters, they cannot SEE the PRECISE and OBVIOUS miracles.

7. By distinguishing those who are regressing and those who are progressing, it opened a new era.. (74:37; by calling oneself progressive one does not become a progressive person!). It distinguishes the real believers from the followers of hadith and sunnah, AND from those who give lip service to the QURAN ALONE message, not because of their strong faith, but because of their political, psychological, personal agendas.

8. ....

But, the most important practical result is: ETERNAL LIFE and HAPPINESS.

Why the "quran" is not a proper name?

June 13, 2004

I read Ayman's article on the issue and I found it unconvincing for the following reasons.

All Arabic proper names have (or at least in their origins) meaning. They reflect the description or what is expected of the object/person named. However, when they are used with the article AL (the) or another definitive article or adjective, and are used to refer a PARTICULAR or SPECIFIC category, concept, item, object, or person, then it also can become a proper name.

The word QURAN, thus, depending on its context, occasionally means quran (any book of recitation) or the Quran (The Book of Recitation). The latter is the name/ism/description of the book given to Muhammad.

Let's reflect on the word Muhammad. This word means THE PRAISED ONE.[30] Well, then using Ayman's logic one can claim that MUHAMMAD was not a proper name but a common name, referring to EVERY PRAISED PERSON!!!

I find some assertion made a DOGMA and passionately promoted by some people here, being half-cooked or outright the product of fallacious thinking. For instance, rejecting the bodily component of Salat prayer, for instance insistence on seeing HELL rather a GREAT PROPHECY AND MIRACLE in chapter 74...

This is my answer to the title of your question. But, your post contains different questions. Perhaps you first meant to ask something else.

As for your question regarding certain things mentioned in the Quran being absolute... A good question! I do not have time right now to deal with it.

Just a quick reminder: verse 16:8 reminds people of God creating other means of transportation, which we came to learn in the last century.

A, B, C and Z of this Forum

June 13, 2004

An interesting fixation! On this forum, some people cannot engage in a decent argument. Here is the caricaturized version of my experience here:

> A: What is the square root of 64?
> Z: Eight.
> A: First, tell me about your Salat. We will NOT listen to anything you say. We are very open-minded!
> B: A is right. Yes, tell us about Salat.
> Z: The word al-Quran means The Book of Recitation.
> A: Tell me about Salat. We will NOT listen to anything you say. We are very open-minded!
> B: I agree with A.
> C: I agree with both A and B.
> Z: Okay, I will write something on this issue. Give me a week or so.
> A: You better write soon.

---

[30] About seven years after this discussion, in late 2011, I noticed the mispronounciation and distorting of the meaning in the name of the Last prophet. His names should be read as Muhammed not Muhammad, and it means "the one praises (God) frequently." See my article titled, *What was the name of the last prophet?*

Z: Ok, don't have a horn.

Z: I posted my criticism to your position under the title: Salat Prayer Both with Mind and Body.

B: You believe in 19; it is a hoax.

Z: 19 is not a hoax. I will write an article on that issue too.

A. You better write soon.

B: He is a very arrogant person. Who does he think he is?

Z: Ok, don't have a noodle.

Z: Have you read my recent article on Salat?

A: You do not believe the Quran is detailed....

Z: My position is the same regarding the detailed nature of the Quran. Based on the details of the Quran I conclude that Salat is performed both mentally and bodily.

B: What about 19; it is about the HELL not a NUMERICAL MIRACLE.

Z: Obviously, you have not read the article yet.

B: What about Salat? Dou you still worship the stone called Kaba?

Z: One at a time. First let's understand each other and correct our errors regarding the performance of Salat, and then we can discuss this issue.

C: We will not listen to you. We are free-minders.

B: Yeah; C is right. We do not hear what some people say. We are free!

A: I agree with both. Bodily bowing down in front of God is shirk. He is a *mushrik*.

Z: I see. You are very free. I see that you in fact have no concern about what the Quran says; but all you are attempting is to justify your lifestyle, to calm your feeling of inferiority, to eliminate any religious restriction or duty, to be free from any ritual. As long as your desire is this, no verse, no evidence and no reason will have value for you. You want to make your own religion according to your own desires and wishful thinking.

You do not hesitate to ignore the facts, you do not hesitate to fabricate lies about your opponents, you do not hesitate to change the subject whenever you are proved to be wrong in one, you do not hesitate to come up with ridiculous speculations to justify your freedom from God's religion.

So, your religion is to you and my religion is to me. Besides, I never expected you to FOLLOW me, and as for my advice, my advice is for those who have ears to hear.

When you are ready to have DIALOGUE then I will engage in discussing with

440

you. But, as long as you start with insults and lies and proudly declare that you do not hear whatever I say, it is futile to engage in a dialogue. It is sad that those whose mind is the least free assume to be free-minds. It is sad that those who have the least respect to the Quran claim to follow the Quran alone. It is sad that those who cannot think beyond an inch pretend to be the promoters of islam.

To Zlatan

Feb 2005

Zlatan, I see that you have been tormented by number 19, as it is prophesized by 74:30-31. You ignorantly depict one of the greatest divine signs as "human made" and thus, like the prototype disbeliever mentioned in chapter 74, you are too committed to the 19: The number 19 is a SAQAR for you and it will torment you forever. To the Question "Which one do you see: Hell or Miracle?" your answer has been Hell, and you are living hell already in this world.

As a one whose faith increased tremendously by witnessing the fulfillment of the prophecy of the mathematical code, both intellectually and spiritually, I will not spend more time with you and your phobia; you deserve it and you will suffer from this divine punishment; sure, unless you repent and change yourself.

And, you never expressed YOUR EVIDENCES why you believe in 9:128-129 being God's words.

Shabbir Ahmed and his Group initially Act like a Pack of Wolves and then Escape like Zeal of Zebras

I received an email informing me about a heavily moderated (read: controlled) forum by a retired medical doctor and author, Shabbir Ahmed. The title of the article was one of the many arrogant attacks to the numerically structured book (Kitabun Marqum), the Quran. After I challenged them for a face-to-face public debate, initially they rejected, then one of them (Ismail Bohra) decided to sponsor the debate, but all those who were posting rants and rages against the number 19 decided not to accept such a challenge, since it was not worth it! Of course, there were some mature and reasonable participants of the forum, such as Ali Noor and Anisur Rahman, and I do not classify them with the rest.

The leading attacker, Shabbir Ahmed, who calls himself always as Dr. Shabbir, came up with the following pathetic reason to avoid my challenge for a public debate based on reason and evidence:

> "Debates fail to accomplish anything. Did any prophet/messenger engage in it? Debate is nothing but a clash of two intellects. The smarter and more eloquent of the two parties will impress the audience more and claim victory. Hence, debate is one of the most futile exercises except for granting fun to the audiences."

And everyone agreed with Shabbir, except a few. For instance, though Adeel Nawaz agreed with Shabbbir about the worthlessness of the 19, he was "confused" by Shabbir's depiction of debate as a mean of finding the truth. Adeel quoted from Shabbir's own translation where Moses engages in a face-to-face debate with magicians.

Ironically, those who rejected my offer to debate the issue face to face were attacking it with monologues in their moderated/controlled forum through hundreds of messages filled with ignorance, false assumptions, straw-men arguments, red-herring, ad-hominem attacks, mockery, distortion, deception, and contradiction! My Interned debates with Lomax, Ayman and many more in both English and Turkish is solid evidence that I do not escape from debating the issue in writing. But, debating against a group of people in a forum strictly moderated by their leader was not a reasonable option. For each of my message they would post dozens of messages filled with plethora of false statement and distortion. I would not be able to handle the attack just because I would be outnumbered. Besides, I would be relying on the mercy of their leader to publish my response.

So, debating the issue in writing would give them an unfair advantage that had nothing to do with the merit of their position. The facts would be lost among the noise, cheers and jeers created by the group. I know well these sorts of environments and I did not wish to spend hours and days responding the members of a group who would not even read my responses, let alone ponder on them. Through my experience, I know the power of Socratic method in exposing contradiction and deception. Thus, I invited them to debate it face-to-face rather than debating them in their arena; they would condemn me like the majority of 501 jurors had done to Socrates!

In sum, according to these doctors and professors, a face-to-face debate was futile but producing volumes of silly arguments, condemnation and mockery behind screens in their bandwagon controlled by Shabbir was productive!

The Quran contains at least 332 cases of arguments between muslims and ingrates. Qalu (they said) and Qul (you say) is repeated in the Quran 332 times. Many messengers and prophets invited their opponents to a reasoned debate or arguments (in philosophical context). Ironically, they were not performing those debates through internet behind computer screens, but face to face, like Socrates, Abraham, Moses, Jesus, Muhammad had done. The Quran instruct us to engage in philosophical (hikma) and reasoned debate (hujja), but warns us against quarrel, putting emotions before reason, group think, red herring, false accusations, personal attack, prejudice, bigotry and mockery. Ironically, these were exactly what Shabbir and his group were engaging in. Many in Shabbir's group did exactly what the Quran warned us against, ignored and even rejected what the Quran asked us to do.

Below you will see only a sample of what many in the group spew against this great prophetical miracle. Just look at the number of exclamation marks used by the first one and infer the torment and *fitna* they are experiencing, exactly in accordance to the prophetic description of ingrates and hypocrites when they are exposed to this number (74:1-56).

(I corrected a few spelling error in the following messages).

443

Edip Yuksel: LIES, DAM LIES, AND STATISTICS

JAWAID AHMED (UK):

31 January 2012

19! Rashad Khalifa [RK] discovered that the Quran is a mathematical miracle based on the number 19; the number of chapters, words, letters, their numerical values, all have a connection to 19. This proves that the Quran is from Allah and no man could have produced a book with a mathematical code that was only discovered using a computer.

Edip Yuksel, an associate of RK, has carried on investigating the 19 code following the murder of RK, which any right thinking human being will unreservedly condemn. Edip has stated that RK made some mistakes with certain counts and as such he has written a book entitled Nineteen, God's Signature in Nature and Scripture, showing the correct proofs and which I am going to give a brief rebuttal to [it has to be brief because the book consists of 620 pages, although only approximately half have any 'evidence' for 19, the rest is personal anecdotes].I am going to comment on those points that I found of particular interest.

> [E.Y.'s Note: Followed by 7 pages repeating exactly some of the criticism directed by Lomax, Phillip and Ayman, which I have responded and refuted here. So they are skipped.]

Edip accuses all who do not believe in 19 as blind, indoctrinated in hadith and falsehood; he says we are ingrates for not accepting the 19 code RK discovered and which he is promoting. I say that such statements are only an attempt at mental persuasion of the gullible; if you are not a 19 coder, you are a hadithist, not a Quran only follower. This is another lame argument.

I will let you work out who are the gullible [lap up everything], and who are Rashad and Edip from the following tale [PS, I am the little boy].

> [E.Y.'s Note: Two pages wasted to tell the details of this well-known story are skipped]

"The boy is right! The Emperor is naked! It's true!"

The Emperor realized that the people were right but could not admit to that. He though it better to continue the procession under the illusion that anyone who couldn't see his clothes was either stupid or incompetent. And he stood stiffly on his carriage, while behind him a page held his imaginary mantle.

Their nonexistent 19 is made visible by their weaving the Quran every which way but truthfully

!!!
!!!!!!!!!!!!!!!

Challenge to Ingrates at Ourbeacon.com

31 January, 2012

EDİP YÜKSEL (Arizona): I would like to challenge the top deniers of the code 19 in this forum to a face-to-face debate on this issue.

I have been answering to the repeated criticisms mostly containing false assumptions and poor research. Responding to each of the same separately is time consuming and unproductive.

I have refuted all the major criticism, yet many who publish their criticism have not even read the book, or my response to similar criticism. It takes days to make any progress just on one issue via writing, and there are millions of people out there who are ready to REPEAT the same criticism without even blinking their eyes.

So, I have learned that the best way to EXPOSE the false ideas is through Socratic Method. A face-to-face debate provides the best means to discuss such an issue and expose the party with false ideas or facts, usually after a few follow-up questions and reasoned dialogue.

So, I would like to give you an opportunity to EXPOSE my lies, distortions, and ignorance!

Pick one person in this forum who you think is the best critic of this "miracle" and I will pay for his trip to Tucson, Arizona or I will fly to where he resides. As the condition of this debate, we will record the debate, and both parties will have the right to publish the video recording severally.

I have debated this issue for about 30 years, and have spent too much time responding each critic one by one. Considering the number of Sunnis, Shiites and others who have problem with accepting a numerical structure in the Quran and their quality of thinking, it is impossible for me to respond each, individually through such a limited media.

So, such a video debate will help others too to hear both sides and make their mind. Face-to-face debate is the best way to discuss such an issue complicated with diverse range of ignorance and bigotry. Peace.

SIDQI: First, if anyone can figure out this one, please let us know, not just a game. http://www.regiftable.com/regiftingrobinpopup.html

TAYLOR STEVENSON: Look forward to seeing who is knowledgeable enough on this to be able to debate. I do not have time to do proper research on 19. I work full-time and study Qur'an. The person would also probably need to read some Edip's books, so time would probably need to be set aside. I heard recently that 19 followers have calculated the end of the world. Can this be serious? Would this be in alignment with Qur'an?

Let me understand one point further. If a person had never read Qur'an could they be Muslim, and serve God (by serving Humanity)? If a person read Qur'an could they be non-Muslim. (not speaking truth, and not upholding Human Rights)? The answer here is obvious. Yes to both, simply be cognizant of Qur'an does not generate Righteousness.

Now; If a person disregards a numerical significance in Creation and Qur'an how does that make them less Muslim? If I never heard of code 19, and only follow God and Messenger through the Qur'an, what is the logical significance of a numerical standard of 19?

To me to say you will be in a great detriment for not knowing 19 is similar to salvation through Jesus Christ. A man in a cave CANNOT be saved, in the Christian viewpoint. A man in a cave without Qur'an is in a great detriment because of an ignorance of the number 19?! Salaamun Alaikum

ASEF ZAHER (Siwan, Bihar, India): Salaam alaikum the method on which that game is working: whenever you subtract the two digit from the number you have already in mind the answer is always 9 by adding the two digits of the result of the new number. Let's take an example

Let the number you think is 25,then according to question you have to do=25-2-5=18.....this 18 yields 1+8=9...and hence on next page you will be given some things name corresponding to each number...and every number which gives 9 after adding their digit will have the same thing i.e. at 9, 18, 27, 36, 45, 54, 63,

446

72, 81, 90, 99 (99 is also an answer because if the addition is greater than 9,then the two digit is taken as separately).and every time after subtraction you get a number in such a way that after adding their digit it is always 9 ... this is the beauty of mathematics ... it works only for 0-99. Example:

54 54-4-5=45=4+5=9
88 88-8-8=72=2+7=9...etc

ADIL SIDQI: Asif. Great work to unfold this math mystery. Thank You. In the same way, i think number 19 is also a man developed thing out of his natural instinct to find mystery. Quran is just a simple divine book on knowing a creator, next life, human orders and civil society management.

EDIP: If this is what you understand from 19, then all I can say: have a nice day -:) Peace.

ADEEL NAWAZ (Scotland, UK): Every answer that you will get, it will be a multiple of 9. In every box that is a multiple of 9, there is the same picture. Check box 9, box, 18, box 27 , box 36 etc and see that they will always have the same gift in them. Check them. The trick is that the gift picture is changed in each 9 box every time to add to the realism of the trick. That is all.

JAWAID AHMED (UK): Please provide the website address of your debate with Ayman, announced in 2008 so that we can see whether there is any point in using this method.

http://free-minds.org/forum/index.php?topic=9597165.0

You had multiples of 19 written posts with Ayman on Freeminds and nothing came from this, so what will a live debate prove?

ALİ NOOR (Atlanta): One very basic question :-- Numerology (Power of words (Harf) given) is man-made and Quran is Allah's Word then how can the man judge veracity of Allah's words from his own Numerology? Do not try to judge Arsh (Heaven) by being on (Farsh) Earth.

He believes it is Allah's numerology so he uses this pretext to remove verses from Allah's Book and change letters! This false challenge he is proposing is like these: Challenge to the deniers of the trinity, OR Challenge to the deniers of idol worshippers. Both of the above have arguments to 'prove' their case; all three are wrong.

447

SHABBIR AHMED (Florida, 1 February 2012): Dear brothers Edip, Jawaid, Taylor and all. All of us know that we should not accept or reject anything without due reflection. I had conducted a thorough, unbiased study on late Dr Khalifa's theories in the 1990s. The flaws were so obvious that I did not find my findings worth publishing. I wish I had the time and energy to reproduce even the highlights of my humble research.

I respect brother Edip. He is indeed brilliant but that does not make anyone infallible. On this score, I find myself in agreement with Azizam Jawaid. Those who wish to analyze Dr Khalifa's work can save much time by going through the appendix of his "Authorized English Version". They will find it comical and know the true caliber of his deluded mind.

Debates fail to accomplish anything. Did any prophet/messenger engage in it? Debate is nothing but a clash of two intellects. The smarter and more eloquent of the two parties will impress the audience more and claim victory. Hence, debate is one of the most futile exercises except for granting fun to the audiences.

MUHAMMAD RAFI (Karachi): The 19 debate is a waste of time. It distracts Muslims from the simple message and guidance of the Quran. I fail to understand what can be gained from this futile debate.

FADIL (Cameroon): Dear brother Shabbir. I am very much satisfied with your post. It has said it all. It will be absolutely foolish entering debates especially on the foolish nonsensical number 19. Please, let's avoid such posts. People who want to know more about Number 19 could refer to other forums. Salam

ADEEL NAWAZ (Scotland, UK): Dear Dr. Shabbir. I completely agree with you and I believe that this 19 debate is absolutely a waste of time. I'd rather spend my time studying the Quran rather than unnecessary and useless books on this topic. However I am a little bit confused with not entering a debate altogether because you said:

> "Did any prophet/messenger engage in it? Debate is nothing but a clash of two intellects. The smarter and more eloquent of the two parties will impress the audience more and claim victory. Hence, debate is one of the most futile exercises except for granting fun to the audiences"

I hope I did not quote out of context. According to QXP - Prophet Moses debated and according to QXP, God also asked him to enter the debate

448

(7:117) Did Prophet Moses not enter a debate to convince them? Is there something I have missed? Please can you or someone else help me to understand please.

7:108 The strength of his conviction shone bright to those present. [20:17-21]

7:109 The chiefs of Pharaoh's nation said, "This is some knowing wizard!

7:110 It appears to us that Moses, after gaining some following, plans to throw you out of your land." Said Pharaoh, "Now, what do you advise?"

7:111 (After consulting) they said, "Put him and his brother off for a while and send to the cities men to summon.

7:112 To bring all master debaters to your court." [Saahir from Sihr = Magician, liar, stunningly eloquent, smart poet, deceitful, trickster, cheater, defrauder, dodger, hoaxer, swindler]

7:113 The master debaters came to Pharaoh. They said, "There will be reward for us if we are victors."

7:114 He answered, "Yes, and you will be among those who are close to me."

7:115 The debaters said, "O Moses! Either you begin or let us be the first to present our case."

7:116 Moses asked them to begin. When they began they cast a spell on the people's eyes, and struck them with awe, and stunned the assembly with their spellbinding eloquence.

7:117 So, We inspired Moses, "Respond with conviction. And then see how it swallows their deceptive rhetoric." ['Asaa = Staff = Cane = Strength = Conviction = Authority = Power = Strong logic = Convincing argument]

7:118 The truth prevailed and their arguments were manifestly proven false.

7:119 Thus, they were defeated right there and humiliated.

7:120 The debaters fell down prostrate.

7:121 They were convinced and said, "We believe in the Lord of the Worlds.

SHABBIR (Florida): Respected brother, You bring up an excellent point. Thank you! I have translated SIHR allegorically in these verses. SIHR = Magic = Lie = Spellbinding eloquence = Seemingly incredible = That stuns the intellect = Apparently beyond reason.

Moses convinces the magicians with reason, not through debate. They present their sly arguments and Prophet Moses eloquently and convincingly responds. A lifelong student

ISMAIL BOHRA (Sweden): Dear Edip & other friends. No disrespect intended. Mr Edip Yuksel, you are evading the forum. I think you have also realized that a debate is an exercise in futility. Dear Sir, you seem to have too much time at hand. Why not invest your time & effort in a better venture? I know five men in Norway and Sweden who can bring you hands down in a debate in minutes, but they are wise enough to focus on issues of real import. Thanks!

ALI NOOR (Atlanta): Assalam Professor Sahib. To resolve the matter once for all please ask your friends to take up this with Edip.

Edip on the run - Challenge!

ISMAIL BOHRA (Sweden, 3 February 2012): In Response To: Re: Edip on the run! (Ali Noor Atlanta) I have spoken to one of those friends adept at debating. He has floored some "expert" 19ers in 15 minutes. Oslo can be a suitable place if Edip would accept the challenge. I will gladly bear all expenses involved for Br Edip Yuksel, a one week trip to Norway with boarding and lodging. Thank you!

EDIP: Finally, a proper response which I expect from any seeker of truth. I could not believe that many people would dismiss my invitation by rejecting a reasoned face-to-face discussion on this issue while they find it morally okay by making mockery of this issue with little or no knowledge at all. So, I gladly accept this invitation by brother Ismail Bohra. And I appreciate his generosity. Please contact me via my email at 19@19.org by including this invitation. Please provide me with your phone number so that we can discuss the best time for this event. This issue has been a controversial issue and very divisive among those who follow the Quran. So, I believe that such a face-to-face real-time debate or Socratic Dialogue will allow many people to hear both sides in a very efficient mode and hopefully attentive and cheerful mode. Peace, Edip

ADEEL NAWAZ: Jawaid Ahmed, I have wasted £16-00 on his book and he is already promoting his new book-Run Like Zebras. This is a debate with someone called Lomax where Edip refuses to accept any criticism of his changing methodologies etc. etc at fixing a 19 answer. Hence, I would not waste my money on an all expenses trip to Norway. We should see if he answers my rebuttal to his book and then judge the man by his cover.

So to understand the significance of 19 in the Quran, we have read Edip's HADITHS to understand it? Have we not heard a similar argument before somewhere? Namely, to understand the Quran you need hadith. Same old same old, Edip doesn't even realise that he is doing exactly what he is asking others not to do... Ridiculous!

ALI NOOR: AOA Professor Let your friends of Sweden & Denmark take up with Edip so that the matter is resolved once for all.

JAWAID AHMED: Dr Saab, I understand your sentiments but when someone presents a challenge the Quran tells us to ask them for proof, if they are truthful. The less than brilliant [I beg to differ with you on this point as well] Brother Edip said that RK had made errors in his calculations [imagine, a rasool making errors!] so we must read his book, *19 God's Signature in Nature and Scripture*, to see the truth behind 19. I did and presented my results on the Edip Yuksel; lies, dam lies and statistics post below, for which I genuinely hope he will answer the specific problems I raised about his methodologies, distortions etc, etc, etc.....

I have no intention of engaging in futile to and fro debates, more so in a live debate where points raised cannot be verified or refuted in the heat of the moment, but in order to help those already caught up in this 19 fiasco and to stop others from being caught, we must engage with these people.

Taylor has already mentioned that this group has worked out when Judgment Day is, 2280 AD, we also know they believe RK was a rasool, age of responsibility is 40 years old so anyone committing a sin and dying before 40 goes straight to heaven, etc, etc, so we must expose this falsehood in the same way we criticize the N2I. At the very least, let me have a little fun!

SHABBIR: Azizam Jawaid, Keep enjoying, no restraints :-) I did my part on RK at the right time when he was becoming known and I could reserve one full month for his work. Ourbeacon (Lighthouse at that time) saved many from falling into SAQAR. Now, bright and young people like you are effectively refuting and pulverizing the "messenger-ship" of RK with all its baggage. His ship is sinking faster than Titanic! With due regards to all, A lifelong student

ANISUR RAHMAN: Yes, debate makes sense. Even our father prophet Ibrahim engaged serious debate with one of the worst disbelievers at that time. It is a good example for us. Why should not we follow Ibrahim's example?

> [The Quran 2:258] Have you noted the one who argued with Abraham about his Lord, though GOD had given him kingship? Abraham said, "My Lord grants life and death." He said, "I grant life and death." Abraham said, "GOD brings the sun from the east, can you bring it from the west?" The disbeliever was stumped. GOD does not guide the wicked.

PS:

EDİP: My invitation to discuss this issue a face-to-face in public (while video-recorded for the world to see) still stands, yet this time my invitation is directed to Shabbir and Jawaid.

18

Other Miracles in the Sura "The Hidden"

From philosopher/theologian Dr. Caner Taslaman's book, *Quran: Unchallengeable Miracle*. Also see www.quranmiracles.com)

While all the verses of the sura "The Hidden" are very short, the 31st verse it that mentions the functions of 19 is quite long. Although the 282nd verse of the sura The Cow is the longest verse of the Quran, the 31st verse is six times longer than it when the verse averages of those two suras are considered. The 20th verse of the sura The Enwrapped has the same feature. Yet this verse is eight times longer than the word average of the verses in that sura. Nevertheless the 31st verse of the sura "The Hidden" is more than twelve times longer than the word average of that sura. . Thus, this verse has the maximum words when the word proportions of the verses are considered in the whole Quran. The 31st verse of the sura "The Hidden", the only verse mentioning the function of number 19, has a special case. When we examine the number of letters and words of this verse, we witness that this verse is encoded with 19, as well as the fact that the verse mentions the functions of 19.

1) This verse consists of 57 words (19×3). Since this verse mentioning the functions of 19 has 57 (19×3) words, the preceding verse "Over it, is 19" consists of 3 words in Arabic. Thus, 3 has a meaning here because the number of words in this verse (57) is 3 times as much as 19 words.

2) In the part of this verse which says that "What does God mean by this symbol", the description of the functions of 19 ends. While this part of the verse consists of 38 (19×2) words, the rest of it consists of 19.

3) It is very interesting that the number of the words of the only verse mentioning the functions of 19 is equal to the number of the words of the first 19 verses in the sura "The Hidden" which consists of 57(19×3) words.

453

4) The first 30 verses of the sura "The Hidden" referring to the functions of 19 consist of 95(19×5) words and this is another proof evidencing the fact God uses the code 19 many times in this Sura.

5) The 30th verse of the sura "The Hidden" is the only verse referring to the number "19". From the beginning of the sura "The Hidden" till the beginning of the word nineteen, there are 361(19×19) letters. How great and detailed God's miracles are!

6) The sura "The Hidden" has 56 numbered verses and an unnumbered Basmalah at the beginning. Thus, the sura "The Hidden" has 56+1=57(19×3) verses.

We would like to call your attention to two points which are not as assertive as the particularities above. The 31st verse of the sura *"The Hidden" is the last verse of the Quran whose words are the multiplie of 19. In addition, the statement "None knows the armies of your Lord, except He"* which is the 31st verse of the Sure Mudessir consists of 19 letters.

How many words are there in the 31st verse of the sura The Hidden which mentions the functions of 19?	57 (19x3)
How many words are in the section which mentions the functions of 19 in the 31st verse of the sura The Hidden?	38 (19x2)
How many words are there in the first 19 verses of the sura The Hidden?	57 (19x3)
How many words are there from the beginning until the 31st verse of the sura The Hidden?	95 (19x5)
How many letters are there until the word nineteen in the sura The Hidden?	361 (19x19)
How many verses are there in the sura The Hidden?	57 (19x3)

We have seen the sura "The Hidden" pointing to the number 19 is the basis of the mathematical miracles in the Quran which includes many mathematical miracles. We would like to point out an interesting sign in this sura. When you think about what was the most important event on the world around 1974, the date 19's discovery, you will see that in 1969, man landed on the moon. This event is one of the most important events in the history of the world. Human beings went to the moon after a short time of the discovery of the mathematical miracle which relates to the lexical concordance and a very short time before the discovery of 19. If we consider this, it is very significant that the Quran calls attention to the moon in the 32nd verse just after the 31st verse of the sura "The Hidden" referring to the functions of 19. (We have examined the statements indicating that one day men would go to the moon in the Quran in the 16th chapter this book)

74:32 Absolutely, by the moon.

19

In Defence of Rashad Khalifa against Slanderers

who try to divert the argument through adhominem attack and defamation

Damnare:

You have found an old "allegation" from a newspaper to slander and defame Rashad Khalifah, whom you hate to your bones. Through your posts on this forum and your chat conversations, it is obvious to me that you hate Rashad because he invited you to follow the Quran alone by rejecting to worship your religious leaders and prophets through associating them to God in law-making and intercession. You could not show sufficient bravery and wisdom to question the religion you have inherited from your parents.

Your adhominem attack is clear and your intention is not clean. You know it well that the allegation you are trying to propagate did not survive the scrutiny of the court. Otherwise, a person committing rape should have been convicted by the court. The court found no evidence of allegation. You appear to be very good in digging dirt about people. But, why you failed to find any conviction about Rashad Khalifa on this matter?

You know that the allegation was false, but you act as if you are an honest person accidentally subjected to these allegations, got confused and needing clarification. In a previous private conversation, I informed you about the real story but you laughed and told me that you would go a head and post defamatory news about Rashad.

Well, I will clarify the issue for those who have good intention, but are expecting a clarification. First, to establish foundation, I will tell a little about my personal relationship with him:

456

After rejecting the polytheistic religion of my parents, I became a target in my home country. I lost my best-selling author title, popularity, and I became radioactive. When I received death threats and experienced physical attack to my life from fanatic people who shared the same faith as you do, I decided to immigrate. Before that, Rashad had sent me an invitation letter together with his promise of helping me regarding boarding and other necessities. So, I had gotten visa with no problem. When I arrived to the airport at New York, the officer there extended my visa without even I asked him.

Initially, I lived in a room adjacent to Masjid Tucson, and then Rashad rented an apartment for my wife and me. I spent one year in Masjid Tucson working with Rashad, until his assassination by a Sunni terrorist group, and during that year, I spent almost every day of the week with him in the Masjid, from morning prayer until night prayer. I also had a round trip together with him driving a car from Tucson to Vancouver, and back. I was working on the Turkish translation, reviewing his revision of his translation, discussing our differences, writing articles for the monthly bulletin, once a while giving Friday sermons, participating in discussions with Sunni and Shiite visitors, interviewed by TV stations or journalists... In 1989, I participated in an international conference held in Chicago by Sunni scholars, on the Finality of Prophethood. They had sent an invitation to Rashad to be their audience! Rashad asked me represent him there. He prepared a special issue of the Muslim Perspective to be distributed there. I wrote the "19 Questions For Muslim Scholars" to hand out to the participating scholars.

The hosts of the conference mistreated me without justification. I submitted the 19 Questions to the scholars on the panel. The young organizers immediately gathered around me, physically pushing me out from the conference room. This is their routine reaction to criticism to their dogmas. So, I spent my time in the lobby talking to their youth, who were curious about what I had to say. This made them furious; but could not ask the police to stop me hanging around in my hotel's lobby. Later, I learned that they had further plans to hurt me in my hotel room. Thank God, a black sunni muslim, whom I did not know personally, came to me in hurry and took me out from the hotel before they performed their plan.

During this one-year period, I had observed great integrity, wisdom, dedication, bravery and camaraderie from Rashad. He was a great friend, teacher and at the same times a humble student of the Quran. I loved him very much, as I loved my father, even more; but I never accepted his opinion on the Quran or on any other issues like some people did; I always checked for myself and occasionally

457

disagreed with him. My discussions and arguments with him, unbeknownst to me, created some enemies from his secret worshippers who were then incubating in their cocoons. In fact, several days before his assassination, I had an intense argument with him in front of the community, and afterwards we did not talk to each other for several days, until he came to me apologizing for his words. Those who know the nature of our relationship know well that I never treated him as a cult leader, but as a friend, a mentor, and a partner in jihad. In fact, in the front page of a copy of his translation, which he gave me as a gift; on top of his signature, he called me his brother in jihad.

I can write many pages regarding his strength, skills, good character, integrity, faith, trust in God, intelligence, charitable acts, etc. However, during my one year working with him I also witnessed some human weaknesses and traits. For instance, I saw tears coming from his eyes and his belly shaking from laughter; I witnessed him making continuous corrections in his understanding of the Quran; I found him usually insisting that his understanding of a particular verse was more accurate than of mine, and occasionally he accepted mine. Once he disappointed me by uttering the four-lettered word S..T, when a police officer stopped me, while we were driving to California (I had temporary driving permit, then; and I am a puritan). Furthermore, he continuously irritated my sense of strict accuracy whenever he was praising every food he was cooking for us in Masjid's kitchen by saying "This is the most delicious food in the world!" (Ironically, years later, I started saying the same thing and now my son is getting irritated of hearing so many MOSTs). One more thing: He liked mangoes very much and when he would eat them, he would make a mess. Surely, these are all trivial errors or perceived errors. But, I think, one error or weakness was outstanding: he was very kind to people around him and he could not tell those who were excited with a so-called discovery of mathematical miracle, "What you have discovered has no mathematical significance," or "Please give up using calculator to discover mathematical miracles; you have no clue about math and probability." I remember criticizing him frequently for his liberal attitude on this issue for not discouraging those innumerate miracle-hunters.

During my one-year close work with him, I never noticed him treating women inappropriately or any flirtatious behavior with anyone. We were very close. One day he told me about his past work in American Muslim student association. He was one of the founders of Islamic Center or Mosque of Tucson. However, in 1970's, he started questioning hadith and consequently discussing this issue with his close friends in that mosque. The straw that broke the camel's back was his rejection of Stoning-to-death penalty for adulterers. He found it anti Quranic and barbaric. His friends immediately ask him to cut his relationship from the Center.

Soon, he purchased an apartment complex on lot 114 on the corner of Euclid and 6th Avenue, by the University of Arizona, about a mile from the Islamic Center. However, soon he found himself been accused by a young Mexican girl. He had no doubt, that she was paid and hired by his former Sunni friends to defame him. I had plenty reason to believe him, since I knew him and also I knew his enemies. Sunnis had fabricated many lies about me too, from being a member of Moon Cult to being paid by CIA or Israel, etc.

Now, you are chewing over the same false accusation, which was fabricated by the believers of stoning-to-death practice, a practice that relies on the authority of hadith books that narrate the most ridiculous stories, such as a group of monkeys stoning an adulterer monkey, or a hungry holy goat eating the stoning verse after Muhammad's death and abrogating from the Quran.... There is no doubt, your animosity towards Rashad is because of his rejection of making those garbage teachings a part of his religion. Throughout history, *Mushriks* have slandered, attacked, tortured, and even killed those who dedicated themselves to God ALONE by rejecting Satan and his polytheistic doctrines. You and your comrades are no different from those Arab *mushriks* who tried everything, including false accusations to defame and deter Muhammad from promoting monotheism. What you brought was rejected by the court but you and your gang help bring some verses of the Quran to life again. No wonder the verses do not specify names:

24:11   A gang among you produced a big lie. Do not think that it was bad for you; instead, it was good for you. Meanwhile, each one of them has earned his share of the guilt. As for the one who initiated the whole incident, he has incurred a terrible retribution.

24:12   When you heard it, the believing men and the believing women should have had better thoughts about themselves, and should have said, "This is obviously a big lie."

24:13   Only if they produced four witnesses (you may believe them). If they fail to produce the witnesses, then they are, according to GOD, liars.

24:14   If it were not for GOD's grace towards you, and His mercy in this world and in the Hereafter, you would have suffered a great retribution because of this incident

24:15   You fabricated it with your own tongues, and the rest of you repeated it with your mouths without proof. You thought it was simple, when it was, according to GOD, gross.

24:16   When you heard it, you should have said, "We will not repeat this. Glory be to You. This is a gross falsehood."

24:17   GOD admonishes you that you shall never do it again, if you are believers.

Addendum

I never approved many of Rashad's mathematical manipulation in the footnotes of his translation regarding his messengership. However, my faith in his mission was primarily based on the following objective and subjective facts:

1. His clarity and sensitivity regarding monotheism.
2. His insight to the Quran.
3. His courage and honesty.
4. His not asking money for his work.
5. His discovery of the prophesized miracle of the Quran.
6. The root of his name been mentioned in the Quran 19 times, and interesting implications in verses where the root of his name is mentioned.
7. My discovery of the time of the end of the world from the Quran without knowing his method.
8. His happy and cheerful personality against all kinds of rejection, false accusations, threats, etc.
9. His patience and dedication in his struggle.
10. My personal interaction and discussions with him.

I can add several more reasons, but now I need to go home.

20

Dogmatic Skepticism, Organized Atheism (I)
Finding Mixed Nuts in James Randi's Amazing Hat, and
Losing Michael Shermer in the Cave of Patternicity

Edip Yuksel vs James Randi
July 2011

Why did they turn away from this reminder?
Like fleeing zebras... Running from the lion?
(74:49-51)

In this chapter I will share with you my communication with the renowned atheist magician, the so-called "Amazing Randi", and my response to the article published by Michael Shermer of Skeptic Society as a reaction to my interview in June 17, 2011.

FROM: Edip Yuksel
TO: Sadie Crabtree
DATE: Tue, Jun 7, 2011 at 9:26 PM
SUBJECT: Re: Interview with James Randi

Dear Sadie, thank you for the prompt response:

Enclosed is the letter I received from the director of the upcoming documentary. The group is mainly focusing on my recently published book, *NINETEEN: God's Signature in Nature and Scripture*. The book will be advertised heavily with the release of the documentary...

We have two different expectation regarding Randi:

1. The producers are interested in sharing Randi's skeptical position against religions, paranormal claims, which we believe is very important and necessary to be heard. They are very appreciative of Randi's work.

2. This involves my personal challenge to his categorical rejection of paranormal phenomena. I would like to respond to his challenge of the so-called 1 million dollar award. Those who know me, know well that I have no interest in getting someone else's money, especially the money of an honest man like Randy, but I will take the challenge since this case is unique. We have an extraordinary evidence for the extraordinary claim regarding God and the divine message.

If Randi is interested, I could order the book to his address or he could just visit the following site and check the book's SEE INSIDE feature.

> http://www.amazon.com/dp/0979671590/

There are also a few articles on this subject at:

> www.19.org

I also can give him access to the electronic version, which will be the fastest way.

The team will leave in the end of this month. So, we are looking forward to have a wonderful meeting with Randi.

Peace, Edip

FROM: James Randi
TO: Edip Yuksel
DATE: Tue, Jun 28, 2011 at 10:06 PM
SUBJECT: 19 -- or any other number

I have no interest in discussing numerology. I'm a grown adult, with far more important things to do with my time.

FROM: Edip Yuksel
DATE: Tue, Jun 28, 2011 at 11:42 PM
TO: James Randi
SUBJECT: Re: 19 -- or any other number

James Randi,

You are right, James. As it appears, I was wrong in expecting a magician and a strawman-puncher to engage in a serious philosophical debate that involves math and sciences. Your response demonstrates your paranoia, which appears to

462

be contagious in your group. I assume that your partner, Michael Shermer, warned you about us: "These guys are not the kinds of stupid believers and charlatans that we have been beating and declaring cheap victories. Stay away from them!"

We have the video records of Michael, acting like a bigoted believer, falling into multiple contradictions and ending up with an incriminating confession.

You have time to discuss the primitive trick of spoon-bending, claims of spiritual-healing, or stories of UFO-believers, but you have no time for discussing the claim about a verifiable and falsifiable numerical system in a book! You are a grown adult producing rabbits from your hat, bending spoons, putting up a million dollars challenging silly paranormal claims! In fact, your statement above shows that you consider our argument not to be in the category of the childish claims that you have devoted yourself to.

Though I support your work in exposing hoaxes and fraud in the name of God and the paranormal, I find you acting like a child when you are challenged by rational monotheists. Unlike Michael, you might have barely escaped this "ambush", but you will one day be caught by monotheistic critical thinkers and your bigotry will be exposed. God willing, we are going to debunk the debunkers. Wait and see. You may encounter us in the least expected places.

Peace (and I mean it)

Edip Yuksel

FROM: James Randi
TO: Edip Yuksel
DATE: Wed, Jun 29, 2011 at 5:02 AM
SUBJECT: Re: 19 -- or any other number

Ooh! Now I'll have sleepless nights waiting for a misinformed, bigoted, naive, religious nut to jump on me from out of the shadows! Enjoy your well-nurtured delusions, and anticipate the crowd of eager virgins who await you when you arrive in Valhalla -- or is it Paradise, Disney World, or some other amusement park...? I can't keep them all in my mind, there are so many imaginary places that you nut-cases have invented...!

The life of a Grubby is your only path, it seems... Enjoy!

Oh, by the way, the ancient "19" delusion has long ago been falsified, didn't you know...? No one cares, you see...

James Randi.

FROM: Edip Yuksel
TO: James Randi
DATE: Jun 29, 2011 at 6:22 PM
SUBJECT: Re: 19 -- or any other number

74:49 Why did they turn away from this reminder?*
74:50 Like fleeing zebras,
74:51 Running from the lion?
6:4 Whenever a sign came to them from their Lord, they turned away from it.
6:5 They have denied the truth when it came to them. The news will ultimately come to them of what they were mocking.
7:146 I will divert from My signs those who are arrogant on earth unjustly, and if they see every sign they do not acknowledge it, and if they see the path of guidance they do not take it as a path; and if they see the path of straying, they take it as a path. That is because they have denied Our signs and were heedless of them.*
15:9 We, indeed We, it is We who have sent down the Reminder, and indeed it is We who will preserve it.*
15:10 We have sent before you to the factions of old.
15:11 Any messenger that came to them, they would mock him.
15:12 We thus let it sneak into the hearts of the criminals.
21:41 Messengers before you have been mocked, but those who mocked were then surrounded by the object of their mockery!
26:5 Not a new reminder comes to them from the Gracious, except that they turn away from it.
26:6 They have denied, thus the news will come to them of what they used to ridicule.
29:47 Similarly, We have sent down to you the Book. Thus, those whom We have given the book will acknowledge it. Also, some of your people will acknowledge it. The only ones who mock Our signs are the ingrates.
45:34 It will be said to them: "Today We will forget you, just as you forgot the meeting of this day. Your abode is the fire, and you will have no helpers."

45:35	"This is because you took **God**'s signs in mockery, and you were deceived by the worldly life." So from this day, they will never exit therefrom, nor will they be excused.
45:36	All praise belong to **God**; the Lord of the heavens, and the Lord of the earth; the Lord of the worlds.
45:37	To Him belongs all majesty in the heavens and the earth. He is the Noble, the Wise.

FROM: James Randi
TO: Edip Yuksel 19@19.org
DATE: Tue, Jul 19, 2011 at 2:05 PM
SUBJECT: Re: Dogmatic Skepticism and Organized Atheism

Sir, you must try to understand. It is very difficult to argue with children and with idiots. They find meaning where there is none, they re-discover patterns that are simply mathematical realities, and they simply react, instead of using reason. Children usually grow out of such ways, idiots may be stuck for life in this trap.

No, we are not fleeing like zebras from a lion. We flee like thinking humans from participating in arguments with a rock, which wastes time and breath. Yes, the truth - facts, evidence, reality - can set you free. Fantasy only traps you more firmly.

James Randi

The Number 19

An attempted ambush interview turns into
a lesson in patternicity and numerology

by Michael Shermer

On Friday, June 17, a film crew came by the Skeptics Society office to interview me for a documentary that I was told was on arguments for and against God. The producer of the film, Alan Shaikhin, sent me the following email, which I reprint here in its entirety so that readers can see that there is not a hint of what was to come in what turned out to be an attempted ambush interview with me about Islam, the Quran, and the number 19:

> Dear Michael!
>
> I am the director of a film crew hired by a non-profit organization, Izgi Amal, from Kazakhstan, which has no connection with the American brat, Borat. We have been working on a documentary film on modern philosophical and scientific arguments for and against God for almost a year. We have been taking shots and interviewed theologians, philosophers and scientists in England, Netherlands, USA, Turkey, and Egypt.
>
> We are planning to finish the film by the end of this year and participate in major film festivals, including Cannes. We will allocate some of the funds to distribute thousands of copies of the film for free, especially to libraries and colleges.
>
> Our crew will once again visit the United States and will spend the rest of June interviewing various people, from layman to artists, from academicians to activists.
>
> Though we are far out there, we know your work and we think that it contributes greatly to the quality of this perpetual philosophical debate. We would like to include perspective and voice in this discussion. We would appreciate if you let us know what days in JUNE would be the best dates to meet you and interview you for this engaging and fascinating documentary film.
>
> Since we are planning to interview about 10 scholars and experts of diverse positions such as atheism, agnosticism, deism, monotheism,

466

and polytheism, it is important to learn all available days in this month of June.

Please feel free to contact us via email or our cell phone numbers, below. If you respond via email and please let us know the best phone number and times to reach you.

Peace,
Alan Shaikhi

In hindsight perhaps I should have picked up on his admission that "we are far out there," which in fact they turned out to be. Present were Mr. Shaikhin, another gentleman named Edip Yuksel, a couple of film crew hands, and a woman videographer who was setting up all the lighting and equipment. Before we began Shaikhin explained that they were actually filming two projects, and that his colleague (Mr. Yuksel) would be interviewing me after he, Shaikhin, was finished. Yuksel, in fact, was very fidgety and throughout the interview with Shaikhin I could see him out of the corner of my eye feverishly taking notes and fiddling around with books whose titles I could not see.

Shaikhin's interview, in fact, included mostly standard faire questions for such documentaries: Do I think there's a conflict between science and religion?, What do I think about this and that argument for God's existence?, Why do I think people believe in God?, etc. He was unfailingly polite and professional. Toward the end he did make some vague reference to Islam and our cover story of *Skeptic* on myths about the Islamic religion (the myth of the Middle East Madman, the myth of the 72 virgins, etc.), but I begged off answering anything about Islam because I haven't studied it much nor have I read the Quran.

My first clue that the interview was about to take a sharp right turn came when Shaikhin acted shocked that I would edit an issue of *Skeptic* on Islam without myself having read the Quran. I explained that I write very few articles in *Skeptic* and that my job as editor is to find writers who are experts on a subject, which was, in fact, the case with this issue when our Senior Editor Frank Miele interviewed the University of California at Santa Barbara Islamic scholar R. Stephen Humphreys. Nonetheless, Shaikhin continued to act surprised, repeating "you mean to tell me that you edited a special issue of *Skeptic* on Islam and haven't read the Quran?" I again explained that editors of magazines are not always (or ever) the world's leading expert on the topics they publish, which is the very reason for contracting with experts to write the articles for magazines.

The BELIEVING BRAIN

From GHOSTS and GODS to POLITICS and CONSPIRACIES — How We Construct Beliefs and Reinforce Them as Truths

MICHAEL SHERMER

Author of Why People Believe Weird Things

With this first part of the interview completed, Edip Yuksel leaped up out of his chair like a WWF wrestler charging into the ring for his big match. He grabbed a chair and pulled it over next to mine, asked for a bottle of water for the match, and instructed the videographer to widen the shot to include him in the interview. Only it wasn't an interview. It was a monologue, with Yuksel launching into a mini-history of how he wrote Carl Sagan back in 1992 about the number 19 (he didn't say if Sagan ever wrote back), how Carl had written about the deep significance of the number π (pi) in his science fiction novel *Contact*, how he is a philosopher and a college professor who teaches his students how to think critically, and that he is a great admirer of my work. However (you knew this was coming, right?), there is one thing we should not be skeptical about, and that is the remarkable properties of the number 19 and the Quran.

At this point I had a vague flashback memory of the Million Man March in Washington, D.C. and Louis Farrakhan's musings about the magical properties of the number 19. The transcript from that speech confirmed my memory. Here are a few of the numerological observations by Farrakhan that day in October, 1995:

> There, in the middle of this mall is the Washington Monument, 555 feet high. But if we put a one in front of that 555 feet, we get 1555, the year that our first fathers landed on the shores of Jamestown, Virginia as slaves.
>
> In the background is the Jefferson and Lincoln Memorial, each one of these monuments is 19 feet high.
>
> Abraham Lincoln, the sixteenth president. Thomas Jefferson, the third president, and 16 and three make 19 again. What is so deep about this number 19? Why are we standing on the Capitol steps today? That number 19—when you have a nine you have a womb that is pregnant.

468

And when you have a one standing by the nine, it means that there's something secret that has to be unfolded.

I want to take one last look at the word atonement.

The first four letters of the word form the foundation; "a-t-o-n" ... "a-ton", "a-ton". Since this obelisk in front of us is representative of Egypt. In the 18th dynasty, a Pharaoh named Akhenaton, was the first man of this history period to destroy the pantheon of many gods and bring the people to the worship of one god. And that one god was symboled by a sun disk with 19 rays coming out of that sun with hands holding the Egyptian Ankh – the cross of life. A-ton. The name for the one god in ancient Egypt. A-ton, the one god. 19 rays.

This is a splendid example of what I call *patternicity*: the tendency to find meaningful patterns in both meaningful and meaningless noise. And Edip Yuksel launched into a nonstop example of patternicity when he pulled out his book entitled ***Nineteen: God's Signature in Nature and Scripture*** (2011, Brainbow Press; see also www.19.org) and began to quote from it. To wit...

- The number of Arabic letters in the opening statement of the Quran, *BiSMi ALL*aĤi AL-RaĤMaNi AL-RaĤYM (1:1) **19**
- Every word in *Bismillah...* is found in the Quran in multiples of **19**
- The frequency of the first word, Name (*Ism*) **19**
- The frequency of the second word, God (*Allah*) **19** x 142
- The frequency of the third word, Gracious (*Rahman*) **19** x 3
- The fourth word, Compassionate (*Rahym*) **19** x 6
- Out of more than hundred attributes of God, only four has numerical values of multiple of **19**
- The number of chapters in the Quran **19** x 6
- Despite its conspicuous absence from Chapter 9, Bismillah occurs twice in Chapter 27, making its frequency in the Quran **19** x 6
- Number of chapters from the missing Ch. 9 to the extra in Ch. 27. **19** x 1
- The total number of all verses in the Quran, including the 112 unnumbered Bismillah **19** x 334
- Frequency of the letter Q in two chapters it initializes **19** x 6
- The number of all different numbers mentioned in the Quran **19** x 2
- The number of all numbers repeated in the Quran **19** x 16
- The sum of all whole numbers mentioned in the Quran **19** x 8534

469

This goes on and on for 620 pages which, when divided by the number of chapters in the book (31) equals 20, which is one more than 19; since 1 is the cosmic number for unity, the first nonzero natural number, and according to the rock group Three Dog Night the loneliest number, we subtract 1 from 20 to once again see the power of 19. In fact, 19 is a prime number, it is the atomic number for potassium (flip that "p" to the left and you get a 9), in the Baha'i faith there were 19 disciples of Baha'u'llah and their calendar year consists of 19 months of 19 days each (361 days), and it's the last year you can be a teenager and the last hole in golf that is actually the clubhouse bar. In point of fact we can find meaningful patterns with almost any number:

- **99**: names of Allah; atomic number for Einsteinium; Agent 99 on TV series Get Smart
- **40**: 40 days and 40 nights of rain; Hebrews lived 40 years in the desert, Muhammad's age when he received the first revelation from the Archangel Gabriel and the number of days he spent in the desert and days he spent fasting in a cave
- **23**: The 23 enigma: the belief that most incidents and events are directly connected to the number 23
- **11**: sunspot cycle in years, the number of Jesus's disciples after Judas defected
- **7**: 7 deadly sins and 7 heavenly virtues; Shakespeare's 7 ages of man, Harry Potter's most magical number
- **3**: number of dimensions; number of sides of a triangle, the 3 of clubs—the forced pick in one of Penn & Teller's favorite card tricks
- **1**: unity; the first non-zero natural number, it's own factorial and it's own square; the atomic number of hydrogen; the most abundant element in the universe; Three Dog Night's song about the loneliest number
- **π (pi)**: a mathematical constant whose value is the ratio of a circle's circumference to its diameter, or 3.14159.... Make of this what you will, but Carl Sagan did elevate π to significance at the end of *Contact*:

The universe was made on purpose, the circle said. In whatever galaxy you happen to find yourself, you take the circumference of a circle, divide it by its diameter, measure closely enough, and uncover a miracle—another circle, drawn kilometers downstream of the decimal point. In the fabric of space and in the nature of matter, as in a great work of art, there is, written small, the artist's signature. Standing over humans, gods, and demons, subsuming Caretakers and Tunnel builders, there is an intelligence that antedates the universe.

At this point in the filming process I interrupted Yuksel and told Shaikhin that the interview was over, that he could use the footage from the first part of the interview but not this monologue mini-lecture that was an undisguised attempt to convince me of the miraculous properties of the number 19. I didn't sign any waiver or permission to use any of the footage shot that day, but just in case I was relieved when the videographer came to me in private to apologize and explain that she had nothing to do with the rest of the crew, that she was just hired to do the filming, and that after I had put an end to the interview she stopped filming.

At some point I asked Edip why he felt so compelled to convince me of the meaningfulness of the number 19 in the Quran, when I told him that I haven't read the Quran and hold that all such numerological searches are nothing more than patternicity. The impression I got was that if he could convince a professional skeptic then there must be something to the claim. I asked him what other Islamic scholars who have read the Quran think of his claims for the number 19, and he told me that they consider him a heretic. He said it as a point of pride, as if to say "the fact that the experts denounce me means that I must be on to something."

P.S. Edip Yuksel did strike me as a likable enough fellow who seemed genuinely passionate about his beliefs, but there was something a bit off about him that I couldn't quite place until I was escorting him out of the office and he said, "I see you are a very athletic fellow. Can I show you something that I learned in a Turkish prison?" With scenes from *Midnight Express* flashing through my mind, I muttered "Uhhhhhh... No."

Patternicity Challenge to Readers

As a test—of sorts—I would like to hereby issue a challenge to all readers to employ their own patternicity skills at finding meaningful patterns in both meaningful and meaningless noise with such numbers and numerical relationships, both serious and lighthearted, related to the number 19 or any other number that strikes your fancy. Post them in the comments section of this *eSkeptic* and we shall publish them in a later feature-length article I shall write on this topic. **S**

—Michael Shermer, Editor, *Skeptic*

21

Dogmatic Skepticism, Organized Atheism (2)
Finding Mixed Nuts in James Randi's Amazing Hat, and Losing Michael Shermer in the Cave of Patternicity

Edip Yuksel vs Michael Shermer
July 2011

The leading American skeptic, Michael Shermer, after getting disoriented in an interview which he calls an "ambush", engages in a pre-emptive strike. He responds with a criticism infected with logical fallacies such as poisoning the well, straw-man, red-herring, hasty generalization and a false innuendo about my prison term in Turkey. After analyzing his article, I propose two options.

"He produced a blog post and he wants to hide the only evidence that is related to the blog post. This placed him in a deep contradiction with his position as a skeptic." Gazi Alankuş

Dear Michael:

You have published an article reacting and commenting on our interview with you on June 17, 2011. You published the article entitled **The Number 19: An attempted ambush interview turns into a lesson in patternicity and numerology,** in three websites:

Your article was an evasive reaction to the 19-based mathematical system in the Quran. Before sharing some examples of 19 in the Quran, you added a few silly words and then you hurled a few punches at a straw-man by quoting the unrelated and nonsensical remarks by a dogmatic clergyman with a 3.5 pound potato for a brain. Thereby, you wished to poison the well and create prejudice against the Quranic examples. After quoting some examples from my book, you continued employing logical fallacies by attempting to confuse them with unrelated, anecdotal, and arbitrary semi-silly examples about dozen of different numbers. Finishing your article, by referring to my prison experience in Turkey with an innuendo, was a clever and creative ad hominem attack. In fact, your article was filled with ad hominem. Instead of dealing with the philosophical debate on the evidences I provided, you indulged in describing me as "very fidgety" "feverishly taking notes and fiddling around with books." "leaped up out of his chair like a WWF wrestler", etc. Interestingly, you forgot to quote your own description of yourself during the cross-examination!

Let me concede. Indeed you are more fluent than me; English is chronologically my fifth language and I have a strong accent, which usually creates a low impression on Americans hooked on American idol or the Kardashian show. Despite your balding head you are more handsome than me. You generate better smiles on your face, at least during the first session with Matthew (not Alan!). You were not very fidgety. You did not feverishly take notes; in fact, you did not take any. You did not fiddle around with books; in fact, you treated them like a bibliophobe. But, towards the middle of my cross-examination, you looked disoriented, disturbed, confused, and utterly lost in your own office. You treated me like a "bad cop" while you sought clarification from my friend, whom you perceived as the "good cop."

It seems that you panicked after you were ambushed intellectually. In front of five people and two recording cameras you confessed twice: "I look like an ***** here!" and you asked us not to release the second part of the interview. All have been recorded *in high definition*!

You lost debate because of your arrogance

No Michael, you are not an *****; you are a bright and educated man. Yet indeed on the afternoon of June 17, 2011 you looked like an *****because you acted like a bigot, demonstrating no interest in an intellectual debate on the topic of your expertise. Though my argument was philosophical and scientific, and my examples were verifiable and falsifiable, you noticed the compelling nature of my argument and you wished to avoid it by declaring your ignorance of the Quran and Islam. Ironically, you had just made the Quran and Islam the cover story of the Skeptic magazine, at which you are the chief-editor and publisher. Bad timing!

In brief:

1. You lost philosophically through your reaction and evasive response to the prophetic mathematical evidence for God's existence.
2. You lost scientifically through your inability to account for the verses of the Quran on cosmology, embryology, physics, and other issues supported by modern science.
3. You lost professionally when you pleaded the Fifth after publishing false information about the Quran through your magazine.
4. You lost your integrity when you published an article mocking our evidences by mixing it with irrelevant remarks; the evidences that you could not handle during the Socratic grilling session, which you described as an "ambush".

As I told you during my cross-examination, you did not need to respond to my statements on theology. Though months before the interview I had mailed to you the **Quran: a Reformist Translation** and later the **NINETEEN: God's Signature in Nature and Scripture**. As it appears, you have interest in everything, including paranormal claims, religious mythologies, and scientific issues in the world, but you have not even looked at the pages of those books that equally challenge the religious establishment and modern atheism. Apparently, you did not even show interest in those books, especially the second one that refers to some of your books, even during the time you were editing articles about and against the Quran.

Then the day of reckoning came, which is much easier than the real day of reckoning in the hereafter. Here I was there at your office, challenging your atheistic dogmas and your claims about the Quran with philosophical arguments supported by verifiable/falsifiable evidences. As a prominent skeptic and atheist, what did you do? You acted like a religious bigot with no interest in philosophical debate. You acted as if you had seen a lion (or a WWF wrestler!) in the open. I was surprised when I provided some examples of numero-semantical system in the Quran, such as the frequency of the word Month being

474

12, and the frequency of the word Day, as you expected, being 365, even so you still did not demonstrate any interest in the subject matter. You acted as if you were sitting on a bunch of nails with sacks of potato chips on your shoulders. You knew that you were being subjected to a Socratic cross-examination and the person next to you was not a religious piñata, a believer of dogmas and contradictory stories whom you could easily punch to death and declare another cheap victory. It was too late when you realized that you made a remark that you regret making. Interestingly, you repeated that remark at least twice.

You lost the debate. No, it was not you who lost. It was your prejudices and arrogant claims about God and the Quran that lost. Instead of considering it as a victory for truth and for yourself, instead of asking me to meet you another time after you study our arguments more carefully, you acted like the prototype ingrate described in 74:16-25 and 7:146 and chose to stay in Saqar, which is described in 74:26-30.

A Hodgepodge of Logical Fallacies

Interestingly, you managed to squeeze in half-a-dozen major logical fallacies in a short article, which may be used as a good material in Logic classes to show examples of logical fallacies. Here is the list of fallacies crept in your article:

1. Ad Homonym (WWF wrestler; false implication by referring to the Midnight Express film)
2. Poisoning the Well (Louis Farrakhan, numerology; Three Dog Night)
3. Straw-man (Lois Farrakhan; numerology)
4. Red-herring (the entire article)
5. Hasty Generalization and Mockery (the entire article, especially through the patternicity label)
6. False Analogy (The paragraph which mentions the number of pages in my book; the mishmash list following that paragraph, which contains 8 different numbers and about 20 different concepts/facts as an analogy to the pattern in the Quran, which is about one number (19) in one book (Quran).

Fearing that the video recording of you with self-incriminating confessions of ****** may be released to the public, you calculated the risks and decided to pre-emptively react, by any means possible. The evasive, the "not-knowing-anything," the "looking-like-an-*****" atheist against me in June 17th 2011 somehow turned into a skeptic hero who was "ambushed" by a lunatic man who had escaped from a Turkish prison! A nice Hollywood movie!

You have my book and you may learn from it that I was imprisoned by a martial court after the military coup in 1980 for two of my articles which were critical of the Turkish government. Many activists, academics, authors and political figures too were imprisoned by the military regime. Then, Recep Tayyip Erdoğan, the current Turkish Prime Minister, my former comrade, too was arrested for participating in a political rally led by me, and we spent a night in jail together with other 300 protesters. This story was publicized in a political biography by his former advisor, Mehmet Metiner, and recently became the focus of controversy in Turkish media. Though I was subjected to torture in prison, I did not experience the kinds of abuse you are implying. I invited you to play a physical game after the interview while the camera operator was packing up. This was to reduce your anxiety and to comfort you since you looked fazed after being grilled by an incarnated Socrates with funny accent and weird demeanor. Though I am uncompromising when it involves theological and philosophical debates, I have compassion even to my ardent enemies, which I do not consider you belonging to that category. Albeit, perhaps I was naïve not to think that it might take you to the Midnight Express and invoke some disturbing scenes. If you really got that impression, I do apologize for that. My friends knows about the game, which is an entertaining balancing and concentration challenge; we invented it when we were in a small military prison where we did not have room to exercise. So, I am still very good at that.

Michael, your critical article on 19 is juvenile. You cannot mock and distort the facts and get away with it forever. I advise you to get yourself out of the panic mode and deal with this challenge as an intellectual and honorable person, which I believe is your essence. As I told you, I support your work and enjoy many of your articles debunking religious lies and mythologies. But, when you were examined by a rational monotheist, you fell apart like the religious figures have done in your investigation.

Patternicty or the pattern of ignoring examples of pattern and design that refute the atheistic dogmas

Dear Michael, the light of truth may initially hurt our eyes. Learning that that there is an intelligent creator might pose problems with your petty vested interest in leading an atheist organization, but it will save you from darkness of atheism and nihilism; it will open the gate of eternal life with God. I had similar experience myself in 1986; for several months, I insisted to stay in the darkness of Sunni religion.

At the end of your article, you have promised your readers to write a feature article and solicited the following:

> "Patternicity Challenge to Readers: As a test—of sorts—I would like to hereby issue a challenge to all readers to employ their own patternicity skills at finding meaningful patterns in both meaningful and meaningless noise with such numbers and numerical relationships, both serious and lighthearted, related to the number 19 or any other number that strikes your fancy. Post them in the comments section of this eSkeptic and we shall publish them in a later feature-length article I shall write on this topic."

This is a good try. But, looking at the meaningless patterns you have listed in your article and the patterns your admirers posted in the comment section, I do not expect a serious response. Your readers will come up with arbitrary and silly lists of some numbers and events or inconsistent and anecdotal examples of the number 19 in a text of their choice, which they will pretend utilizing a similar pattern. I am still looking forward to your article, since it might be better than the current one.

If you look at the book I gave to you and autographed upon your request, you will see that I have several articles in the Appendices statistically analyzing the so-called patterns fabricated by the knee-jerk critics and innumerate numerologists. For instance, I recommend you reading the following articles:

- Appendix 2, The Gullible, the Blind, and Boxes of Diamonds and Glass, pp. 388-391.
- Appendix 3, Diluting the Miracle, pp. 394-399.
- Appendix 5, Diamond vs. Glass, by Prof. Richard Voss, pp. 402-429.
- Appendix 6, The Ingrates React, pp. 430-433.

As it seems, you failed again to study the subject of your critical article, which has been the common reaction of religious people. For instance, you wonder about whether Carl Sagan responded to my mail. If you had browsed the book, you would see the copy of his second mail on page 126 and my response in the following pages. You could learn a few things from my response, or at least, you could criticize my argument, rather than punching a straw man like Louis

477

Farakhan, whose nonsensical remarks have nothing to do with the subject of the book.

Atheist Sam Harris: "We will continue to spill blood"

Perhaps, you will understand my point better if I treat you with your own medicine. No, I will not mock you as a person as you and your friend Randi have done (See: http://19.org/641/randi/). Trust me I can do a fine job in making a mockery of you; but it would be a disservice to such an important subject. Besides, you are a good-natured person and I have no personal vendetta against you. I will only respond to your straw-man argument, transforming into an educational boomerang. To mirror your twist, I will use a few sentences from your introductory paragraph about Louis Farakhan:

> At this point I had a vague flashback memory of another famous atheist, Sam Harris, who justified the USA-Inc's invasions and massacres around the world, killing more than a million in Iraq, and tens of thousands in Afghanistan. A little Internet search confirmed my flashback. Below is a quote from Sam Harris, a best-selling atheist author whose remarks comparing rape to religion in an interview with ABC Radio host Stephen Crittendon in 2006 made the headlines. Though Sam Harris did not condone rape, his statement that "there is nothing more natural than rape" stirred controversy. Here is the Sam's argument to spill more blood of the "innocents abroad":
>
>> "The link between belief and behaviour raises the stakes considerably. Some propositions are so dangerous that it may even be ethical to kill people for believing them. This may seem an extraordinary claim, but it merely enunciates an ordinary fact about the world in which we live. Certain beliefs place their adherents beyond the reach of every peaceful means of persuasion, while inspiring them to commit acts of extraordinary violence against others. There is, in fact, no talking to some people. If they cannot be captured, and they often cannot, otherwise tolerant people may be justified in killing them in self-defense. This is what the United States attempted in Afghanistan, and it is what we and other Western powers are bound to attempt, at an even greater cost

to ourselves and to innocents abroad, elsewhere in the Muslim world. We will continue to spill blood in what is, at bottom, a war of ideas." (Sam Harris, *The End of Faith*, pp. 52-53.)

The warmongering argument above justifies all the terror, atrocities and tortures committed by USA-Inc and Zionists against civilian population of poor countries. The prominent atheist sides with jingoism and imperialism against gang terrorism, which is ironically the byproduct of imperialism and its policy of invasions, covert operations, supporting tyrants and authoritarian regimes in countries with rich natural resources. When in power, atheists are ruthless. To promote their ideologies, atheists like Stalin and Pol Pot did not hesitate to kill millions of people.

Now, would you be happy if this were my response to any of your articles promoting atheism, especially, if you disagree with the warmongering position of Sam Harris or with the policy of Stalin and Pol Pot? Interestingly, your philosophical position is much closer to that of Sam Harris than is my philosophical and theological position of Louis Farrakhan. But, you did not care about that. In haste to poison the well and manipulate your readers, you just wanted to find a silly remark made by a "Muslim" clergyman. This would lead your readers to punch a straw man through a hasty generalization. This is not even "guilt by association," since I do not consider Louis Farrakhan as a rational monotheist. I feel much closer to an atheist than to a professional religious leader who has no inhibition in peddling false stories about God. Louis is a Sunni demagogue, a clergyman who lives a lavish life through donations made by his poor congregation. Louis does not engage in critical thinking; to the contrary, like all religious preachers, he invites people to believe on faith and join his bandwagon!

Two options

I challenge you for a real face-to-face debate that will not make you feel as if you are being ambushed by a fidgety wrestler. In fact, you may ambush me at any time in any location you wish! I have no problem with truth. If you falsify my claims I am ready to change my position! You may even ask your magician friend, the so-called "Amazing Randi", to join you for help. Though he is a hard nut to crack, he may produce rabbits out of his magician's hat to distract or entertain the audience! He may even call me nuts, peanuts, coconuts, walnuts as much as he wishes. I am not offended by the insults of a *moruk* whose intellectual capacity is limited by a bowl of nuts.

So, here is my proposal:

1. Let's set up a date for a two-hour debate with me on the subject of NINTEEN in front of a public audience before the end of this coming October. We may pick a University in California or New York.

OR:

2. Give permission to the producing company to release the video recordings of my cross-examination of you which you describe as an "ambush," within their documentary film, so that people could see the nature of the ambush you are complaining about.

Why I prefer face-to-face debate? Well, I have debated this issue in writing for more than two decades. I have already published a book, Running Like Zebras, containing a lengthy debate between me and Abdurrahman Lomax. I have also extensive debate in writing with another critic who calls himself Ayman. For complicated and comprehensive issues like this, I prefer a face-to-face debate, since it is immediate, and allows follow-up questions and answers, thereby enabling parties to focus on a particular issue and expose contradictions, logical fallacies, and false assumptions in much efficient way.

So, it is no surprise that anyone who engaged in a face-to-face debate with me on this issue lost the debate. The former head of the Turkish Religious Affairs, Professor Suleyman Ateş, literally escaped from the TV studio during a live program watched by millions of Turkish audience. Twice! Each time he was brought back by the TV host who received concessions from me to be softer on him. In order not to disturb the chips on his shoulder, I did not even respond to him when he added two imaginary letters to the first statement of the Quran, increasing the numbers of its letters from 19 to 21! The third guest on the panel, Pofessor Haluk Oral, a mathematician at Bosporus University, could not understand the phobic reaction demonstrated by the prominent Sunni cleric.

By now, you should know that you cannot get help from your comrade, Randi. (See the first part of this article). I do not recommend David Silverman, the president of American Atheist Organization, either. I had an hour-and-half debate with him in the Atheist Organization's New Jersey headquarter. Unlike you, he was more comfortable and even enthusiastic during the debate; yet he found himself in a foreign zone when I brought the issue of code 19. Instead of reacting reflexively like you did, he wanted to outsource me to Randi. Though I told him that a mathematician, not a magician, would be more appropriate for this issue, he insisted me to contact a magician who happened to be nut-caster.

You should also know that, before publishing the book NINETEEN, I had contacted 50 mathematicians who signed a paper rejecting the claims made by Drosnin in the Bible Code. None responded, except one from Israel. If you would like to contact him, I would be glad to put you in contact with him. There have been serious critics of code 19, whom I had the opportunity to debate via Internet. For instance, Abdurrahman Lomax, Abdullah Ayman are among them. Just a week before interviewing you, I contacted Dave Thomas of New Mexicans for Science and Reason. He had written a critical article against 19, and I was going to cross-examine him and expose the manifold problems with his criticism. Unfortunately, citing a family issue, Dave declined the interview. In brief, you could invite the two Muslim critics of the code, Abdurrahman Lomax and Adullah Ayman together with Dave Thomas to join you in that debate.

Peace,
Edip Yuksel

Note: The person who interviewed you before me was not Alan Shaikhin, who hardly speaks English and this was his first visit to the USA. The interviewer was Matthew Capiello, the spokesperson of Muslims for Peace, Justice and Progress (MPJP). Matthew was not aware of my "ambush." You may also be interested in knowing that Alan Shaikhin prefers Dr. Pepper, since back in Kazakhstan he had seen Tom Hanks drinking it in the film, Forrest Gump. (The last information is meant to provide you with an irrelevant material so that you might use in your next article! In case you might need an ingredient for distraction! ☺)

22

A Short Discussion With
Skeptic Author James J. Lippard

19-21 September 1995

JAMES: Got your pamphlet. I have a few questions:

1. What is the significance of the fact that the words "satan" and "angel" each occur 88 times (not a multiple of 19)?
2. What is the significance of the fact that the words "this world" and "hereafter" each occur 115 times (not a multiple of 19)?
3. How many times do the words for "morning" and "night" occur? The same number each? A multiple of 19?

EDIP: They are not multiple of 19.

JAMES: You didn't answer my questions in 1 and 2. What is the significance? You list them in your booklet as though it's something remarkable that was found, but why? Looks arbitrary to me.

JAMES:

4. How many chapters are there from the beginning to the missing basmalah? How many from the extra basmalah to the end? Multiples of 19?

EDIP: I thought I have answered this question; it was asked by Sagan. Have you read the entire booklet?

JAMES: Not carefully, no.

JAMES: 5. How many times does the word "disbelievers" occur? A multiple of 19?

EDIP: Probably not.

JAMES: I believe that Sagan's a priori/a posteriori question gets right to the heart of the matter. My speculation is that approximately 1/19 of the miraculous seeming numerical correspondences that could be found are found, and that the other 18/19 don't hold and are ignored. That is, given 19 times the effort that you and others have put into finding what you've found, 19 times as many *non*-correspondences could be found. Unfortunately, nobody is going to be interested in putting that kind of effort into the search. Do you have any suggestions on alternative ways of refuting my speculation? Perhaps if you could be more specific about where multiples of 19 must be found, so that others could propose as-yet-unchecked tests for the hypothesis?

EDIP: With a blind a priori statistical approach you may not find any significance of 19 in the Quran. I believe an entirely blind a priori test is not valid in this context, since it treats every element equally.

JAMES: Well, you need some objective criteria of significance, and I suspect you cannot produce it except on a post-hoc basis.

EDIP: For instance, it treats the number of chapters, number of verses, key verses, and key words equally with the number of insignificant words. If important elements of the Quran are determined a priori, then it would be a meaningful test. We do not claim that everything in the Quran is the multiple of the number 19, and believe that it contains information. However, there is a system which has pattern and predictability. Please reflect on the examples I have provided.

JAMES: I don't think they are that impressive in light of the observations I've made above.

> **JAMES**: By the way, I have seen exactly the same sorts claims made for the Hebrew scriptures (Tanakh) and for the Protestant Bible, with multiples of 7. In response to that claim, I took a piece of Homer's _The Odyssey_ (the introduction and the story of the lotus eaters) in Greek and applied the same kind of numerical analysis to it. With only a few minutes of playing around, I came up with this list:
>
> The first sentence has 14 words (7 x 2). The first three words have 14 letters (7 x 2).

EDIP: If the first sentence was consisted of three words then it would be significant. Here the "three" is arbitrary.

483

JAMES: This is a subjective judgment. The criteria of significance need to be put on the table up front, in advance of tests. If this can't be done, then you're just building sandcastles in the clouds.

JAMES: The last four words have 28 letters (7 x 4).

EDIP: Again there is no reason or a pattern to justify your counting of "four" words. Arbitrary.

JAMES: The total number of words is 231 (7 x 33).

EDIP: The total number of words in what?

JAMES: The passage.

JAMES: This is also 7 x 11 x 3; adding the factors produces a sum of 21, which is 7 x 3.

EDIP: There is no pattern to justify this calculation.

JAMES: I saw some factor adding or digit adding in your numerology, what's the justification there?

JAMES: The sum of the values of all the words in the passage is 120,169, which is 7 x 17,167.

EDIP: I did not understand this.

JAMES: Greek letters are used to represent numbers, too. If you take the values of all the words in the passage I described and add them up, you get the above sum

JAMES: The number of words in the vocabulary is 186; 1 + 6 = 7, 8 - 1 = 7, the number between 8 and 6 is 7.

EDIP: Again another arbitrary calculation without a pattern.

JAMES: I don't understand this. There clearly is a pattern. I agree that it is arbitary, but that's the point.

JAMES: Of the words in the vocabulary, 25 have values divisible by 7; 222 + 5 = 7.

484

EDIP: Another different method of arriving to 7. Arbitrary! Not supported by other examples.

> **JAMES**: The number of proper nouns in the vocabulary is 7.

EDIP: A simple coincidence.

JAMES: Of course it is.

> **JAMES**: Granted, these are less impressive than yours--but I produced them from only a small piece of text in a very short period of time. Given the entirety of _The Odyssey_ and many hundreds of hours, I have no doubt that I could find all manner of interesting correspondences.

EDIP: No Lipard, the examples you have given are unrelated and in almost each case you have employed a different method of calculation (in fact manipulation of numbers). I am surprised that you think that the examples I have presented in the "Prime Argument" have the same characteristics.

JAMES: Just as I am surprised that you think otherwise.

EDIP: Thank you for your answer.

JAMES: You're welcome.

23
Diluting the Miracle

In this chapter, you will find some of my articles and
email/internet communications on various numerical
assertions, manipulations, and distortions.

Inferring contradictory results from numerical associations, or claiming an
arbitrary and utterly absurd/subjective numerical relationship to be an
extraordinary sign or a miracle is a dangerous path.

There are some interesting findings in your calculations. However, I find your
adding ALLAHU AKBAR after each of your calculation to be unnecessary. I
have seen many ridiculous calculations, especially by the false prophets that
started popping up after Rashad's departure. For instance, the "Doomsday
Prophets", a gang of exceptionally sincere and yet equally credulous people led
by Kay Emami, Behrouz Mofidi, Feroz Karmally, and Douglas Brown, ended
their silly numerical manipulations regarding "May 19, 1990 Doomsday"
exactly with the same exclamation. Many of the conjurers of arbitrary
calculations and absurd speculations have done the same. This is my first
suggestion.

Second, you are using a good program and you occasionally come up with
interesting discoveries. That is beautiful. But, it seems that you are unaware that
you are exceeding the limits and jumping from one relation to another one.

For instance, you check whether a number is divisible by 19 or not. If yes, you
declare it to be a miracle. If not, you check with the gematrical value of a
particular word. If that one fits, you declare it a hit. If not, you check it whether
it matches a related number of a chapter. If it does, you claim another hit. If it
doesn't, you add a verse number or numbers. If it fits, then you celebrate the
discovery of another example of code 19. If it doesn't, you look for another, any
relationship.

It seems, whichever number you land, you have a way to relate it to the number nineteen, to another number, another verse, another chapter, or to another gematrical value. You do not care, as it seems, about the consistency in the system, predictability, objectivity, or the laws of probability.

Just check your recent claim regarding the letter N. You claim that the verses starting and ending with letter N contain 50 Ns in them. It is not 133 or any other multiple of 19. But, you find a connection between 50 and the gematrical value of N. But, to reach this conclusion you need more than one example. You should check the verses that start and end with Qaf. If those verses too contain 100 Qafs, equaling to the gematrical value of letter Qaf, then you got a support. You should check other initial letters. If none exhibit such a connection, you should consider it as mere chance, your own wishful association. You cannot dismiss other letters by saying N is the most important one, since you have no evidence for such a claim. Even if you had reasons for such a claim, still you needed more consistent examples to justify your inference.

Perhaps you might get angry with me, like many of those who have indulged in this number-hunting, who have confused the numerical system with arbitrary numerology.

But, I urge you to study the dozens of pages of calculations presented by Kay and other false prophets of doom, in the last few years, and learn a lesson. As long as you are not able to see the methodological and logical problems with their calculations, as long as you are not able to distinguish yours and theirs mathematically, then please stop your calculations and speculations on the subject and try to learn something from them.

In brief: you may continue observe relationship among numbers of the Quran, but do not haste to attribute them to God by calling it a miracle or implying it by capitalized ALLAHU AKBAR after eac, with exclamation marks.

****

Dear Brother,

I am involved in studying the numerical structure of the Quran since 1980, and I know that the biggest attack to this great miracle has come not from the ardent critics like Lomax or Ayman, but from overzealous and gullible "miracle hunters." Though, I do not want to discourage you in your search for more details of the numerical system, I want you not to be one of those who present a bunch of arbitrary calculations and personal speculations, akin to pieces of glasses, and then presenting those worthless pieces of glasses as diamonds.

When I see pieces of glass in your hand screaming Allahu Akbar, I am not offended by your praise of God. I am offended by your attributing your false perception to God. Do you understand what I mean?

Those who sent letters to the media and gave interviews to the local TV stations prophesizing that an asteroid would hit Saudi Arabia in May 19, 1990, were very angry with me when I opposed to their childish calculations. They considered me as a disbeliever. Ironically, they continued their personal hatred towards me when it became obvious that their calculations were the figment of their imagination.

Though you are not producing false prophecies, you are attributing to God some arbitrary and useless calculations, and you should be glad that I am reminding you that. I take my time and give you an example of manipulation or arbitrary calculation you are presenting here. Instead of giving up from them or defending them based on the patterns of the mathematical structure, you are showing emotional reaction. Don't be carried away by some friends who are cheering for whatever numerical "miracle" you are coming up with. They cannot lead you to the truth. Beware the cheers of the gullible. Peace.

Code Seven?

25 July 2005

This is indeed an excellent example shared by brother Bahnam. The fact that there are only 114 verses containing ALL 14 initial letters/numbers is a very clear example that fits the jigsaw puzzle.

But, I have not yet seen a systematic pattern in your claims about the role of the number 7 in the mathematical structure of the Quran. You are not the first one though. In 1980's a Turkish medical doctor wrote a book asserting such a thing, which I found a series of disconnected, arbitrary and inconsistent calculations.

First of all, the number Seven is relatively small number; so according to the laws of probability you should find about three times more examples to create similar effect created by examples of code 19. Besides, when you use occurrences of the number 7 TOGETHER with 19, the probability of chance increases. Yes, it becomes less significant. If we wish to dilute the miracle even further, we could add the number 3 in the calculations, and we would end up

488

with hundreds of more examples, examples that would mathematically be worthless in terms of claim of the "extraordinariness" of the alleged pattern. Trust me, it is very easy to add another number in the game and come up with many more examples.

So, before getting too excited about your discoveries, first reflect on whether it decreases or increases the probability and the quality of system. But, having discussed this issue so many times with so many people who are so deeply indulged into discovering their OWN codes or miracles, my words of caution to you might be another meaningless try. Perhaps by now, you have fallen in love with number 7 and nothing will make you wake up and break up your sweet relationship.

However,

Knowing your honesty, high intelligence, and devotion, I want to give one more chance to myself to see what you are claiming about. So, help me out brother Bahnam. Please do not ask me to go through numerous mixed, jumbled, and disconnected calculations dispersed in dozens of pages on internet, but present me the STRONGEST examples of the so-called code Seven. I would appreciate if you keep it concise and provide me about twenty or thirty examples, without much explanation. I am familiar with the Quran and one or two statement for each example should be sufficient.

You know, I can provide anyone a list of twenty examples about 19 that would be sufficient to show the "miracle" for an objective audience. So, you should be able to do the same.

Please do not mix your example with 19, since you will be in trouble with all other examples of 19 that do not meet the number 7.

**

Yes Behman, I do believe that most of those so-called evidences listed in the link you have given have no statistical merit. There are 513 direct references to the word Rasul (messenger) in the Quran and one every out of 19 of them would be multiple of 19. Add to this, literally millions of possible manipulations when you take them alone, OR put them before OR after chapters, OR add their digits OR add the entire numbers, OR add the gematrical value of RK before OR after the number OR add it to the digits, OR take the gematrical value of the first name, OR etc., EVERY SINGLE PERSON on this planet should be able to find MANY MORE example of multiples of 19 in connection to their names in verses containing the word messenger when they apply similar arbitrary method.

Rashad, especially towards his last days, lost his objectivity and mathematical intuition. I know some of the reasons for this anomaly, but I will leave it to another time.

Binary Symmetric Book

9/13/2010

Dear Erdem,

I spent about two hours to study the observations and claims made in *Symmetric Book* by Halis Aydemir, the Turkish engineer. I tried my best to remain open-minded despite early signs of arbitrary calculations and innumeracy. I could not continue reading after reaching page 202. As you will see in the attached document, I made the necessary corrections on the verse numbers of chapter 9 and the related calculations, which did not make much change.

We should first consider the following statistical facts:

From 1 to 114, there are 30 prime numbers. In other words, in that zone, one every four numbers are prime. Within the first thousand numbers, the ratio of primes to composites decreases to one in six. Since most of his calculations are done over small numbers, we can assume the probability of hitting a prime once every four numbers.

To find an either prime or composite number, excluding 0 and 1, is hundred percent. Interestingly, our author uses both of them! Now, let's add the following facts to the facts above:

1. Some observations are too subjective or esoteric; they are specifically picked among dozens of similar calculations.

2. Some of them happen to be prime numbers and others composite, and yet the author pulls our attention to each. Fifty out of hundred similar calculations should produce the symmetry our author is seeking. Occasionally, he encounters some numbers that do not fulfill his expectations and he finds them interesting too! Sometimes the first number is prime the other not or the first is composite and the second is prime. Both permutations transform into interesting and even beyond interesting in the eyes of the author.

3. I noticed only a few interesting observations. However, whoever works with many numbers in such a fashion for moths, perhaps years, should be able to find many more interesting coincidences.

4. In most cases, changing the number of verses in chapter 9 to 127 does not create a problem with Mr. Aydemir's observations. Knowing that the author would not include the unwanted numbers, I can claim that the verse numbers of Chapter 9 being 127 would not change a thing.

I cannot speak for the rest of the book, but with the exception of the borrowed table in the beginning of the book, the first 200 pages have no mathematical value. You may find similar even more interesting, yet arbitrary numerical calculations in the Appendices section of Dr. Rashad Khalifa's translation of the Quran. You may even find more interesting ones in the calculations of copycat messengers proliferated after Rashad Khalifa. If you wish, I could check my computer and send you a sample from the interesting calculations of those who deluded themselves to be the messengers after Rashad Khalifa.

I find Mr. Aydemir's work as a waste of time. I hope that you will not waste your skills, energy and time on this. You should not contribute to a false claim that distorts the mathematical aspect of the Quran.

Selam, Edip

AbuJamil, who is a business professor friend of mine, wrote an excellent evaluation on the probability of the so-called "positional partitionability" promoted by my friends Milan Sulc and Dr. Ali Fazely. Before reading his evaluation of the claims, what about starting from a brilliant remark from his article: "Rephrased for our purposes, one might say that the more ways we find to show significance in our findings, the more we destroy that very significance."

Positioning Partitionability?

AbuJamil
03/25/98,

Salaam!

Sorry about the long delay, but my second field exams are fast approaching and I've been neglecting just about everything else.

Just for the curious, I've "discovered" ... a *sign*!!! The letters in "Lomax" add up to a multiple of 17!!! Aaaaarrrrggghhz! And to make matters worse, so do the letters in "AbdulraHman." They are 17 x 43 and 17 x 31, respectively. That means ... he's the devil!! Aaaaarrrrggghhz!

Unfortunately, if you concatenate the GV [Geometric Value] of his original first name (Daniel) with the GV of "Lomax," you get 99731, which is 19 x 5249. That means ... he's a messenger!! Aaaaarrrrggghhz!

(Not that I expect too many people are actually reading this, but just for the record I hope no one takes this too seriously. For anyone who likes recreational numerology, which can admittedly be mildly fun if you have nothing better to do, I used the Roman qua Arabic system in which $h = 8$, $i = 9$, $j = 10$, $k = 20$, etc.)

Seriously though, re: "Lomax finally read your series on my web site and stated that it was one of the most impressive set of statistical tests he has seen thus far, but of course he had to spoil it by saying 'However, it is still deceitful.'"... it sure is easy to call things "deceitful" when you don't have to prove it. But that's just like him. He doesn't know enough about statistics to make an informed judgment, but he's already decided he knows all the answers in spite of his ignorance.

On the other hand, maybe that's okay. By comparison, I hear there are actually people who play with numbers until something comes out right and then try to hawk their absurd "findings" as "miracles" from God! How blasphemous! At least Lomax doesn't do that.

(I have just confirmed that my full name adds up to 1625, which is neither a multiple of 17 nor a multiple of 19 -- evidently, I am neither the devil nor a messenger ... I am relieved.)

Part of my brain has been casually pondering the positional partitionability question -- how to test it -- for several months. I have tried mapping complicated "trees" from the various rules involved, but without a fancy computer program to do this is not feasible. Its intent was to provide a complete database of outcomes from a single number thought to be significant and then assess whether the chains of relationships do indeed tend toward the same numbers again and again rather than fly off into random space. To the extent that I have been able to map "trees," I have found such an explosion of data points because of the variety of rules involved that they have simply deluged the precious few "significant" data that emerged from that process. But after a lot of thought on this, I have concluded the following:

1. My suggestion a long time ago that statistical tests of positional partitionability should define their domains in terms of the number of digits in the data points is a good idea.

2. Because of the property of fixed (non-sample-specific) relationships between significant and ordinarily non-significant numbers, the latter are rendered significant by association. To clarify, once it is shown that a particular number is associated, via any kind of chain, with the number 19, 9127, etc., that number itself must be declared "significant." This is because every subsequent occurrence of that number will necessarily yield the same result. This means that such numbers must be listed as permanently "significant," just as though they were multiples of 19 or otherwise meaningful in and of themselves.

The implications of #2 above can be seen clearly in the following example:

    a.   The probability of getting a multiple of 19 in a random selection of two- digit numbers may be computed by listing all such multiples and dividing by the total number of two-digit numbers in existence. Thus, "significant" numbers include 19, 38, 57, 76, and 95, and the total set of possibilities consists of 10, 11, ... , 99, which equals 99 - 9 numbers = 90. So the probability of getting a multiple of 19 in that set is 5/90 = p = .0556.

b. The probability of getting *either* a multiple of 19 *or* a number that is positionally related to the number 19 in a random selection of two-digit numbers may be computed by starting with multiples of 19 (i.e., 19, 38, 57, 76, 95), and then adding to these the following: 30 (the 19th composite), 67 (the 19th prime). That gives us a total of seven numbers out of 90, or 7/90 = p = .0778.

c. If we wish to consider *anything* positionally related to 19 to be significant, no matter how long the chain, then we need to add the 30th composite (45), the 45th composite (64), and the 64th composite (88). We also need to add the 67th composite (92). The 30th prime is greater than 100, so we must stop there. Now we have the following set of significant numbers: 19, 30, 38, 45, 57, 64, 67, 76, 88, 92, 95. That makes 11, so our probability has now gone up to 11/90 = p = .122.

d. It is easy to see how the addition of a few more numbers thought to be meaningful will inflate that probability even more. For example, 66 is the GV of "Allah," so that is obviously significant. There are 14 sets of Qur'anic initials, so that is significant. There are 29 initialed Suras, so that is significant. The number 90 might be considered to stand for Sad, which takes us to Surah Sad, etc., so that might be considered significant. The two qaf- initialed surat are numbers 42 and 50, so those might be considered significant. Lastly, the last initialed surah is numbered 68, so that might be considered significant. To these we must add the numbers that are positionally related, and in addition to these, we might as well add anything that is positionally related to a multiple of 19. So our final data set is composed of: 10, 11, 12, 13, 14, 15, 17, 18, 19, 20, 21, 22, 24, 25, 27, 28, 29, 30, 31, 32, 33, 34, 36, 37, 38, 40, 41, 42, 43, 45, 46, 47, 49, 50, 52, 54, 55, 57, 58, 59, 60, 61, 62, 64, 65, 66, 67, 68, 69, 70, 71, 74, 76, 77, 79, 80, 81, 82, 84, 85, 86, 87, 88, 89, 90, 92, 93, 95, 96, 97, 98, 99. Thus, our total is now 72 out of 90, or p = .800.

It is clear that our definition of "significant" is critical. The more freely we mathematically associate ordinary numbers with meaningful ones to show the intricacy of the chain, the more we reduce the significance of our findings. In a manner of speaking, this is like an interesting case of the Heisenberg principle, whereby our observation of a set of dynamics influences the dynamics being observed. Rephrased for our purposes, one might say that the more ways we find to show significance in our findings, the more we destroy that very significance.

All this is not to say that there is no significance in positional partitionability. It merely points up the fact that the consistency with which numbers are shown to be associated with meaningful values must be just about perfect; otherwise, the level of significance shown in a statistical analysis will virtually always be uninteresting. Peace!

A composite example!

Milan Sulc and Ali Fazely have hundreds of dizzying observations like the one in the message below:

Date: 28 Aug 97 02:48:15 EDT
From: Milan Sulc <100102.204@CompuServe.COM>
To: City of Light <CITYOFLIGHT@automailer.com>
Subject: Curious findings
Sender: owner-cityoflight@automailer.com Precedence: bulk

Salaamun alaykum, Richard

The 381st composite is 473, which is 11x43, and we know that 1143 is 9x127, and that 1143rd composite is 1362 and that 1362nd numbered ayat is 9:127.

If we add the composite 473 to its rank 38, we get 854, which is 307th "|C+R|C" type composite and 854 plus its subordinated rank of 307 adds up to 1161, which is of course 9x129.

Salaam, Milan

Theometer or Sectometer

(First conducted on the participants of my lectures at Oxford University in November 3-5, 2008)

Edip Yuksel

Name: _____

Email Address: _____

Phone: _____ Age: _____

Occupation: _____

Nationality: _____

Have you read the Manifesto for Islamic Reform? _____

Favorite Books/Authors: _____

Your Sect: (a) Sunni (b) Shiite (c) Salafi (d) Another sect (d) No sect

Please put a CIRCLE around the letter of your choice:

1. According to the Quran, which one of these is not and cannot be idolized by people?
a. Prophet Muhammad
b. Desires or Wishful thinking (Hawa)
c. Crowds or peers
d. Ancestors or children
e. Reasoning (Aql)

2. Which one of these is a true statement?
a. The Quran is not sufficient to guide us; in addition we need Hadith and Sunna.
b. The Quran is not sufficient to guide us; we need Hadith, Sunna and follow the teaching of a Sunni sect.
c. The Quran is not sufficient to guide us; we need Hadith, Sunna and follow the teaching of a Shiite sect.
d. The Quran is not sufficient to guide us; we need Hadith, Sunna, follow the teaching of a sect and join a religious order.
e. The Quran is sufficient to guide us when we understand and follow it through the light of reason.

3. Which one of these hadiths narrated by Bukhari, Muslim and other "authentic" hadith books, do you think are fabricated?
a. Muhammad was illiterate until he died.
b. Muhammad married Aisha at age 54 while she was only 9 or 13 years-old.
c. Muhammad dispatched a gang of fighters (sariyya) to kill a woman poet secretly during night in her home, for criticizing him publicly through her poems.
d. Muhammad slaughtered 400 to 900 Jews belonging to Ben Qurayza for violating the treaty.
e. All of the above.

4. Which one of these laws or rules does not exist in the Quran?
a. Stone the married adulterers to death
b. Do not play guitar
c. Men should not wear silk and gold
d. Men are superior to women
e. All of the above

5. The Quran instructs us to follow the messengers. Following the messenger means:
a. Follow Hadith and Sunna; Bukhari, Muslim, Ibn Hanbal, etc.
b. Follow his Ahl-al-Bayt.
c. Follow hadith, sunna, consensus of sahaba, ijtihad of imams and fatwas of ulama.
d. Follow Muhammad.
e. Follow the message he was sent with, which was Quran alone.

496

6. The Quran is God's word, because:

   a. There are verses of the Quran stating that it is God's word.

   b. The Quran is a literary miracle. None can bring a sura like it surpassing its literary qualities.

   c. I do not need to have a reason. Reason is not reliable. I have faith in the Quran.

   d. The moral teaching of the Quran is the best for individual and humanity.

   e. The Quranic signs (aya) do not have internal contradiction nor does it contradict the signs in nature. Besides, it is numerically coded book with an extraordinary mathematical structure integrated with its composition and Arabic language.

7. Which one of the following is correct for Muhammad:

   a. Muhammad was the final messenger and prophet.

   b. Muhammad had the highest rank above all humans.

   c. Muhammad demonstrated many miracles such as splitting the moon, healing the sick, and crippling a child

   d. All of the above´

   e. Muhammad was a human messenger like other messengers.

8. In what year he Bukhari started collecting hadith for his hadith collection known as the Sahih Bukhari, the most trusted Sunni hadith collection?

   a. During the life of Muhammad in Medina

   b. Ten years after Muhammad's death.

   c. 130 years after Muhammad's death.

   d. 200 years after Muhammad's death

   e. 230 years after Muhammad's death.

9. According to Bukhari himself, he collected the 7,275 hadith among the 600,000 hadiths he collected. If each hadith, together with its *isnad* (the chain of reporters) and *sanad* (the text that was attributed to Muhammad) took about half a book page, how many volumes of books with 500 pages would they take to record all those 600,000 hadith allegedly collected by Bukhari?

   a. 7 volumes

   b. 10 volumes

   c. 70 volumes

   d. 100 volumes

   e. 700 volumes

10. What are the last statements in the Farewell Sermon (Khutba al-Wada) which was reportedly witnessed by more than 100,000 sahaba, making it by far the most authentic hadith among the thousands of hadiths?

   a. I leave you Abu Bakr; you should follow him.

   b. I leave you my sahaba; you may follow any of them.

   c. I leave you the Quran and Sunna; you should follow both.

   d. I leave you the Quran and Ahl-al-Bayt (my family); you should follow them.

   e. I leave you the Quran, you should follow it.

11. According to some "authentic hadith" found in Bukhari and other hadith books, there was a verse instructing muslims to stone the married adulterers to death: "Al-shayhu wal-shayhatu iza zanaya farjumuhuma nakalan..." According to hadith reports, what happened to those verses?

   a. After the Prophet Muhammad's death, Umayyad governor Marwan burned the pages where those verses were written.

   b. Angle Gebrail came down and deleted it from the scripture.

   c. Ibni Abbas forgot it yet Abu Hurayra never forgot it.

   d. There is no reference to such a verse in any authentic hadith books.

   e. After the Prophet Muhammad's death, the skin which the verse was written on was protected under Aisha's bed. A hungry goat ate it. Thus, it was abrogated literally yet kept legally.

12. According to both Bukhari and Muslim, when Muhammad was in his death bed, he asked his comrades around to bring him a paper and pen to write something for them so

497

that they would not divert from the right path. According to the same "authentic" Sunni hadith books, Omar bin Khattab stopped a sahaba who was hurrying for a paper and pen and said the following: "The prophet is sick and has fever. He does not know what he is saying. God's book is sufficient for us." According to the hadith, all the prominent comrades (sahaba) agreed with Omar and Muhammad passed away without writing down his advice. What do you think about this hadith?

a. If it is narrated by both Bukhari and Muslim, then it must be true
b. If it is true, then, Omar and all other Sahaba must have betrayed Muhammad and committed blasphemy.
c. If it is true, then, Omar and all prominent Sahaba were followers of the Quran alone.
d. If it is false then all other hadith too should be rejected.
e. C and D must be true

13. Do we need to SAY "sallallahu alayhi wasallam" after Muhammad's name?

a. Yes, every time Muhammad is mentioned we have to praise his name.
b. Yes, but we need to say only once in our lifetime.
c. Yes, the more we say the better.
d. Yes, and those who do not say it after Muhammad's name disrespect him and they will not receive his intercession.
e. No, the Quran does not ask us to say anything after Muhammad's name; muslims were asked (salli ala) to support him, as he was also asked to support them (salli alayhim).

14. What is the correct Testimony (shahada) according to the Quran:

a. I bear witness that there is no god but the God and the Quran is God's word.
b. I bear witness that there is no god but the God and Muhammad is His messenger.
c. I bear witness that there is no god but the God and Muhammad is His messenger and His servant.

d. I bear witness that there is no god but the God and Abraham, Jesus, Moses and Muhammad are His messengers.
e. I bear witness that there is no god but the God.

15. Should Muslims who do not observe daily prayers be beaten in public?

a. Yes.
b. No.

16. Should Muslims who are caught for consuming alcohol for the fourth time be killed?

a. Yes.
b. No.

17. Did the prophet give permission to kill women and children in the war?

a. Yes.
b. No.

18. According to the Quran, are women banned from reading Quran and pray during their menstruation periods?

a. Yes
b. No.

19. In the daily Sala prayers, do you recite "attahiyyatu lillahi wassalawatu .... as salamu alayka ayyuhannabiyyu wa rahmatullahi wa barakatuhu"?

a. Yes
b. No

20. Does the Quran justify taxing Jewish and Christian population under Muslim authority with extra or different taxation called Jizya?

a. Yes
b. No.

21. Does the Quran instruct women to cover their hair?

a. Yes.
b. No.

22. Are woman restricted from leading congregational prayers?

a. Yes.
b. No.

23. Are women mentally and spiritually inferior to men?

a. Yes.
b. No.

24. Does the Quran restrict women from initiating divorce?
  a. Yes.
  b. No.

25. Is polygamy with previously unmarried women allowed?
  a. Yes, up to four women.
  b. No, polygamy is allowed only with the widows who have orphans.

26. Do pilgrims need to cast stones at the devil?
  a. Yes.
  b. No.

27. Is the black stone near Kaba holy?
  a. Yes.
  b. No.

28. May a muslim own slaves?
  a. Yes.
  b. No.

29. Is circumcision a required or encouraged practice in Islam?
  a. Yes.
  b. No.

30. Should converts change their names to Arabic names?
  a. Yes.
  b. No.

31. How much charity one should give away?
  a. 2.5%
  b. As much as one can afford, without making themselves needy.

32. Are those who break their fast during Ramadan before the sunset required to fast 60 consecutive days as a punishment for not completing the day?
  a. Yes.
  b. No.

33. Is leadership the right of Quraish tribe?
  a. Yes.
  b. No.

34. Is drawing pictures or making three dimensional statutes a sin?
  a. Yes.
  b. No.

35. Are there more dietary prohibitions besides pork, carcass, running blood, and animal dedicated to idolized names?
  a. Yes.
  b. No.

36. Is displaying Muhammad's name and the names of his closest companions next to God's name in the mosques idol-worship?
  a. Yes.
  b. No.

37. Did Muhammad advise some sick people to drink camel urine?
  a. Yes.
  b. No.

38. Did Muhammad gauge people's eyes with hot nails?
  a. Yes.
  b. No.

39. After following the advice of Moses, did Muhammad bargain with God about the number of prayers, lowering down from 50 *times* a day to 5 times a day?
  a. Yes.
  b. No.

40. Does Muhammad have the power of intercession?
  a. Yes.
  b. No.

41. Was Muhammad sinless?
  a. Yes.
  b. No.

42. Did God create the universe for the sake of Muhammad?
  a. Yes.
  b. No.

43. Did Muhammad have sexual power of 30 males?
  a. Yes.
  b. No.

44. Was Muhammad bewitched by a Jew?
  a. Yes.
  b. No.

45. Do some verses of the Quran abrogate other verses?
  a. Yes.
  b. No.

Here is the story and the answer of this test:

Between November 3 and 10 of 2008, I traveled to UK and Turkey to deliver four lectures; first two at Oxford University, the third at Muslim Institute in London and the fourth one in Istanbul Book Fair. I had prepared a test containing 45 multiple choice questions just the night before my travel. I duplicated them on both sides of a single sheet and I distributed to the audience before the lecture... They were asked to write their name, age, occupation, email address, favorite authors, and their sectarian affiliation. It was a bit awkward to test an audience that consisted of students and professors at one of the world's top universities. The multiple-choice test proved to be a powerful instrument to deliver the message of Islamic Reform under the light of the Quran. The correct answer for each multiple choice question was the E option, and for the Yes or No questions was the B option. So, it would take me a few seconds to evaluate the tests after they were returned to me.

The Sunni or Shiite test-takers found themselves in quagmire of contradiction with their own sectarian teachings. They learned that they were thirty, forty or even more than fifty percent infidels or heretics. Some of those who marked Sunni as their sectarian affiliation contradicted the Sunni teachings on most of the issues. According to their own confessed sects, their lives were worthless; they deserved to be killed! I did not let this mirror or sect-o-meter remain an individual experience; I publicly declared the overall results. Many got all answers correct, including Eric, a monotheist from Unitarian church who already had a copy of the *Quran: a Reformist Translation* in his possession. Eric knew the original message of islam better than all the mullahs and the so-called "ulama" combined.

If you have chosen the wrong option for any of the questions and you are wondering why you have contradicted the Quran, please visit **www.islamicreform.org** and read the full version of the *Manifesto for Islamic Reform*. If you prefer to have it in a book form, you may order it by visiting **www.brainbowpress.com**

SELECTED BIBLIOGRAPHY:

Abdulbaqi, Fuad: *al-Mucam ul-Mufahras li-Alfaz al-Quran al-Karim*

Asfahani: *Al-Mufradat fi Gharib al-Quran*

Ateş, Süleyman: *Kuran Ansiklopedisi /Encyclopedia of the Quran*

Baz, Sheikh Ibnul: *El-Edilletün Naqliyyetu vel Hissiyatu Ala Cereyaniş Şamsi ve Sukunil Ardi ve Imkanis Suudi ilal Kavakibi/The Traditional and Empirical Evidences for the Motion of the Sun and Stillness of Earth, and Possibility of Ascending to Planets*, University of Medina, 1975.

Bazargan, Mahdi: *Sayr-i Tahawul-i Quran/Process of Quranic Evolution*, Book Distribution Center, Houston, 1974.

Caldwell, Chris K. and Honaker, G.L. Jr., *Prime Curios! The Dictionary of Prime Number Trivia*, CreateSpace, 2009.

Clifford, A. Pickover, *The Math Book*, Sterling, New Yok, 2009.

Çelakıl, Ömer: *Kuran'ı Kerim Şifresi*, (Kelepir/Düş, Istanbul, 2002-2005)

Dan, Joseph: *Studies In Jewish Mysticism*, Proceedings of Regional Conferences Held at the University ofCalifornia, Los Angeles and McGill University in April, University of California Press, 1978.

Deedat, Ahmad: *Al-Quran: the Ultimate Miracle*, Islamic Propagation Centre, Durban, 1979-1986.

Drosnin, Michael: *The Bible Code*, Simon & Schuster, 1997.

Dudley, Underwood: *Numerology or What Phythagoras Wrought*, Mathematical Association of America, 1997.

Firuzabadi, Muhammad ibn Ya'qub, *Al-Qamus al-Muhit/Comprehensive Dictionary*.

Gardner, Martin: *Scicentific American*, September, 1980.

Guy, Richard K., *Unsolved Problems in Number Theory*, Springer-Verlag, 1994,

Haddad, Yvonne Yazbeck and Smith, Jane Idleman: *Mission to America: Five Islamic Sectarian Communities in North America*, University Press of Florida, 1993

Hamidullah, Muhammad: *Le Saint Coran/The Glorious Quran*.

Ibn Manzur: *Lisan ul-Arab/The Arabic Language*.

Ifrah, Georges: *The Universal History of Numbers*, John Wiley & Son, 2000.

Khalifa, Rashad: *Miracle Of The Quran: Significance Of The Mysterious Alphabets*, Islamic Productions, St. Louis, Missouri, 1973.

Khalifa, Rashad: *The Computer Speaks: God's Message To The World*, Renaissance Productions, Tucson, Arizona, 1981.

Khalifa, Rashad: *Qur'an: The Final Scripture, Islamic Productions*, Tucson, Arizona, 1981.

Khalifa, Rashad: *Quran: Visual Presentation Of The Miracle*, Ibid, 1982.

Khalifa, Rashad: *Qur'an,* Hadith *and Islam*, Ibid, 1982.

Khalifa, Rashad: *Quran: The Final Testament*, Ibid, 1989.

Livio, Mario: Is God a Mathematician, Simon & Schuster, 2009.

Lucas, Jerry and Wasburn, *Del: Theomatics: God's Best Kept Secret Revealed.* Stein and Day 1977-1986.

Majul, Adib: *The Names of Allah in Relation to the Mathematical Structure of Quran*, Islamic Productions, Tucson, 1982.

Majul, Adib: *Various personal letters to Edip Yuksel*, 1991-1998. Personal collection.

Nawfal, Abdurrazzaq: al-*I'jaz al-'Adadi fi al-Qur'an al-Karim/Numerical Miracles in the Holy Quran*, Dar-ul Kitab-il Arabiy, Beirut, 4th edition, 1983.

Paulos, John Allen: *Innumeracy: Mathematical Illiteracy and its Consequences*, Vintage Books, New York, 1990

Posamentier, Alfred S. and Lehmann, Ingmar, *Mathematical Amazements and Surprizes: Fascinating Figures and Noteworthy Numbers*, Prometheus Books, 2009.

Sagan, Carl: *The Demon-Haunted World: Science as a Candle in the Dark*, Ballantine Books, 1st edition, 1997.

Sagan, Carl: *Contact*, Simon and Schuster. New York: 1985

Shimmel, Annemarie: *The Mystery of Numbers*, Oxford University Press, 1993.

Taslaman, Caner, *The Big Bang, Philosophy and God*, Nettleberry, 206.

Taslaman, Caner: *Kuran Hiç Tükenmeyen Mucize/Quran: Unchallengeable Miracle*, Istanbul Yayinevi, Istanbul, 2002.

Toptaş, Mahmut; Zeyveli, Hikmet; Kutman, Orhan; Yüksel, Sadreddin: 19 Efsanesi/The Myth of 19, Inkilab, Istanbul, 1988-2005

Yuksel, Edip: *The Prime Argument,* Monotheist Productions Int, Tucson, 1995, 64 p.

Yuksel, Edip: *Running Like Zebras... (74:50): An Internet Debate*, Monotheist Productions Int, Tucson, 1995, 120 p.

Yuksel, Edip: *Which one do you See: Hell or Miracle?* (19.org, 2004)

Yuksel, Edip: *Running Like Zebras 2... (74:50)*, (19.org, 2005)

Yuksel, Edip, et al: *Quran: a Reformist Translation*, Brainbow Press, 2007-10.

Yuksel, Edip: *NINETEEN: God's, Signature in Nature and Scripture*, Brainbow Press, 2011.

Yüksel, Edip and Deedat, Ahmad: Kuran En Büyük Mucize /Quran, the Greatest Miracle (Inkilab, Fatih-Istanbul, 1983-88, 16 editions, 204 p.)

Yüksel, Edip: *Kuran'da Demirin Kimyasal Esrari/Chemical Secrets of Iron in the Quran*. Timaş, Çağalfoğlu-Istanbul, 1984, 48 p.

Yüksel, Edip: *Kuran Görülen Mucize/Quran: The Visual Miracle,* Timaş, Çağaloğlu -Istanbul, 1985, 308 p.

Yüksel, Edip: Ilginç *Sorular-1/ Interesting Questions-1,* Inkilab, Fatih-Ist., 1985-1987, 8 editions, 214 p. Beyan, Cağaloğlu-Istanbul, 1988, 9th edition, 214 p.

Yüksel, Edip: *Ilginç Sorular-2/ Interesting Questions-2,* Yüzondört, Fatih-Istanbul, 1987, 190 p. Beyan, Cağaloğlu-Istanbul, 1988, 2nd and 3rd editions, 190 p.

Yüksel, Edip: *Üzerinde Ondokuz Var / On It Is Nineteen,* Ay Yayıncılık, 1997, 300p. Ozan, 2005, 320 p.

Yüksel, Edip: *Mesaj, Kuran Çevirisi/The Message, Translation of the Quran* (Ozan/19.org, 1999-2005).

CPSIA information can be obtained
at www.ICGtesting.com
Printed in the USA
BVHW081657120620
581249BV00001B/6